Current and Future Developments in Nanomaterials and Carbon Nanotubes

(Volume 4)

Synthesis and Applications of Semiconductor Nanostructures

Edited by

Karamjit Singh Dhaliwal

Department of Physics
Punjabi University
Patiala, Punjab, India

Current and Future Developments in Nanomaterials and Carbon Nanotubes

(Volume 4)

Synthesis and Applications of Semiconductor Nanostructures

Editor: Karamjit Singh Dhaliwal

ISSN (Online): 2589-2207

ISSN (Print): 2589-2193

ISBN (Online): 978-981-5080-11-7

ISBN (Print): 978-981-5080-12-4

ISBN (Paperback): 978-981-5080-13-1

First published in 2023.

need for a court order if at any point you breach any terms of this License Agreement. In no event will any delay or failure by Bentham Science Publishers in enforcing your compliance with this License Agreement constitute a waiver of any of its rights.

3. You acknowledge that you have read this License Agreement, and agree to be bound by its terms and conditions. To the extent that any other terms and conditions presented on any website of Bentham Science Publishers conflict with, or are inconsistent with, the terms and conditions set out in this License Agreement, you acknowledge that the terms and conditions set out in this License Agreement shall prevail.

Bentham Science Publishers Pte. Ltd.
80 Robinson Road #02-00
Singapore 068898
Singapore
Email: subscriptions@benthamscience.net

BENTHAM SCIENCE

CONTENTS

FOREWORD

It gives me immense pleasure to write this foreword for the book titled *"Synthesis and Applications of Semiconductor Nanostructures"*, which is the fourth volume of the book series titled *"Current and Future Developments in Nanomaterials and Carbon Nanotubes"*. For the last four decades, semiconductor nanostructures have been the subject of scientific interest because of their size-tunable optical and electrical properties caused by unique quantum confinement effect and augmented surface-to-volume ratio at nanoscale size dimensions. A thorough study of these materials constitutes a new perspective for basic and applied research in nanophotonics, opto-electronics, nanobiotics, nanocosmetics, nanomedicine, and photo-chemistry. This book provides adequate knowledge about the fundamentals of semiconductor nanostructures, synthesis and characterization of nanostructures of various morphologies, as well as metal-organic frameworks and nanostructure-impregnated metal-organic frameworks. Multidisciplinary smart applications of these materials are described in a lucid way in this volume. All the chapters have been appositely chosen to justify the theme of the book. I think this book would be of immense use to academicians, researchers and technocrats, who may like to get a better understanding of synthesis and applications of semiconductor nanostructures.

Prof. Arvind
Vice Chancellor
Punjabi University
Patiala, Punjab, India

PREFACE

Nanotechnology is a multidisciplinary field benefitting from efforts and developments in the various spheres of science and technology, such as applied physics, materials science, interface and colloid science, supramolecular chemistry, chemical engineering, biological engineering, mechanical engineering, electrical engineering and computer engineering. Its theme is the design, fabrication, characterization and application of structures, devices and systems by controlling shape and size at the nanometer scale. This book focuses on the synthesis, characterization and multifaceted potential applications of semiconductor nanostructures, metal-organic frameworks (MOFs) and nanostructure-impregnated MOFs, along with some glimpses of doped glasses, functionalized carbon nanotubes, doped graphene and graphene nanoribbons. Numerous bottom-up and top-down synthesis techniques opted for the synthesis of various morphologies of semiconductor nanostructures have been discussed at length, along with details of synthetic metal-organic frameworks and nanostructure-impregnated MOFs constructed from a variety of inorganic and organic components. The photocatalytic activity potential of synthetic materials for hydrogen production, wastewater treatment, carbon dioxide reduction, environmental pollution monitoring and management, as well as oxidation of alcohols has been described comprehensively. The sensing ability of nanostructures and MOFs has been thoroughly illustrated to utilize these materials as chemical sensors, gas sensors and biosensors. Many other smart applications in gas separation, gas storage, drug delivery, nanomedicine, nano-cosmetics as well as antibacterial uses have been reported lucidly in this book. This book also deals with the geometric and electronic properties of doped graphene and graphene nanoribbons. Spin-polarized density functional study results have been included to describe the structural, electronic and magnetic properties of strained graphene nanoribbons.

Although this book is not edited with any special course in mind, it may serve as a unique source of knowledge for postgraduate students, academicians, researchers and technocrats, who want to carry out R&D activities with nanomaterials and metal-organic frameworks.

Karamjit Singh Dhaliwal
Department of Physics
Punjabi University
Patiala, Punjab, India

DEDICATION

Dedicated to My Father Late S. Major Singh Dhaliwal.

List of Contributors

Aman Grover	Department of Chemistry, Punjabi University, Patiala-147002, Punjab, India
Aman Mahajan	Department of Physics, Materials Science Laboratory, Guru Nanak Dev University, Amritsar-143005, Punjab, India
Amanpreet Kaur	Department of Microbiology, Mata Gujri College, Fatehgarh Sahib-140 407, Punjab, India
Ankush Vij	Department of Physics, University of Petroleum and Energy Studies, Dehradun-248 007, Uttarakhand, India
Anshul Kumar Sharma	Centre for Sustainable Habitat, Guru Nanak Dev University, Amritsar-143005, Punjab, India
Anup Thakur	Department of Basic and Applied Sciences, Advanced Materials Research Lab, Punjabi University, Patiala-147 002, Punjab, India
Ashok Kumar	Department of Applied Sciences, National Institute of Technical Teachers Training and Research, Chandigarh-160019, India
Ashok Kumar Malik	Department of Chemistry, Punjabi University, Patiala, Punjab, India
Babita Rani	Department of Physics, Punjabi University, Patiala-147002, Punjab, India
Baby	Department of Microbiology, Mata Gujri College, Fatehgarh Sahib-140 407, Punjab, India
Balwinder Kaur	Department of Chemistry, Punjabi University Patiala-147002, Punjab, India
Dinesh Kumar	Department of Physics, Punjabi University, Patiala-147002, Punjab, India
Gulshan Dhillon	Chitkara University Institute of Engineering and Technology, Chitkara University, Rajpura, Punjab, India
Inderjeet Singh Sandhu	Chitkara University Institute of Engineering and Technology, Chitkara University, Rajpura, Punjab, India
Inderpreet Singh Grover	Department of Chemistry, Public College Samana, Patiala, Punjab, India
Irshad Mohiuddin	Department of Chemistry, Punjabi University, Patiala-147002, Punjab, India
Isha Mudahar	Department of Basic and Applied Sciences, Punjabi University, Patiala-147 002, Punjab, India
Jaspreet Kocher	Department of Physics, National Institute of Technology, Kurukshetra-136119, Haryana, India
Jatinder Singh Aulakh	Department of Chemistry, Punjabi University, Patiala-147002, Punjab, India
Jiri Pechousek	Regional Centre of Advanced Technologies and Materials, Department of Experimental Physics, Faculty of Science, Palacky University, Olomouc, Czech Republic
Kavita	Department of Physics, Multani Mal Modi College, Patiala, Punjab, India
Kuldeep Kaur	Department of Chemistry, Mata Gujri College, Fatehgarh Sahib, Punjab, India

Manju	Department of Basic and Applied Sciences, Advanced Materials Research Lab, Punjabi University, Patiala-147 002, Punjab, India Department of Physics, Punjabi University, Patiala-147 002, Punjab, India
Manreet Kaur Sohal	Department of Physics, Materials Science Laboratory, Guru Nanak Dev University, Amritsar-143005, Punjab, India
Mansi Chitkara	Chitkara University Institute of Engineering and Technology, Chitkara University, Rajpura, Punjab, India
Megha Jain	Department of Basic and Applied Sciences, Advanced Materials Research Lab, Punjabi University, Patiala-147 002, Punjab, India Department of Physics, Punjabi University, Patiala-147 002, Punjab, India
Nancy	Department of Physics, Punjabi University, Patiala-147002, Punjab, India
Pooja Rani	Department of Physics, Multani Mal Modi College, Patiala, Punjab, India
Rajeev Sharma	Department of Chemistry, Multani Mal Modi College, Patiala, Punjab, India
Rajesh Kumar	Department of Chemistry, Government Degree College, Sugh-Bhatoli-176 047, Kangra, Himachal Pradesh, India
Sandeep Kaur	Department of Basic and Applied Sciences, Punjabi University, Patiala-147 002, Punjab, India
Sandeep Kumar	Department of Chemistry, Punjabi University, Patiala-147 002, Punjab, India
Sanjay Kumar	Department of Chemistry, Multani Mal Modi College, Patiala-147 001, Punjab, India
Saurabh Gupta	Department of Microbiology, Mata Gujri College, Fatehgarh Sahib-140 407, Punjab, India
Seema Maheshwari	Department of Chemistry, Mata Gujri College, Fatehgarh Sahib, Punjab, India
Sharmila Kumari Arodhiya	Department of Physics, National Institute of Technology, Kurukshetra-136119, Haryana, India Department of Physics, Rajiv Gandhi Government College for Women, Bhiwani-127021, Haryana, India
Shashank Priya	Materials Research Institute, Penn State University, PA 16801, USA
Shikha Bhogal	Department of Chemistry, Punjabi University, Patiala, Punjab, India
Shyam Sundar Pattnaik	Media Engineering, National Institute of Technical Teachers Training and Research, Chandigarh-160019, India
S.M. Rao	Institute of Physics, Academia Sinica, Taipei-11529, Taiwan
Subhash Chand	Department of Chemistry, L.R.D.A.V College, Jagraon-142 026, Punjab, India
Supreet Pal Singh	Department of Physics, Punjabi University, Patiala-147002, Punjab, India

List of Abbreviations

AC	Acetylacetone
AD	Alzheimer's Disease
AF	Antiferromagnetic
AFM	Atomic Force Microscope
AGNRs	Armchair Graphene Nanoribbons
ALP	Alkaline Phosphatase
AOPs	Advanced Oxidation Processes
API	Active Pharmaceutical Ingredients
AR	Analytical Reagent
ATP	Adenosine Triphosphate
AZT-Tp	Azidothymidine Triphosphate
Aβ	Amyloid-β
BAU	Bangladesh Agricultural University
BDC	Benzene-1 4 Dicarboxylic Acid
BE	Binding Energy
BET	Brunauer Emmett and Teller
BPA	Bisphenol A
BSG	Box-shaped Graphene
Bu	Busulfan
CB	Conduction Band
CDV	Cidofovir
CGNRs	Chiral Graphene Nanoribbons
CL	Chemiluminescence
CNF	Cellulose Nanofiber
CNS	Central Nervous System
CNTs	Carbon Nanotubes
COF	Covalent Organic Framework
CoPc	Cobalt-phthalocyanine
CQDs	Carbon Quantum Dots

CRET	Chemiluminescence Resonance Energy Transfer
CSF	Cerebrospinal Fluid
CTAB	Hexadecyltrimethyl Ammonium Bromide
CVD	Chemical Vapour Deposition
CuPc	Copper-phthalocyanine
D	Dimensional
DBI	Dihydroxybenzene isomers
DDF	Doctor's Dermatologic Formula
DEF	N,N-diethylformamide
DFT	Density Functional Theory
DMF	Dimethylformamide
DNA	Deoxyribonucleic Acid
DOE	Department of Energy
DOS	Density of States
Doxo	Doxorubicin
DRIFTS	Diffuse Reflectance Infrared Fourier Transform Spectroscopy
DSM	Double Solvent Method
2DEG	Two-dimensional Electron Gas
EDS	Energy-dispersive X-ray Spectroscopy
ERGO	Electrochemically Reduced Graphene Oxide
Erα	Estrogen Receptor Alpha
FC	Field Cooled
FCC	Face Centred Cubic
FDA	Food and Drug Administration
FFT	Fast Fourier Transform
fGO	Functionalized Graphene Oxide
f-MWCNTs	Functionalized Multi-walled Carbon Nanotubes
FRET	Fluorescence Resonance Energy Transfer
FTIR	Fourier Transform Infrared
FWHM	Full Width at Half Maximum
5-FU	5- ourouacil
GAs	Graphene Aerogels
GC-MS	Gas Chromatography Mass Spectrometry
GMS	Glycerol a-monostearate
GN	Graphene Nanosheet

GNPs	Gold Nanoparticles
GNRs	Graphene Nanoribbons
GO	Graphene Oxide
HA	Hyaluronic Acid
HER	Hydrogen Evolution Reaction
HKUST	Hong Kong University of Science and Technology
HOMO	Highest Occupied Molecular Orbital
HPLC	High-performance Liquid Chromatography
HPLC-DAD	High-performance Liquid Chromatography with Diode Array Detector
HPLC-PDA	High-performance Liquid Chromatography with Photodiode Array Detector
IBU	Ibuprofen
ICP-MS	Inductively Coupled Plasma Mass Spectrometry
ICP-OES	Inductively Coupled Plasma Optical Emission Spectroscopy
IMI	Imidacloprid
IONPs	Iron Oxide Nanoparticles
LBNP	Lipid-based Nanoparticle
LC	Liquid Chromatography
LMM	Localized Magnetic Moment
Ln-MOF	Lanthanide Organic Framework
LSPR	Localized Surface Plasmon Resonance
LUMO	Lowest Un-occupied Molecular Orbital
M	Metal
MB	Methylene Blue
MBB	Molecular Building Blocks
MBE	Molecualr Beam Epitaxy
MDR	Multiple Drug Resistant
MEMS	Micro-electromechanical Systems
MIPs	Molecularly Imprinted Polymers
miRNA	Micro Ribonucleic Acid
MM	Magnetic Moment
MNase	Micrococcal Nuclease
MNP	Magnetic Nanoparticles
MNPs	Metal Nanoparticles
MO	Methyl Orange
MOCVD	Metal Organic Chemical Vapor Deposition

MOFDC	Metal Organic Framework Derived Carbon
MOFs	Metal Organic Frameworks
MOVE	Methane Opportunities for Vehicular Energy
MOXs	Mixed Oxides
MP	Methyl Parathion
MPA	Mercaptopropionic Acid
MPcs	Metallophthalocyanines
MW	Microwave
MWCNTs	Multi-walled Carbon Nanotubes
NBOs	Non-bridging Oxygens
NCs	Nanocrystals
NEMSs	Nano-electromechanical Systems
NESD	Nano Enabled Scaffold Device
NHE	Normal Hydrogen Electrode
NMR	Nuclear Magnetic Resonance
NPs	Nanoparticles
PAHs	Polycyclic Aromatic Hydrocarbons
PBUs	Primary Building Blocks
PCA	Photo-catalytic Activity
PCFs	Porous Carbon Fibers
PCL	Poly (ε-caprolactone)
PD	Parkinson's Disease
PEC	Photoelectrochemical
PET	Photo-induced Electron Transfer
PL	Photoluminescence
PVD	Physical Vapour Deposition
PVP	Polyvinyl Pyrrolidone
PXRD	Powder X-ray Diffraction
QDs	Quantum Dots
QE	Quantum Efficiency
QW	Quantum Well
RE	Rare Earth
RET	Radiative Energy Transfer
rGO	Reduced Graphene Oxide
RhB	Rhodamine B

RNA	Ribonucleic Acid
RON	Research Octane Number
ROS	Reactive Oxygen Species
rpm	Revolutions per minute
RT	Room Temperature
SAW	Surface-acoustic Waves
SBUs	Secondary Building Units
SCs	Semiconductors
SCF	Solid Carbon Fiber
SCNSs	Semiconductor Nanostructures
SEM	Scanning Electron Microscope
SHE	Standard Hydrogen Electrode
SLNs	Solid Lipid Nanoparticles
SNPs	Silver Nanoparticles
SPE	Solid Phase Extraction
SPIONs	Superparamagnetic Iron-oxide Nanoparticles
SPR	Surface Plasmon Resonance
SQDs	Semiconductor Quantum Dots
SQUID	Superconducting Quantum Interference Device
ss-DNA	Single-stranded Deoxyribonucleic Acid
SS-NMR	Solid-state Nuclear Magnetic Resonance
STM	Scanning Tunneling Microscope
SWCNTs	Single-walled Carbon Nanotubes
TB	Tuberculosis
TEM	Transmission Electron Microscope
TFA	Trifluoroacetic Acid
TGA	Thermogravimetric Analysis
TMM	Total Magnetic Moment
TNase	Thermonuclease
TNP	Trinitrophenol
TNR	Titanium Nanorod
TNT	Trinitrotoluene
TNW	Titanium Nanowire
tRNA	Transfer Ribonucleic Acid
TYL	Tylosin

UV	Ultraviolet
VB	Valence Band
VOCs	Volatile Organic Compounds
VPc	Vanadium-phthalocyanine
WF	Week Ferromagnetic
XPS	X-ray Photoelectron Spectroscopy
XRD	X-ray Diffraction
ZFC	Zero Field Cooled
ZIF	Zeolite Imidazolate Framework
ZGNRs	Zigzag Graphene Nanoribbons

CHAPTER 1

Semiconductor Nanostructures and Synthesis Techniques

Kavita[1] and **Pooja Rani**[1,*]

[1] *Department of Physics, Multani Mal Modi College, Patiala, Punjab, India*

Abstract: Semiconductor nanostructures show different properties compared to their bulk counterparts due to quantum confinement effects and enhanced surface-to-volume ratio with the reduction in particle size on nanoscale dimensions. This chapter introduces the nanomaterials, especially semiconductor nanostructures of various morphologies, quantum nanostructures (quantum dots, quantum wires and quantum wells) along with conventional 3D nanostructures. The present time is the introductory era of nanoscience and nanotechnology; synthesis of highly monodisperse nanostructures for device applications is a challenge for researchers and technocrats. This chapter discusses at length fascinatingly the bottom-up and top-down synthesis approaches along with the commonly used nanomaterial synthesis techniques, such as mechanical milling, lithography, electrospinning, template synthesis, chemical precipitation, sol-gel method, hydrothermal/solvothermal method, laser ablation, and other vapour processing methods.

Keywords: Electrospinning, Lithography, Laser ablation, Milling, Sol-gel, Semiconductor nanostructures, Template synthesis, Vapour deposition techniques.

INTRODUCTION

In 1959, Nobel laureate Richard P. Feynman, in his famous talk "There is plenty of room at the bottom [1, 2]", contemplated a remarkable technology on the scale of a few nanometers. This was the beginning of a new branch of science and technology, which nowadays is known as nanotechnology. The length of a nanometer can be understood through the example of ten hydrogen atoms lined up in a row, which is one nanometer (nm). Materials are defined as nanomaterials if their size or at least one dimension of the structure is in the range of 1 to 100 nm. Nanotechnology is an interdisciplinary paradigm, which conjoins diverse fields of science and engineering at the nanoscale. Numerous research disciplines like physics, chemistry, biotechnology, *etc.*, and technology and industry sectors like

* **Corresponding author Pooja Rani:** Department of Physics, Multani Mal Modi College, Patiala, Punjab, India; E-mail: pgoyal0510@gmail.com

Karamjit Singh Dhaliwal (Ed.)

information technology, energy, environmental science, medicine/medical instrumentation, homeland security, food safety, and transportation, among many others, are to be revolutionized by this interesting science. Many techniques which Feynman envisioned are now well-developed [3, 4]. Multifunctional nanomaterials and devices can behave in extremely different ways, and their physical and chemical properties change drastically if subjected to different external parameters, resulting in variations due to their large surface area to volume ratio and quantum size effect. Technological development in recent years has led to the development of crystalline materials with assured desired qualities, facilitating their applications in areas like electronics (optoelectronics), spintronics, medicine, superconductivity, nuclear and electron resonance, molecular structure investigation, photonics, photocatalysis and photovoltaics. Amongst these crystalline materials, semiconducting nanomaterials are of immense interest because their properties can be easily modified or improved through doping, thus enabling them to cover a wide range of promising applications. Semiconductor nanostructures form a class of materials with a large degree of freedom to design optoelectronic properties through variations in composition, size and dimensionality. Currently, semiconductor nanomaterials are still in the research phase; they are promising contributors to the development of technology in a disciplined manner, for instance, in lighting and displays, laser technology, telecommunication, quantum information processing and sensing. Scaling down feature sizes into the nanometer range is a common trend in advanced compound semiconductor devices, and the progress of nano-fabrication technology has opened up exciting possibilities for constructing novel quantum devices for which the operations are directly based on quantum mechanics. Size tunable physical and chemical properties of semiconductor nanocrystals make these materials very attractive both from a scientific view and optical device application aspect [5 - 11]. This was made possible by the availability of semiconducting materials of unusual purity and crystalline excellence. Such materials can be structured to contain a thin layer of highly mobile electrons. Motion perpendicular to the layer is quantized so that the electrons are confined to move in a single plane. This two-dimensional electron gas (2DEG) combines several required properties not shared by thin films. It has a low electron density, which may be readily varied by employing an electric field. The low density implies a large Fermi wavelength (typically 40 nm), comparable to the dimensions of the smallest structures (nanostructures) that can be fabricated today. Quantum transport is conveniently studied in a 2DEG because of the combination of a large Fermi wavelength and a large mean-free path. Quantum interference becomes more important as the dimensionality of the conductor is decreased [12]. The quantum confinement effects in low dimensional semiconductor systems were studied two decades ago with the stress on the optical properties, including absorption and luminescence [13, 14]. The confinement of an electron and a hole in nanocrystals significantly depends on the material properties, namely the exciton Bohr radius (a_B). For most of the commonly studied semiconductor nanostructures, such as ZnO, ZnS, CdS, CdSe, ZnSe, Cu_2O, SnO_2, TiO_2, Cu_2S, exciton Bohr radius values are less than 10 nm.

This chapter focuses on the basic understanding of semiconductor nanostructures and nanomaterial synthesis techniques. The target is to present upcoming and potentially leading nanomaterials and structures, highlighting various stages of their applications and synthetic methods. Most semiconducting materials, such as the II-VI or III-V compound semiconductors, show quantum confinement behaviour in the 1-20 nm size range. Size reduction affects most of the physical properties (structural, magnetic, optical, dielectric, thermal, *etc.*) due to surface effects, enhanced surface-to-volume ratio, and quantum size effects. Owing to the extremely small dimensions, these materials exhibit properties, which are fundamentally different from, and often superior to those of their conventional bulk counterparts. Optical spectroscopy, being the non-contact method, has proven to be the most suitable technique to monitor the size evolution of the electronic structure [15 - 18]. Due to morphology dependent properties of semiconductor nanostructures, it becomes much more important to understand different nanostructure morphologies and their synthesis.

SEMICONDUCTOR NANOSTRUCTURES

Electronic nanostructures include 2D materials, nanowires, and quantum-confined heterostructures, which reveal fascinating properties from traditional quantum transport to correlated effects, including spintronics.

The fabrication of nanostructures for quantum information is a flourishing field that looks to control electrons' degrees of freedom by local and wide-range interactions. In many cases of both optical and electronic nanostructures, surfaces and interfaces and their control play a significant role in deciding the nature of the properties.

Recently, there has been significant interest in the construction, characterization, and implementation of semiconductor nanoparticles that play a substantial role in numerous novel technologies. The conductivity of the semiconductor and its optical properties (absorption coefficient and refractive index) can be manipulated. Semiconductor nanomaterials and devices are still in the research stage, but they are promising for applications in many fields, such as solar cells, nanoscale electronic devices, light-emitting diodes, laser technology, waveguide, chemical and biosensors, packaging films, parts of automobiles, and catalysts.

Further development of nanotechnology will certainly result in remarkable milestones in the semiconductor industry, such as many kinds of diodes, including the light-emitting diode, the silicon-controlled rectifier, and digital and analog integrated circuits. Some of the semiconductor nanomaterials, such as Si, Si-Ge, GaAs, AlGaAs, InP, InGaAs, GaN, AlGaN, SiC, ZnS, ZnSe, AlInGaP, CdSe, CdS, and HgCdTe, *etc.*, exhibit excellent applications in computers, laptops, cell phones, pagers, CD players, TV remotes, mobile terminals, satellite dishes, fiber networks, traffic signals, car tail lights, and airbags [15, 16].

Classifications of Semiconductor Nanostructures

Semiconductor nanocrystals (NCs) are made from a variety of different compounds. They are referred to as II-VI, III-V or IV-VI semiconductor nanocrystals based on the periodic table groups from which these are formed. For example, silicon and germanium are group IV elements/semiconductors, GaN, GaP, GaAs, InP and InAs are III-V, while those of ZnO, ZnS, CdS, CdSe and CdTe are II-VI semiconductors. In nanocrystalline materials, the electrons are confined to regions having one, two or three dimensions when the relative dimension is comparable with the de Broglie wavelength. For a semiconductor like CdSe, the de Broglie wavelength of a free electron is around 10 nm. In other words, one can say that the quantum confinement effect comes into play when at least one, two or all three dimensions of a semiconductor nanostructure become comparable to the exciton Bohr radius of the semiconductor material.

Heterostructures are semiconductor structures in which chemical composition changes with position [1]. Altering the composition gives spatially varying semiconductor properties. The change in composition performed during the growing process takes place in one dimension, producing homogeneous "layers" of semiconductors in the other two dimensions. This is a so-called quantum well. The restriction on the movement of the electron into this plane affects the energy of the electron. Quantization effects will result in allowed energy bands, whose energy positions are dependent on the height and width of the barrier and can be calculated by solving the one-dimensional Schrodinger equation.

Nanomaterials are often categorized as to how many of their dimensions include in the nanoscale. A nanoparticle is defined as a nano-object with all three external dimensions on the nanoscale, whose longest and shortest axes do not differ significantly. A low-dimensional system is one where the motion of microscopic degrees of freedom, such as electrons, phonons, or photons, is restricted from exploring the full three dimensions. There has been tremendous interest in low-dimensional quantum systems during the past twenty years, fuelled by a constant stream of striking discoveries and also by the potential for and realization of new state-of-the-art electronic device architectures.

Three-dimensional Nanostructures

Some bulk materials contain features on the nanoscale, including nanocomposites, nanocrystalline materials, nanostructured films, nanoporous membranes and nanotextured surfaces [19]. Box-shaped graphene (BSG) nanostructure is an example of a 3D nanomaterial [20]. BSG nanostructure has appeared after the mechanical cleavage of pyrolytic graphite.

Semiconductor nanomaterials with at least one dimension less than 100 nm but larger than the exciton Bohr radius are named 3D nanostructures.

Two-dimensional Nanostructures

The nanostructures of semiconductor crystals having the z-direction below the critical value (critical size value is exciton Bohr radius of the material) are defined as 2D nanostructures (quantum well). 2D materials are crystalline materials consisting of a two-dimensional layer of atoms. Thin films with nanoscale thicknesses are considered nanostructures; however, sometimes, they are not considered nanomaterials because they do not exist separately from the substrate [19]. Every quantum well is a nanofilm, but every nanofilm cannot be named a quantum well.

One-dimensional Nanostructures

When the dimensions both in the x and z direction are below a critical value (critical size value is exciton Bohr radius of the material), the nanostructures are defined as 1D (quantum wire, linear chain structure). The smallest possible crystalline wires with a cross-section as small as a single atom can be engineered in cylindrical confinement [16-18]. Carbon nanotubes, a natural semi-1D nanostructure, can be used as a template for synthesis. Confinement provides mechanical stabilization and prevents linear atomic chains from disintegration; other structures of 1D nanowires are predicted to be mechanically stable even upon isolation from the templates [17, 18].

Semiconductor quantum wires are produced by, for example, micro-structuring the quantum wells or growing the wells at an edge or a groove. These systems involve confinement and consequently, quantization of carriers in two dimensions. Such confinement allows the free-electron behavior in one dimension only. The energy dispersion for a quantum wire is given by:

$$E_{n,k_x,k_y} = \frac{\hbar^2 \pi^2 n^2}{8m^* d^2} + \frac{\hbar^2}{2m^*}(k_x^2 + k_y^2) \ n = 1, 2 \tag{1}$$

where d is the width of the quantum well, n is the quantum number corresponding to the confinement along the z-direction, and k_x and k_y are the wave vectors along the x and y directions, respectively.

Zero-dimensional Nanostructures

When the y-direction is also below the threshold, it means that all three dimensions are below the critical size value; the resulting structures are referred to as 0D (quantum dot).

Quantum dots (QDs) are characterized by the confinement and quantization of carriers in the three dimensions [1]. These can be produced by several techniques, from the same procedure used to develop quantum wells to chemical processes. The different techniques lead to the production of quantum dots with different shapes and sizes. Let us now discuss further the energy and DOS for the quantum dots before turning to the different manufacturing processes.

In this case, confinement takes place in all directions, the energy spectrum is quantized, and there are three quantum numbers n_x, n_y, n_z, each associated with one spatial direction. The density of states (DOS) of a discrete spectrum is merely a number of delta functions at each energy level En_x, n_y, n_z. The DOS for QDs are even sharper than those for other cases, making them especially suitable for photonic applications. The self-assembled quantum dots are naturally fabricated under special conditions of the molecular beam epitaxy (MBE) growing process. If there is a major lattice mismatch between the substrate and the material that is being grown, islands nucleate at the interface, forming quantum dots [21]. This type of dot is more suitable for optoelectronic applications, like lasers [22]. These are being used in commercial products like blue lasers for data reading. Laterally confined quantum dots can be produced by the electrical confinement of a 2D electron gas (formed in the interface of a QW, for example) [23]. For instance, as these dots can be easily integrated into electrical circuits, they can be used in single-electron transistors [24]. Quantum dots can also be produced by colloidal synthesis [25]. These dots are smaller than the others, and they can be attached to proteins or part of the DNA and be used in medical diagnosis applications for the detection of tumors and other medical conditions [26].

Relatively new and sophisticated material growing techniques allow the manufacturing of low-dimensional systems based on semiconductors nanostructures with great accuracy [1]. A few years ago, the molecular beam epitaxy technique was restricted to research institutes; today, it is widely used in large-scale manufacturing of semiconductor-based devices. Using advanced techniques, such as MBE and metal-organic chemical vapor deposition (MOCVD), it is possible to produce a semiconductor structure with a precision of a monoatomic layer and develop devices that constrain the flow of the carriers to low dimensions. These devices, depending on the material and intrinsic properties related to carrier confinement, have many distinct technological applications, especially in optoelectronics.

NANOMATERIAL SYNTHESIS APPROACHES

The diverse methods which are being used to fabricate various nanostructure morphologies are categorized under two approaches; the Top-down approach and the Bottom-up approach. Top-down synthesis methods give appealing ways to approach the nanoscale by starting with the bulk scale materials and then slicing or cutting them to the nanometer-level dimensions. Physical breaking of the bulk material through high-energy processes is involved in these strategies. The widely used top-down approaches to synthesize the nanoparticles include mechanical milling, lithography, electrospinning, laser ablation *etc*. Whereas the Bottom-up approach involves building material from the bottom: atom by atom, molecule by molecule, or cluster by cluster. Bottom-up approaches are also classified into two broad groups, wet chemical methods and gas-phase methods, depending on the medium *via* which nanostructures are formed. Chemical and physical vapor deposition, sol-gel method, solvothermal/hydrothermal methods, *etc*., are the commonly used bottom-up synthesis methods. The schematic flow chart of the duo is shown in (Fig. **1**). Both the top-down and bottom-up approaches have advantages and disadvantages in different aspects. The top-down approach utilizing lithography and etching techniques can be advantageously used to create required nanostructures in a spatially controlled way, which is an important property for the interconnection and integration of nanomaterials into circuit elements and to design other specific applications, while the bottom-up approach is a very powerful technique for the synthesis of monodisperse nanostructures with atomic precision. This precise synthesis is significant for applications that need well-defined nanostructures. The bottom-up approach provides ease of doping; structures with lesser defects and good control over morphology can be easily fabricated under ambient conditions *via* eco-friendly synthesis methods. In addition, the machinery and the costs of both approaches are considerably different. For the top-down techniques, expensive machinery and careful maintenance are generally required. While the bottom-up techniques involve the reactions to be carried out generally in a test tube, and the cost of reagents is a lot cheaper compared to the costs of machinery used in the top-down approach.

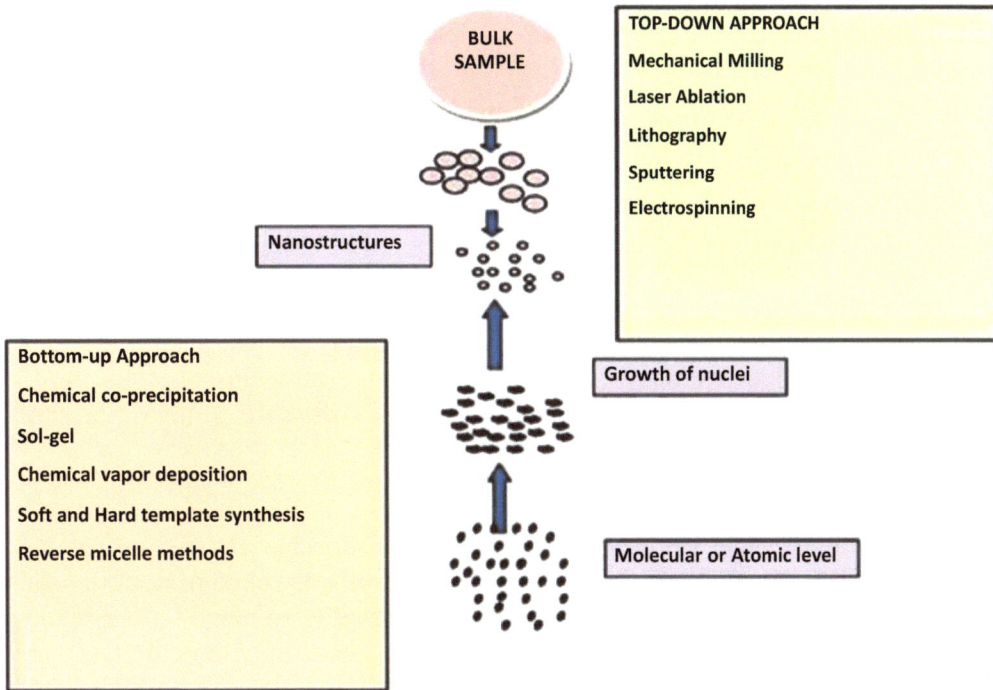

Fig. (1). Schematic flow chart of top-down and bottom-up approaches for the synthesis of nanomaterials.

Commonly used Synthesis Techniques

Mechanical Milling

An appropriate procedure to grind a bulk material into nano-dimensions is mechanical milling. It is a simple and low-cost process and therefore has found huge suitability, especially in industrial nanomaterial preparation environments. In this technique, milling provides a great force to grind materials to the nanoscale. The most commonly used attrition devices for this purpose are a shaker mill, attrition mill, and planetary ball mill [27].

In the shaker mill, the bulk sample to be milled is charged into a vial with spherical balls called "milling balls" made up of hard material. Denser materials (steel or tungsten carbide) are advantageous, as the kinetic energy of the milling balls is a function of their mass and velocity, resulting in the optimization of the size and size distribution of the desired product for the given mill [28]. Afterward, the bulk material is kept in the shaker, and back-and-forth motion starts energetically for numerous thousand cycles per minute. Milling balls collide with each other and with the vial wall throughout this shaking process. During this process, large impact forces are produced, which, in turn, grind the solid bulk material down and thoroughly mix it [29]. The packing of balls should not be too dense because it reduces the mean free path of the ball motion; rather, the packing should be diluted to the distribution, which diminishes the frequency of collisions [30].

The planetary ball mill's name is coined due to the vial motion in the device. In this process, the vials are kept in a disk that rotates about its axis. Each vial rotates about an axis of its own in the opposite direction to the rotating disk. The entire system rotates at many thousands of rpm and hence resulting centrifugal and acting acceleration forces lead to strong grinding effects. Furthermore, the intensive impact and frictional forces grind the material to low dimensions [31]. Planetary Ball Mills are used for fine grinding of soft, hard to brittle or fibrous materials [29].

Attrition milling is a simple and effective method of milling. It is similar to a ball mill in which spherical grinding balls loaded to a horizontal tank are rotated to perform the milling action. However, in attrition mills, the feed material is placed in a stationary tank with the grinding media. The tank is attached with a sequence of cautiously positioned impellers within the mill. These impellers are fixed at the right angle to each other. In contrast to the ball mill, the attrition mill is rotated at high speeds with the impellers, which results in very high shear and impact forces. For the most effective grinding action, these competent forces must be present, but it is not possible to obtain them with conventional ball mills. Thus, size reduction, as well as homogenous particle dispersion with very little wear on the tank walls, is achieved [32 - 35].

Lithography

The word "lithography" is a combination of two Greek words, *lithos* (meaning stone) and *graphein* (meaning to write) [36, 37]. It is a method of printing originally based on the immiscibility of oil and water [38]. It refers to the process invented in 1796 by Alloys Senefelder, where patterns of desired designs were transferred onto a base substrate, mostly using masks [39 - 41]. Lithography originally used an image drawn with oil, fat, or wax onto the surface of a smooth, level lithographic limestone plate. The stone was then treated with a mixture of acid and gum arabic, etching the portions of the stone that were not protected by the grease-based image. This traditional technique is still used in some fine art printmaking applications. In modern lithography, the image is made of a polymer coating applied to a flexible plastic or metal plate [42]. The image can be printed directly from the plate (the orientation of the image is reversed), or it can be offset by transferring the image onto a flexible sheet (rubber) for printing and publication.Sophisticated lithographic techniques can be broadly classified into two categories as microlithography and nanolithography. Microlithography and nanolithography refer specifically to lithographic patterning methods resulting in structuring material on a fine scale [36]. Typically, features smaller than 10 micrometers are considered under microlithographic patterns, whereas features smaller than 100 nanometers are considered nanolithographic features. Photolithography is one of these methods, often applied to semiconductor device fabrication. Photolithography is also commonly used for fabricating microelectromechanical systems (MEMSs).

The microelectronics industry is quite advanced in the mass manufacturing of miniaturized devices, such as integrated circuits or [41] nanoelectromechanical systems (NEMSs). For this purpose, the industry always looks forward to developing advanced lithography procedures. Photolithography is one of the well-known lithographic methods used in the semiconductor industry. Nowadays, lithography has many types, *i.e.*, nanolithography using proximal probe nanolithography, including STM- and AFM-based nanolithographic methods. A variety of other lithography procedures, such as nanoimprint (mold) lithography, plasmonic-assisted lithography, electron beam lithography, laser interference lithography, nanosphere lithography, chemistry-based nanofabrication, local electro etching, dip pen lithograph with AFM are also used.

Photolithography

The term "photolithography" refers to the use of photographic images in lithographic printing, whether these images are printed directly from a stone or a metal plate, as in offset printing. "Photolithography" is used synonymously with "offset printing". Since the 1960s, photolithography has played an important role in the fabrication and mass production of integrated circuits in the microelectronics industry [43 - 45].

Photolithography generally uses a pre-fabricated photomask or reticle as a master from which the final pattern is derived. Basically, photolithography is a photon-based technique comprised of projecting, or shadow casting, an image into a photosensitive emulsion (photoresist) coated onto the substrate of choice. The general sequence of steps for a typical optical lithography process is as follows: substrate preparation, photoresist spin coating, prebake, photon exposure, post-exposure bake, development, and post-bake [46, 47]. Resist stripping is the final operation in the lithographic process after the resist pattern has been transferred into the underlying layer *via* etching or ion implantation. This sequence is generally performed on several tools linked together into a contiguous unit called a lithographic cluster. Photolithography finds numerous applications in nanoelectronics, metrologies, and single-molecule biology. Light diffraction sets a fundamental limit on optical resolution, posing a critical challenge to the down-scaling of nano-scale manufacturing.

Electron Beam Lithography

Although photolithographic technology is the most employed form of nanolithography, more techniques for nanoscale precise miniaturization are explored. Electron beam lithography results in much greater patterning resolution (sometimes as small as a few nanometers). Electron beam lithography is also important commercially, mainly for its use in the manufacture of photomasks. Electron beam lithography is a usually practiced form of maskless lithography, *i.e.*, a mask is not required to generate the final pattern in electron lithography [45]. Instead, the final pattern is created directly from a digital representation on a computer by controlling an electron beam as it scans across a resist-coated substrate. The main drawback of electron beam lithography is that it is much slower than photolithography.

In addition to the above-mentioned commercially well-established techniques, a large number of promising microlithographic and nanolithographic technologies exist or are being developed, including nanoimprint lithography, interference lithography, X-ray litho graphy, extreme ultra violet litho graphy, magneto-lithography and scanning probe lithography. Some of these new techniques have been used successfully for small-scale commercial and important research applications. Surface-charge lithography, in fact, plasma desorption mass spectrometry, can be directly patterned on polar dielectric crystals *via* the pyroelectric effect [46, 47].

Electrospinning

Electrospinning, a word derived from "electrostatic spinning", is a process for producing ultrathin fibers with diameters between 100 nm and a few microns. This process is governed by the electrohydrodynamic phenomena, where electric force is used to draw charged threads from polymer solutions or polymer melts. The basic setup for this technique consists of a polymer fluid contained in a positively charged capillary (typically a syringe) and tipped with a blunt needle (for needle-based electrospraying), a pump, a high voltage power supply and a grounded collector. This method uses a high-voltage electric field applied to polymer fluid derived through the capillary. Fig. (**2**) shows a schematic representation of the electrospinning method. When the electrostatic repulsion is higher than the surface tension, the liquid meniscus is deformed, and a liquid polymer jet is formed and elongated into a conically shaped structure known as the Taylor cone, as shown in Fig. (**2**). Once the Taylor cone is formed, the charged liquid jet is ejected towards the grounded collector, which serves as an electrode and deposits on it [48, 49].

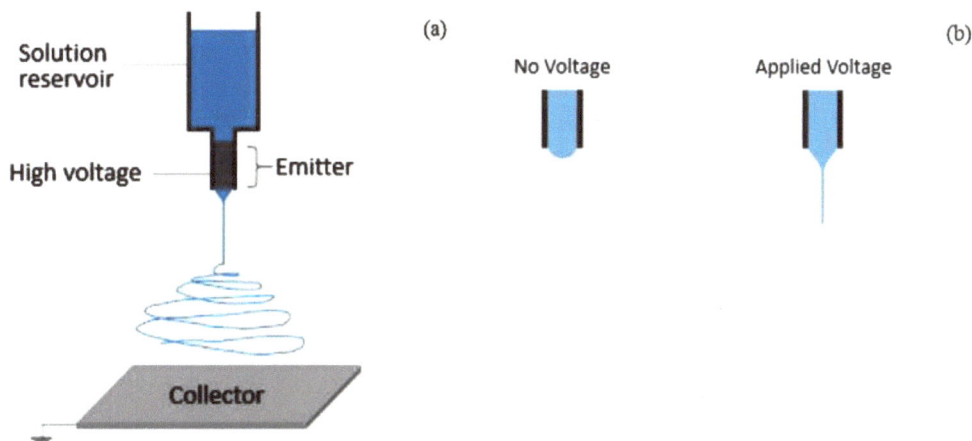

Fig. (2). (a) Typical electrospinning setup **(b)** Diagram of a Taylor cone with no voltage/ applied voltage [*Reprinted with permission from ref. Nanoscience Instruments. 10008 S. 51ˢᵗ Street, Ste 110 Phoenix, AZ 85044 USA*].

There are following two ways of nanofiber fabrication through the electrospinning technique:

a. Needle-less

b. Needle-based.

In needle-less electrospinning, a stationary or rotating platform is used to generate fibers. The initiating polymer solution is transferred to an open vessel. One of the major benefits of the needle-less electrospinning process is the mass production of materials, but there are many disadvantages. The morphology of fibers and their quality cannot be specifically controlled; the raw materials that can be employed are limited, hence reducing the versatility of the process, and process parameters like flow rate cannot be controlled.

On the other hand, in needle-based electrospinning, to reduce and prevent solvent evaporation, the polymer solution is usually contained in an air-tight closed reservoir. Due to this difference, a wide variety of materials, including highly volatile solvents, can be processed effortlessly. Needle-based electrospinning has many advantages, including flexibility to develop different architectures like core-shell and multi-axial fibres. This particular advantage of needle-based electrospinning is beneficial to incorporating active pharmaceutical ingredients (API) within a fibre. Other advantages of needle-based electrospinning are a firmly controlled flow rate, minimized solution waste and a number of jets [50, 51]. One of the major advancements in electrospinning is coaxial electrospinning. The spinneret comprises two coaxial capillaries in coaxial electrospinning. Two viscous liquids, or a viscous liquid as the shell and a non-viscous liquid as the core, can be used to form core-shell nanostructures in these capillaries under the electric field. Core-shell ultra-thin fibres on a large scale can be produced by coaxial electrospinning. The lengths of these ultrathin nanomaterials can be extended to several centimeters. This method has been used for the development of core-shell and hollow polymer, inorganic, organic, and hybrid materials [34, 52].

Laser Ablation

The laser ablation technique generates the nanostructures using a powerful laser beam that hits a solid target material (or occasionally liquid). These adequate energy short pulses result in the vaporization of the target material, which condenses as nano-dimensional material on the substrate. This method is primarily used for preparing nanostructures of noble metals. However, it can also be used to generate a wide range of nanomaterials, such as metal nanoparticles [53], carbon nanostructures [54, 55], oxide composites [56], and ceramics of other metals [57]. The utilization of short pulses in this technique allows the ablation process to be carried out in different mediums, including very volatile solvents to highly reactive monomers. Another unique advantage of this method is the ability to make nanoparticles with very high purity since no by-products or residual chemicals are produced in the process. Pulsed laser ablation in liquids is a promising approach for producing monodisperse colloidal nanoparticle solutions without using surfactants or ligands.

The typical setup generally consists of a pulsed laser, a set of focusing optics and a container containing a metal target. The metal target is placed close to the focus of the laser beam. The average size and distribution of the nanoparticles can be controlled *via* pulse duration, wavelength and the intensity of the laser pulses. Typically, the laser pulses are of extremely short pulse width generally of the order of femtoseconds (10^{-15} s), picoseconds (10^{-12}s) or nanoseconds (10^{-9}s) which are preferred for nanoscale precision, but the pulses up to milliseconds time duration can also be utilized for photoablation. Generally, deep UV, visible and IR lasers such as excimer laser, Nd: YAG laser, ruby laser, and CO_2 laser are used for laser ablation processes [58]. The distinctive advantages of laser ablation techniques over other deposition techniques are the capability of producing multicomponent materials with well-controlled stoichiometry. The laser ablation method is basically an evaporation-based deposition technique, which is a special type of physical vapour deposition (PVD). Details of PVD are discussed in the vapour deposition techniques section.

Template Methods

Template methods involve the fabrication of the desired material within the pores or channels of a nonporous template. Track-etched membranes, porous alumina, and other nonporous structures have been characterized as templates. Soft templating ("endotemplating") and hard templating or nanocasting ("exotemplating") are two categories of templating approaches for the synthesis of ordered mesoporous materials together with several transition and complex metal oxides [59]. Basically, the following three successive steps are involved in template syntheses of nanomaterials: template preparation, template-directed synthesis of the target materials (using sol-gel, precipitation, hydrothermal synthesis, and so on) and then removal of the template. To produce nanoporous materials, soft and hard template methods can be classified roughly into deposition onto colloidal suspensions and emulsions, respectively. In the deposition, hard templates, such as polystyrene nanoparticles, are used as the core for a target material deposition, whereas in the second, emulsions are a clear-cut method, whereby material is deposited onto micelles or phases within a solution. The fundamental principle for emulsion systems is the phase separation of two media from which preferential micelles or nanostructures will form. Nanostructures will be formed when the material gets deposited on the micelles (oil or water) of one phase. As a result, colloidal solutions can act as emulsion templates, such as those comprised of polystyrene latex, a common template for this method [60, 61].

The soft templating method depends on the combined self-assembly of the surfactant and the precursor to form a mesoporous structure. The process is centered on the interactions between inorganics and surfactants, which assemble into inorganic-organic mesostructured composites [62 - 67].

Hydrogen bonding, van der Waals forces and electrostatic forces are the notable interactions between the soft templates and the precursors [68]. Finally, after the removal of the pore-templating surfactant by low-temperature calcination (up to 600 °C) or by different washing techniques, the structure of the desired mesoporous product is obtained. The soft template method is a simple conventional method for the generation of nanostructured materials. The soft templating method is an advantageous technique because of the simplicity of the one-step synthetic procedure for obtaining mesostructured materials, relatively mild experimental conditions, and simple removal of the templating surfactant as compared to the more complex hard templating methods, where removal of silica mold is required [61, 69 - 71]. This method is a suitable technique for the creation of a variety of morphologies of the materials [72]. In the soft templating method, different kinds of soft templates, such as block copolymers, flexible organic molecules, and anionic, cationic, and non-ionic surfactants, can be employed to develop nanoporous materials [73]. Numerous factors can affect mesoporous material structures generated through this method, such as the surfactant and precursor concentrations, surfactant-to-precursor ratio, surfactant structure, and environmental conditions [72].

In the hard template method, solid materials are well-designed to utilize as templates for specific applications [68]. The solid template pores are filled with precursor molecules to obtain nanostructures. The selection of a hard template plays a significant role in developing well-ordered mesoporous nanomaterials. Therefore, one should keep in mind that the hard templates should maintain a mesoporous structure during the precursor conversion process and be easily detachable without disrupting the obtained nanostructure. A series of materials is the choice as hard templates, such as carbon black, silica, carbon nanotubes, colloidal crystals, porous polycarbonate membranes, nanoporous anodic alumina membranes and wood shells [74].

To achieve the well-ordered mesoporous replica structure by the nanocasting, the following points must be taken into consideration:

i. A precursor material has to be preferred that does not chemically react with the template (typically n-hydrates of metal salts, such as nitrates and chlorides);
ii. To avoid the accumulation of the precursor, strong interactions between the precursor and the template should be avoided. Since accumulation near the pore entrance can lead to the blockage of pores.
iii. The solvent choice can help in pore filling; for instance, ethanol is a polar solvent and exhibits the amphiphilic property, which is suitable to the silica pore wall surface, which results in the increase of capillary force and hence eases the diffusion of precursor across the pores.
iv. Moderate filling of the pore is necessary since a very high or extremely low percentage of pore filling can lead to the structural collapse of metal oxide after the removal of the template. For proper cross-linking of the material inside the pore system of the template, an adequate amount of the precursor is needed (usually pores filling percentage: 5 - 15 %).

It is clear that the infiltration of the precursor into the porous template matrix is the most significant factor. Therefore, wet impregnation, incipient wetness, and dual-solvent or solid-liquid methods are usually used. The soft templating synthesis process relies on sol-gel chemistry in which the final product is dependent on the laboratory conditions (temperature, pH, humidity) [75 - 79] and hence possesses a significant disadvantage because of its poor control over the condensation reaction. In the sol-gel process, the limited temperature range during synthesis favors low crystallinity resulting in an amorphous or semi-crystalline phase of the synthesized product. The presence of amorphous phases can have unwanted effects on the specific application of the synthesized mesostructured product. For instance, easing the recombination of electrons and holes decreases the catalytic efficiency and/or reduces the conductivity of the material; in this way, the electronic applications of the material get restricted. Still, the amorphous phase can be eradicated during the process of calcination, and the mesoporous structure cannot be well-preserved after the high-temperature thermal treatment. Contrary to the soft templating method, the hard template method has many advantages, such as controllability, pore regularity, and crystallinity [80 - 82]. A variety of types of mesoporous MOXs have been successfully prepared using the nano-casting method [80, 81, 83], and a lot of them have been developed using mesoporous silica templates [84 - 90].

Chemical Precipitation Method

Chemical precipitation (wet chemical synthesis) is a simple eco-friendly facile bottom-up synthesis technique. Generally, in this method, precursor solutions are mixed stoichiometrically under ambient conditions; the rapid reactions among the precursors cause the formation of nanostructures *via* precipitation. The products of precipitation reactions are generally moderately soluble species formed under conditions of high supersaturation. When precipitation begins, numerous small crystallites initially form (nucleation), but they tend to quickly aggregate together to form larger, more thermodynamically stable particles (growth). Thus, the processes of nucleation and growth govern the particle size and morphology of products in precipitation reactions. As compared to all the above-referred methods, this method has been found to have a number of advantages, including high yield, easy processability at ambient conditions, the possibility of doping of different kinds of impurities with high doping concentrations even at room temperature, good control over the chemistry of co-doping particularly when different impurities are incorporated simultaneously in the host lattice, easiness of surface capping with variety of reagents (organic as well as inorganic) *etc.*

Sol-Gel Method

The sol-gel method is a wet chemical route that is promisingly used for the fabrication of nanomaterials. A variety of high-quality metal-oxide-based nanomaterials can be developed through this method. It is a facile synthesis procedure for the metal-oxide nanoparticles; it consists of the development of inorganic networks through the formation of a colloidal suspension (sol) and gelation of the sol to form a network in a continuous liquid phase (gel) [91]. The sol-gel technique presents a low-temperature method for synthesizing materials that are either inorganic in nature or both inorganic and organic. It is based on the hydrolysis and condensation reactions of organometallic compounds in alcoholic solutions. This method is advantageous for the fabrication of coatings along with an outstanding control on the stoichiometry of precursor solutions, easiness of compositional modifications, customizable microstructure/nanostructure, simplicity of introducing various functional groups or encapsulating sensing elements, relatively low annealing temperatures; the possibility of coating deposition on large area substrates and above all it needs only simple and inexpensive processing equipment. Metal alkoxides are the conventional precursors for the generation of nanomaterials using this method. The synthesis process of nanomaterials *via* the sol-gel method can be completed in a series of distinct steps [92 - 95]:

Step 1: Hydrolysis of the metal oxides (formation of the sol), *i.e.*, the stable solutions of the alkoxide or solvated metal precursor.

Step 2: Condensation; gelation resulting from the formation of an oxide- or alcohol-bridged network (the gel) through a polyesterification or polycondensation reaction that results in a remarkable increase in the viscosity of the solution.

Step 3: Aging of the gel, for the duration in which the polycondensation reactions continue until the gel transforms into a solid mass, together with contraction of the gel network and expulsion of solvent from the gel pores. The aging process of gels can exceed a week or so.

Step 4: Drying of the gel, when water and other volatile liquids are removed from the gel network. The drying process set hurdles due to fundamental changes in the structure of the gel.

Step 5: Dehydration, during which surface-bound M-OH groups are removed, thus stabilizing the gel against rehydration. This is usually achieved by calcining the sample at temperatures up to 800°C.

Step 6: Densification and decomposition of the gels at high temperatures (T > 800°C). The pores of the gel network collapse and the remaining organic species are volatilized. This step is normally reserved for the preparation of dense ceramics or glasses.

Solvothermal and Hydrothermal Methods

Hydrothermal and solvothermal syntheses are branches of inorganic synthesis. The term "hydrothermal" is of geologic origin [96]. It is believed that Schafhäutl described the first results of a laboratory-based hydrothermal reaction in 1845 [97]. He observed the formation of quartz microcrystals from silicic microcrystals. Hydrothermal synthesis refers to the synthesis through heterogeneous chemical reactions in aqueous solution above the boiling point of water, while solvothermal synthesis takes place in a non-aqueous solution at relatively high temperatures. On the other hand, the reactions are referred to as alcohothermal and glycothermal, respectively, when alcohols and glycerol are used as the reaction media. These synthesis strategies are significant for the fabrication of nanocrystals with good crystalline properties. Both synthesis methods are performed in an autoclave, which can bear high temperatures and pressures [98]. Hydrothermal and solvothermal methods are promising methods for the creation of various nano-geometries of materials, such as nanowires, nanorods, nanosheets, and nanospheres [99 - 101]. Usually, the course of action involves loading precursors and other reagents, including the solvent, in proper ratios into an autoclave, where the solution is heated at a set temperature for a fixed time period. The sample is washed with water and alcohol to get rid of impurities, followed by vacuum drying prior to the final product. This method is advantageous as almost any material can be dissolved in the solvent by increasing the temperature and pressure to its critical point [102]. The microwave-assisted hydrothermal synthesis of nanostructures has proven significant for engineering nanomaterials due to the incorporation of the merits of both hydrothermal and microwave methods [103].

Vapour Deposition Techniques

Vapour deposition techniques can be categorized as physical vapour deposition (PVD), chemical vapour deposition (CVD), aerosol-based processes, and flame-assisted deposition methods. Vapour deposition techniques are the most broadly used nanofabrication approaches in nanotechnology research and industry.

The physical vapour deposition process involves the creation of vapor phase species through the following processes: evaporation, sputtering, and ion plating. During transportation in the PVD chamber towards the substrate, the vapour phase species undergo collisions and ionization and subsequently condense onto a substrate where the formation of thin films/nanomaterials occurs *via* nucleation and growth processes. There are three basic physical vapor deposition techniques: (a) evaporation, (b) sputtering, and (c) ion plating, as classified based on various ways of generating the gaseous species. In PVD, the vaporization of material through heating, electron beam, ion beam, plasma, or laser is followed by solidification of the material back at the surface of the substrate that needs to be coated. To enhance the effectivity of the coating, PVD is generally carried out in a high vacuum environment so that interaction with other atoms can be minimized. This technique also has applications in metal coating, glass and semiconductor industries. The laser ablation technique already discussed in this chapter is also a PVD process.

One of the PVD methods, *i.e.*, sputtering, has been discussed for a comprehensive understanding of the evaporation process. Sputtering is a PVD vacuum process. It is used to deposit very thin films onto a substrate for a broad range of commercial and scientific uses. Sputtering involves the deposition of nanomaterials *via* the bombardment of solid surfaces with high-energy particles like plasma or gas. An ionized gas molecule is used to displace small atomic clusters of a particular material depending upon the incident gaseous-ion energy. These atoms then bond at the atomic level to a substrate and create a thin film. Sputtering is considered to be an effective method for producing thin films of nanomaterials [104 - 106]. Several types of sputtering processes exist, such as ion beam, diode, and magnetron sputtering [106]. Generally, sputtering is performed in an evacuated chamber, to which the sputtering gas is introduced. A high voltage is applied to the cathode target in order to create high-energy plasma, and free electrons collide with the gas to produce gas ions. The electric field then accelerates these positive ions towards the cathode target; these ions continuously hit the cathode target, resulting in the ejection of atoms from the surface of the target [107]. The sputtering technique is very interesting as it is cost-effective compared with electron-beam lithography and the sputtered nanomaterial composition remains similar to the target material with fewer impurities [108, 109].

The chemical vapor deposition process is based on the thermal dissociation and/or chemical reaction of the gaseous reactants on or near a heated surface to form stable solid products. A CVD coating is a material applied to a surface by a method called chemical vapor deposition. Chemical vapor deposition methods are significantly used in the generation of carbon-based nanomaterials, semiconductors, and ceramics. The CVD process carries out the reaction of a volatile precursor, which is injected into a chamber (usually under a vacuum). The coating chamber is filled with precursor gases that can undergo reactions or break down into the preferred coating, which in turn bonds to the material surface (substrate) that needs to be coated. Suitability of precursor for CVD is characterized by ample volatility, high chemical purity, fine stability during evaporation, low cost, non-hazardous nature, and a long shelf-life. Besides these factors, the decomposition of the precursor should not result in leftover impurities [110, 111]. The choice of catalyst plays a significant role in the morphology and quality of the nanomaterial obtained. In the CVD-based preparation of graphene, Ni and Co catalysts give multilayer graphene, while a Cu catalyst offers monolayer graphene [112]. CVD is extended by exploiting several variants like different heating methods *e.g.*, thermally activated, photothermal, plasma-assisted, *etc.*), the type of precursor used or the choice of catalyst for specific applications, such as spray-assisted vapor deposition, aerosol-assisted chemical vapor deposition, metal-organic chemical vapor deposition, *etc.*

Aerosol-based processing techniques involve the atomization of a liquid precursor into finely divided submicrometer liquid droplets (aerosol) distributed throughout gas medium. Aerosol-based processing techniques can be subdivided into spray pyrolysis, electrostatic-assisted vapor deposition, *etc.*, based on the different aerosol generation methods used.

By and large, CVD is recognized for the creation of two-dimensional nanomaterials [113] and is an outstanding method for producing high-quality nanostructures [114, 115]. In the same way, electrospraying-assisted deposition is classified as corona spray pyrolysis, electrostatic spray pyrolysis/electrostatic spray deposition, electrospray organometallic chemical vapor deposition, gas-aerosol reactive electrostatic deposition, electrostatic spray-assisted vapor deposition, *etc.*

Flame-assisted deposition (also known as flame synthesis or combustion flame synthesis) involves the combustion of liquid or gaseous precursors injected into diffused or pre-mixed flames. The flame-assisted deposition can be grouped into various categories as counterflow diffusion flame synthesis, combustion flame chemical vapor condensation, and sodium/halide flame deposition with *in situ* encapsulation process [116].

Among all the vapour processing techniques, molecular beam epitaxy (MBE) and metal-organic chemical vapour deposition (MOCVD) are two prominent methods that are used for the fabrication of high-quality, defect-free, highly monodisperse nanostructures for optoelectronic device applications. Molecular beam epitaxy is a refined form of vacuum evaporation. It was first developed for the controlled growth of III-V semiconductor epitaxial layers by Arthur [117] and Cho and Cheng [118]. This method involves neutral thermal energy molecular or atomic beams (Ga, Al, *etc.*) impinging on a hot crystalline substrate maintained in an ultrahigh vacuum environment that provides the required source of growth. On the other hand, MOCVD is the CVD process that uses metal-organic gas or liquid precursor systems.

CONCLUSION

Semiconductor nanostructures show remarkable properties at nano length scale, making them model systems for researchers and technocrats. Due to these amazing properties, semiconductor nanostructures find various applications in electronic devices, optoelectronics, sensors, catalysis and telecommunication. Semiconductor nanomaterials and devices are still in the exploration phase, but they are promising for applications in many fields. In a low-dimensional system, degrees of freedom are restricted to less than three, leading to astonishing revelations. The present era challenge is to devise facile eco-friendly synthesis techniques for the fabrication of highly monodisperse nanostructures for device applications. However, both the top-down and bottom-up synthesis approaches have pros and cons in different aspects. The top-down approach utilizing lithography and etching techniques can advantageously be used to create required nanostructures in a spatially controlled way, while the bottom-up approach seems a powerful tool for the synthesis of defect-free, highly pure monodisperse nanostructures with atomic precision. For futuristic research explorations, the synthesis of highly monodisperse nanostructures requires more attention to modify the existing synthesis methods.

REFERENCES

[1]   Rappoport, T.G. Semiconductors: Nanostructures and applications in spintronics and quantum computation. InAIP Conference Proceedings 2006 Jan 6 (Vol. 809, No. 1, pp. 326-342). American Institute of Physics.

[2]   Available from: http://www.its.caltech.edu/ feynman/plenty.html

[3]   Roukes, M. Plenty of room, indeed. *Sci. Am.,* **2001**, *285*(3), 48-57, 54-57.
        [http://dx.doi.org/10.1038/scientificamerican0901-48] [PMID: 11524969]

[4]   Peercy, P.S. The drive to miniaturization. *Nature,* **2000**, *406*(6799), 1023-1026.
        [http://dx.doi.org/10.1038/35023223] [PMID: 10984060]

[5]   Soref, R.A. Silicon-based optoelectronics. *Proc. IEEE,* **1993**, *81*(12), 1687-1706.
        [http://dx.doi.org/10.1109/5.248958]

[6]   Beenakker, C.W.J.; van Houten, H. Quantum transport in semiconductor nanostructures. *Solid State Phys.,* **1991**, *44*, 1-228.
        [http://dx.doi.org/10.1016/S0081-1947(08)60091-0]

[7]   Singh, V.A.; John, G.C. Processes in Nanocrystalline Silicon. *Physics of Semiconductor Nanostructures*; Jain, K.P., Ed.; Narosa Publishing House: New Delhi, **1997**, p. 186.

[8]   John, G.C.; Singh, V.A. Model for the photoluminescence behavior of porous silicon. *Phys. Rev. B Condens. Matter,* **1996**, *54*(7), 4416-4419.
        [http://dx.doi.org/10.1103/PhysRevB.54.4416] [PMID: 9986385]

[9]   Itoh, T.; Iwabuchi, Y.; Kataoka, M. Study on the Size and Shape of CuCl Microcrystals Embedded in Alkali-Chloride Matrices and Their Correlation with Exciton Confinement. *Phys. Status Solidi, B Basic Res.,* **1988**, *145*(2), 567-577.
      [http://dx.doi.org/10.1002/pssb.2221450222]

[10] Bhargava, R.N.; Gallagher, D.; Welker, T. Doped nanocrystals of semiconductors - a new class of luminescent materials. *J. Lumin.,* **1994**, *60-61*, 275-280.
      [http://dx.doi.org/10.1016/0022-2313(94)90146-5]

[11] Wang, Y.; Herron, N. Nanometer-sized semiconductor clusters: materials synthesis, quantum size effects, and photophysical properties. *J. Phys. Chem.,* **1991**, *95*(2), 525-532.
      [http://dx.doi.org/10.1021/j100155a009]

[12] Geller, MR Quantum phenomena in low-dimensional systems. In: *Fundamentals of Physics,* **2001**.

[13] Sahu, M.K. Semiconductor nanoparticles theory and applications. *Int. J. Appl. Eng. Res.,* **2019**, *14*(2), 491-494.

[14] Sahu, S.N.; Nanda, K.K. Nanostructure semiconductors: physics and applications. *Proceedings-Indian National Science Academy Part A.,* **2001**, *67*(1), 103-130.

[15] https://en.wikipedia.org/wiki/Nanomaterials

[16] Senga, R.; Komsa, H.P.; Liu, Z.; Hirose-Takai, K.; Krasheninnikov, A.V.; Suenaga, K. Atomic structure and dynamic behaviour of truly one-dimensional ionic chains inside carbon nanotubes. *Nat. Mater.,* **2014**, *13*(11), 1050-1054.
      [http://dx.doi.org/10.1038/nmat4069] [PMID: 25218060]

[17] Medeiros, P.V.C.; Marks, S.; Wynn, J.M.; Vasylenko, A.; Ramasse, Q.M.; Quigley, D.; Sloan, J.; Morris, A.J. Single-atom scale structural selectivity in Te nanowires encapsulated inside ultranarrow, single-walled carbon nanotubes. *ACS Nano,* **2017**, *11*(6), 6178-6185.
      [http://dx.doi.org/10.1021/acsnano.7b02225] [PMID: 28467832]

[18] Vasylenko, A.; Marks, S.; Wynn, J.M.; Medeiros, P.V.C.; Ramasse, Q.M.; Morris, A.J.; Sloan, J.; Quigley, D. Electronic structure control of sub-nanometer 1D SnTe *via* nanostructuring within single-walled carbon nanotubes. *ACS Nano,* **2018**, *12*(6), 6023-6031.
      [http://dx.doi.org/10.1021/acsnano.8b02261] [PMID: 29782147]

[19] Lojkowski, W; Turan, R; Proykova, A; Daniszewska, A. Eight Nanoforum Report: Nanometrology. European Nanotechnology Gateway (Nanoforum. org,). **2006**.

[20] Lapshin, R.V. STM observation of a box-shaped graphene nanostructure appeared after mechanical cleavage of pyrolytic graphite. *Appl. Surf. Sci.,* **2016**, *360*, 451-460.
      [http://dx.doi.org/10.1016/j.apsusc.2015.09.222]

[21] Petroff, P.M.; Lorke, A.; Imamoglu, A. Epitaxially self-assembled quantum dots. *Phys. Today,* **2001**, *54*(5), 46-52.
      [http://dx.doi.org/10.1063/1.1381102]

[22] Kirstaedter, N.; Grundmann, M.; Richter, U.; Ustinov, V.M.; Kop'ev, P.S.; Bimberg, D.; Werner, P.; Ruvimov, S.S.; Ledentsov, N.N.; Gösele, U.; Alferov, Z.I.; Heydenreich, J.; Maximov, M.V. Low threshold, large To injection laser emission from (InGa)As quantum dots. *Electron. Lett.,* **1994**, *30*(17), 1416-1417.
      [http://dx.doi.org/10.1049/el:19940939]

[23] Kouwenhoven, L.; Marcus, C. Quantum dots. *Phys. World,* **1998**, *11*(6), 35-40.
      [http://dx.doi.org/10.1088/2058-7058/11/6/26]

[24] Devoret, M.H.; Glattli, C. Single-electron transistors. *Phys. World,* **1998**, *11*(9), 29-34.
      [http://dx.doi.org/10.1088/2058-7058/11/9/26]

[25] Alivisatos, A.P. Semiconductor clusters, nanocrystals, and quantum dots. *Science,* **1996**, *271*(5251), 933-937.

[http://dx.doi.org/10.1126/science.271.5251.933]

[26]   Bentolila, L.A.; Weiss, S. Biological quantum dots go live. *Phys. World,* **2003**, *16*(3), 23-24.
       [http://dx.doi.org/10.1088/2058-7058/16/3/35]

[27]   Koch, C.C.; Whittenberger, J.D. Mechanical milling/alloying of intermetallics. *Intermetallics,* **1996**, *4*(5), 339-355.
       [http://dx.doi.org/10.1016/0966-9795(96)00001-5]

[28]   Zhang, D.L. Processing of advanced materials using high-energy mechanical milling. *Prog. Mater. Sci.,* **2004**, *49*(3-4), 537-560.

[29]   Baheti, V.; Abbasi, R.; Militky, J. Ball milling of jute fibre wastes to prepare nanocellulose. *World J. Eng.,* **2012**, *9*(1), 45-50.
       [http://dx.doi.org/10.1260/1708-5284.9.1.45]

[30]   Ward, T.S.; Chen, W.; Schoenitz, M.; Dave, R.N.; Dreizin, E.L. A study of mechanical alloying processes using reactive milling and discrete element modeling. *Acta Mater.,* **2005**, *53*(10), 2909-2918.
       [http://dx.doi.org/10.1016/j.actamat.2005.03.006]

[31]   Abdellaoui, M.; Gaffet, E. The physics of mechanical alloying in a planetary ball mill: Mathematical treatment. *Acta Metall. Mater.,* **1995**, *43*(3), 1087-1098.
       [http://dx.doi.org/10.1016/0956-7151(95)92625-7]

[32]   https://ninithi.wordpress.com/topdown_methods/

[33]   Prasad Yadav, T.; Manohar Yadav, R.; Pratap Singh, D. Mechanical milling: a top down approach for the synthesis of nanomaterials and nanocomposites. *Nanoscience and Nanotechnology,* **2012**, *2*(3), 22-48.
       [http://dx.doi.org/10.5923/j.nn.20120203.01]

[34]   Baig, N.; Kammakakam, I.; Falath, W. Nanomaterials: a review of synthesis methods, properties, recent progress, and challenges. *Materials Advances,* **2021**, *2*(6), 1821-1871.
       [http://dx.doi.org/10.1039/D0MA00807A]

[35]   Sadler, L.Y., III; Stanley, D.A.; Brooks, D.R. Attrition mill operating characteristics. *Powder Technol.,* **1975**, *12*(1), 19-28.
       [http://dx.doi.org/10.1016/0032-5910(75)85004-2]

[36]   https://en.wikipedia.org/wiki/Lithography

[37]   Chisholm, H.J. *Lithography. Encyclopædia Britannica,* 11th ed; Cambridge University Press, **1911**, pp. 785-789.

[38]   Weaver, P. *The technique of lithography*; BT Batsford: London, **1964**.

[39]   Meggs, P. *History of Graphic Design*; John Wiley & Sons, Inc, **1998**.

[40]   Carter, R. *Ben day, and Philip Meggs. Typograhic Design: Form and Communication,* 3rd ed; John Wiley & Sons, **2002**, p. 11.

[41]   Subramani K, Ahmed W. Fabrication of peg hydrogel micropatterns by soft-photolithography and peg hydrogel as guided bone regeneration membrane in dental implantology. In Emerging Nanotechnologies in Dentistry **2012**. pp. 171-187.
       [http://dx.doi.org/10.1016/B978-1-4557-7862-1.00011-0]

[42]   Hill, J. Digital & Photographic. St Barnabas Press.

[43]   Hannavy, J., Ed. *Encyclopedia of nineteenth-century photography*; Routledge Taylor & Francis Group, **2008**, p. 865.

[44]   Mansuripur, M. Classical optics and its applications. *Cambridge University Press,* **2002**.

[45]   https://en.wikipedia.org/wiki/Electron-beam_lithography

[46]   Paturzo, M.; Grilli, S.; Mailis, S.; Coppola, G.; Iodice, M.; Gioffré, M.; Ferraro, P. Flexible coherent diffraction lithography by tunable phase arrays in lithium niobate crystals. *Opt. Commun.,* **2008**, *281*(8), 1950-1953.
[http://dx.doi.org/10.1016/j.optcom.2007.12.056]

[47]   Mack, CA Field guide to optical lithography Bellingham. *WA: SPIE Press,* **2006**.
[http://dx.doi.org/10.1117/3.665802]

[48]   Gopanna, A; Rajan, KP; Thomas, SP; Chavali, M *Polyethylene and polypropylene matrix composites for biomedical applications. In Materials for Biomedical Engineering.,* **2019**, 175-276.

[49]   Sabantina, L. *Nanocarbons-based textiles for flexible energy storage. In Nanosensors and Nanodevices for Smart Multifunctional Textiles.,* **2021**, 163-188.

[50]   Begum, H.A.; Khan, K.R. Study on the various types of needle based and needleless electrospinning system for nanofiber production. *International Journal of Textile Science.,* **2017**, *6*, 110-117.

[51]   https://www.nanoscience.com/techniques/electrospin/

[52]   Kumar, P.S.; Sundaramurthy, J.; Sundarrajan, S.; Babu, V.J.; Singh, G.; Allakhverdiev, S.I.; Ramakrishna, S. Hierarchical electrospun nanofibers for energy harvesting, production and environmental remediation. *Energy Environ. Sci.,* **2014**, *7*(10), 3192-3222.
[http://dx.doi.org/10.1039/C4EE00612G]

[53]   Zhang, J.; Chaker, M.; Ma, D. Pulsed laser ablation based synthesis of colloidal metal nanoparticles for catalytic applications. *J. Colloid Interface Sci.,* **2017**, *489*, 138-149.
[http://dx.doi.org/10.1016/j.jcis.2016.07.050] [PMID: 27554172]

[54]   Ismail, R.A.; Mohsin, M.H.; Ali, A.K.; Hassoon, K.I.; Erten-Ela, S. Preparation and characterization of carbon nanotubes by pulsed laser ablation in water for optoelectronic application. *Physica E,* **2020**, *119*, 113997.
[http://dx.doi.org/10.1016/j.physe.2020.113997]

[55]   Chrzanowska, J.; Hoffman, J.; Małolepszy, A.; Mazurkiewicz, M.; Kowalewski, T.A.; Szymanski, Z.; Stobinski, L. Synthesis of carbon nanotubes by the laser ablation method: Effect of laser wavelength. *Phys. Status Solidi, B Basic Res.,* **2015**, *252*(8), 1860-1867.
[http://dx.doi.org/10.1002/pssb.201451614]

[56]   Duque, J.; Madrigal, B.; Riascos, H.; Avila, Y. Colloidal metal oxide nanoparticles prepared by laser ablation technique and their antibacterial test. *Colloids and Interfaces,* **2019**, *3*(1), 25.
[http://dx.doi.org/10.3390/colloids3010025]

[57]   Su, SS; Chang, I Review of production routes of nanomaterials. Commercialization of Nanotechnologies– A Case Study Approach. **2018**, 15-29.

[58]   Available from: https://ninithi.wordpress.com/topdown_methods/

[59]   Savic, S.; Vojisavljevic, K.; Počuča-Nešić, M.; Zivojevic, K.; Mladenovic, M.; Knezevic, N. Hard template synthesis of nanomaterials based on mesoporous silica. *Metallurgical and Materials Engineering,* **2018**, *24*(4), 225-241.
[http://dx.doi.org/10.30544/400]

[60]   Dalapati, G.K.; Masudy-Panah, S.; Moakhar, R.S.; Chakrabortty, S.; Ghosh, S.; Kushwaha, A.; Katal, R.; Chua, C.S.; Xiao, G.; Tripathy, S.; Ramakrishna, S. Nanoengineered advanced materials for enabling hydrogen economy: functionalized graphene–incorporated cupric oxide catalyst for efficient solar hydrogen production. *Glob. Chall.,* **2020**, *4*(3), 1900087.
[http://dx.doi.org/10.1002/gch2.201900087,] [PMID: 32140256]

[61]   Xie, Y; Kocaefe, D; Chen, C; Kocaefe, Y Review of research on template methods in preparation of nanomaterials. *J. Nanomaterials,* **2016**, (8), 1-10.
[http://dx.doi.org/10.1155/2016/2302595]

[62]   Gu, D.; Schüth, F. Synthesis of non-siliceous mesoporous oxides. *Chem. Soc. Rev.,* **2014**, *43*(1), 313-

344.
[http://dx.doi.org/10.1039/C3CS60155B] [PMID: 23942521]

[63]   Ren, Y.; Ma, Z.; Bruce, P.G. Ordered mesoporous metal oxides: synthesis and applications. *Chem. Soc. Rev.,* **2012**, *41*(14), 4909-4927.
[http://dx.doi.org/10.1039/c2cs35086f] [PMID: 22653082]

[64]   Deng, Y.; Wei, J.; Sun, Z.; Zhao, D. Large-pore ordered mesoporous materials templated from non-Pluronic amphiphilic block copolymers. *Chem. Soc. Rev.,* **2013**, *42*(9), 4054-4070.
[http://dx.doi.org/10.1039/C2CS35426H] [PMID: 23258081]

[65]   Prieto, G.; Zečević, J.; Friedrich, H.; de Jong, K.P.; de Jongh, P.E. Towards stable catalysts by controlling collective properties of supported metal nanoparticles. *Nat. Mater.,* **2013**, *12*(1), 34-39.
[http://dx.doi.org/10.1038/nmat3471] [PMID: 23142841]

[66]   Wan, Y.; Zhao, D. On the controllable soft-templating approach to mesoporous silicates. *Chem. Rev.,* **2007**, *107*(7), 2821-2860.
[http://dx.doi.org/10.1021/cr068020s] [PMID: 17580976]

[67]   Soler-Illia, G.J.A.A.; Sanchez, C.; Lebeau, B.; Patarin, J. Chemical strategies to design textured materials: from microporous and mesoporous oxides to nanonetworks and hierarchical structures. *Chem. Rev.,* **2002**, *102*(11), 4093-4138.
[http://dx.doi.org/10.1021/cr0200062] [PMID: 12428985]

[68]   Poolakkandy, R.R.; Menamparambath, M.M. Soft-template-assisted synthesis: a promising approach for the fabrication of transition metal oxides. *Nanoscale Adv.,* **2020**, *2*(11), 5015-5045.
[http://dx.doi.org/10.1039/D0NA00599A] [PMID: 36132034]

[69]   Lee, J.; Kim, J.; Hyeon, T. Recent progress in the synthesis of porous carbon materials. *Adv. Mater.,* **2006**, *18*(16), 2073-2094.
[http://dx.doi.org/10.1002/adma.200501576]

[70]   Wan, Y.; Shi, Y.; Zhao, D. Designed synthesis of mesoporous solids *via* nonionic-surfactan--templating approach. *Chem. Commun. (Camb.),* **2007**, (9), 897-926.
[http://dx.doi.org/10.1039/B610570J] [PMID: 17311122]

[71]   Wan, Y.; Yang, H.; Zhao, D. "Host-guest" chemistry in the synthesis of ordered nonsiliceous mesoporous materials. *Acc. Chem. Res.,* **2006**, *39*(7), 423-432.
[http://dx.doi.org/10.1021/ar050091a] [PMID: 16846206]

[72]   Liu, Y.; Goebl, J.; Yin, Y. Themed issue: Chemistry of functional nanomaterials. *Chem. Soc. Rev.,* **2013**, *42*, 2610-2653.
[http://dx.doi.org/10.1039/C2CS35369E] [PMID: 23093173]

[73]   Li, W.; Zhao, D. An overview of the synthesis of ordered mesoporous materials. *Chem. Commun. (Camb.),* **2013**, *49*(10), 943-946.
[http://dx.doi.org/10.1039/C2CC36964H] [PMID: 23249963]

[74]   Szczęśniak, B.; Choma, J.; Jaroniec, M. Major advances in the development of ordered mesoporous materials. *Chem. Commun. (Camb.),* **2020**, *56*(57), 7836-7848.
[http://dx.doi.org/10.1039/D0CC02840A] [PMID: 32520012]

[75]   Crepaldi, E.L.; Soler-Illia, G.J.A.A.; Grosso, D.; Cagnol, F.; Ribot, F.; Sanchez, C. Controlled formation of highly organized mesoporous titania thin films: from mesostructured hybrids to mesoporous nanoanatase TiO_2. *J. Am. Chem. Soc.,* **2003**, *125*(32), 9770-9786.
[http://dx.doi.org/10.1021/ja030070g] [PMID: 12904043]

[76]   Scolan E, Louis A, Albouy PA, Sanchez C. Design of meso-structured titanium oxide based hybrid organic–inorganic networks. *New J. Chem.,* **2001**, *25*(1), 156-165.
[http://dx.doi.org/10.1039/b006139p]

[77]   Grosso, D. de AA Soler-Illia GJ, BabonneauF, Sanchez C, Albouy PA, Brunet-Bruneau A, Balkenende AR. Highly organized mesoporous titania thin films showing mono-oriented 2D

hexagonal channels. *Adv. Mater.,* **2001**, *13*(14), 1085-1090.
[http://dx.doi.org/10.1002/1521-4095(200107)13:14<1085::AID-ADMA1085>3.0.CO;2-Q]

[78]   Soler-Illia, G.J.A.A.; Louis, A.; Sanchez, C. Synthesis and characterization of mesostructured titania-based materials through evaporation-induced self-assembly. *Chem. Mater.,* **2002**, *14*(2), 750-759.
[http://dx.doi.org/10.1021/cm011217a]

[79]   Crepaldi, E.L. de AA Soler-Illia GJ, Grosso D, Sanchez C. Nanocrystallisedtitania and zirconia mesoporous thin films exhibiting enhanced thermal stability. *New J. Chem.,* **2003**, *27*(1), 9-13.
[http://dx.doi.org/10.1039/b205497n]

[80]   Tian, B.; Liu, X.; Yang, H.; Xie, S.; Yu, C.; Tu, B.; Zhao, D. General synthesis of ordered crystallized metal oxide nanoarrays replicated by microwave-digested mesoporous silica. *Adv. Mater.,* **2003**, *15*(16), 1370-1374.
[http://dx.doi.org/10.1002/adma.200305211]

[81]   Jiao, F.; Harrison, A.; Jumas, J.C.; Chadwick, A.V.; Kockelmann, W.; Bruce, P.G. Ordered mesoporous Fe_2O_3 with crystalline walls. *J. Am. Chem. Soc.,* **2006**, *128*(16), 5468-5474.
[http://dx.doi.org/10.1021/ja0584774] [PMID: 16620119]

[82]   Lai, X.; Li, X.; Geng, W.; Tu, J.; Li, J.; Qiu, S. Ordered mesoporous copper oxide with crystalline walls. *Angew. Chem. Int. Ed.,* **2007**, *46*(5), 738-741.
[http://dx.doi.org/10.1002/anie.200603210] [PMID: 17146817]

[83]   Shi, Y.; Guo, B.; Corr, S.A.; Shi, Q.; Hu, Y.S.; Heier, K.R.; Chen, L.; Seshadri, R.; Stucky, G.D. Ordered mesoporous metallic MoO2 materials with highly reversible lithium storage capacity. *Nano Lett.,* **2009**, *9*(12), 4215-4220.
[http://dx.doi.org/10.1021/nl902423a] [PMID: 19775084]

[84]   Sun, X.; Hao, H.; Ji, H.; Li, X.; Cai, S.; Zheng, C. Nanocasting synthesis of In_2O_3 with appropriate mesostructured ordering and enhanced gas-sensing property. *ACS Appl. Mater. Interfaces,* **2014**, *6*(1), 401-409.
[http://dx.doi.org/10.1021/am4044807] [PMID: 24308308]

[85]   Imperor-Clerc, M.; Bazin, D.; Appay, M.D.; Beaunier, P.; Davidson, A. Crystallization of β-MnO_2 nanowires in the pores of SBA-15 silicas: in situ investigation using synchrotron radiation. *Chem. Mater.,* **2004**, *16*(9), 1813-1821.
[http://dx.doi.org/10.1021/cm035353m]

[86]   Dickinson, C.; Zhou, W.; Hodgkins, R.P.; Shi, Y.; Zhao, D.; He, H. Formation mechanism of porous single-crystal Cr_2O_3 and Co_3O_4 templated by mesoporous silica. *Chem. Mater.,* **2006**, *18*(13), 3088-3095.
[http://dx.doi.org/10.1021/cm060014p]

[87]   Ryoo, R.; Joo, S.H.; Jun, S. Synthesis of highly ordered carbon molecular sieves *via* template-mediated structural transformation. *J. Phys. Chem. B,* **1999**, *103*(37), 7743-7746.
[http://dx.doi.org/10.1021/jp991673a]

[88]   Yu, C.; Fan, J.; Tian, B.; Zhao, D.; Stucky, G.D. High-yield synthesis of periodic mesoporous silica rods and their replication to mesoporous carbon rods. *Adv. Mater.,* **2002**, *14*(23), 1742-1745.
[http://dx.doi.org/10.1002/1521-4095(20021203)14:23<1742::AID-ADMA1742>3.0.CO;2-3]

[89]   Kleitz, F.; Hei Choi, S.; Ryoo, R. Cubic Ia3d large mesoporous silica: synthesis and replication to platinum nanowires, carbon nanorods and carbon nanotubesElectronic supplementary information (ESI) available: TEM images of mesoporous cubic silica and Pt networks, XRD patterns during formation of the cubic phase. See . *Chem. Commun. (Camb.),* **2003**, (17), 2136-2137.
[http://dx.doi.org/10.1039/b306504a] [PMID: 13678168]

[90]   Deng, X.; Chen, K.; Tüysüz, H. Protocol for the nanocasting method: Preparation of ordered mesoporous metal oxides. *Chem. Mater.,* **2017**, *29*(1), 40-52.
[http://dx.doi.org/10.1021/acs.chemmater.6b02645]

[91]   Danks, A.E.; Hall, S.R.; Schnepp, Z. The evolution of 'sol–gel' chemistry as a technique for materials synthesis. *Mater. Horiz.,* **2016**, *3*(2), 91-112.
[http://dx.doi.org/10.1039/C5MH00260E]

[92]   Tseng, T.K.; Lin, Y.S.; Chen, Y.J.; Chu, H. A review of photocatalysts prepared by sol-gel method for VOCs removal. *Int. J. Mol. Sci.,* **2010**, *11*(6), 2336-2361.
[http://dx.doi.org/10.3390/ijms11062336] [PMID: 20640156]

[93]   Parashar, M.; Shukla, V.K.; Singh, R. Metal oxides nanoparticles *via* sol–gel method: a review on synthesis, characterization and applications. *J. Mater. Sci. Mater. Electron.,* **2020**, *31*(5), 3729-3749.
[http://dx.doi.org/10.1007/s10854-020-02994-8]

[94]   Znaidi, L. Sol–gel-deposited ZnO thin films: A review. *Mater. Sci. Eng. B,* **2010**, *174*(1-3), 18-30.
[http://dx.doi.org/10.1016/j.mseb.2010.07.001]

[95]   de Coelho Escobar, C.; dos Santos, J.H.Z. Effect of the sol–gel route on the textural characteristics of silica imprinted with Rhodamine B. *J. Sep. Sci.,* **2014**, *37*(7), 868-875.
[http://dx.doi.org/10.1002/jssc.201301143] [PMID: 24478149]

[96]   Kaflé, BP Chemical Analysis and Material Characterization by Spectrophotometry. *Elsevier,* **2019**.

[97]   Parker, G. Encyclopedia of Materials: Science and Technology. **2001**.

[98]   Chen, A.; Holt-Hindle, P. Platinum-based nanostructured materials: synthesis, properties, and applications. *Chem. Rev.,* **2010**, *110*(6), 3767-3804.
[http://dx.doi.org/10.1021/cr9003902] [PMID: 20170127]

[99]   Dong, Y.; Du, X.; Liang, P.; Man, X. One-pot solvothermal method to fabricate 1D-VS4 nanowires as anode materials for lithium ion batteries. *Inorg. Chem. Commun.,* **2020**, *115*, 107883.
[http://dx.doi.org/10.1016/j.inoche.2020.107883]

[100]   Jiang, Y.; Peng, Z.; Zhang, S.; Li, F.; Liu, Z.; Zhang, J.; Liu, Y.; Wang, K. Facile in-situ Solvothermal Method to synthesize double shell $ZnIn_2S_4$ nanosheets/TiO_2 hollow nanosphere with enhanced photocatalytic activities. *Ceram. Int.,* **2018**, *44*(6), 6115-6126.
[http://dx.doi.org/10.1016/j.ceramint.2017.12.244]

[101]   Chai, B.; Xu, M.; Yan, J.; Ren, Z. Remarkably enhanced photocatalytic hydrogen evolution over MoS 2 nanosheets loaded on uniform CdS nanospheres. *Appl. Surf. Sci.,* **2018**, *430*, 523-530.
[http://dx.doi.org/10.1016/j.apsusc.2017.07.292]

[102]   Andrews, D; Scholes, G; Wiederrecht, G Comprehensive nanoscience and technology. *Academic Press,* **2010**.

[103]   Meng, L.Y.; Wang, B.; Ma, M.G.; Lin, K.L. The progress of microwave-assisted hydrothermal method in the synthesis of functional nanomaterials. *Mater. Today Chem.,* **2016**, *1-2*, 63-83.
[http://dx.doi.org/10.1016/j.mtchem.2016.11.003]

[104]   Ayyub, P.; Chandra, R.; Taneja, P.; Sharma, A.K.; Pinto, R. Synthesis of nanocrystalline material by sputtering and laser ablation at low temperatures. *Appl. Phys., A Mater. Sci. Process.,* **2001**, *73*(1), 67-73.
[http://dx.doi.org/10.1007/s003390100833]

[105]   Son, H.H.; Seo, G.H.; Jeong, U.; Shin, D.Y.; Kim, S.J. Capillary wicking effect of a Cr-sputtered superhydrophilic surface on enhancement of pool boiling critical heat flux. *Int. J. Heat Mass Transf.,* **2017**, *113*, 115-128.
[http://dx.doi.org/10.1016/j.ijheatmasstransfer.2017.05.055]

[106]   Wender, H.; Migowski, P.; Feil, A.F.; Teixeira, S.R.; Dupont, J. Sputtering deposition of nanoparticles onto liquid substrates: Recent advances and future trends. *Coord. Chem. Rev.,* **2013**, *257*(17-18), 2468-2483.
[http://dx.doi.org/10.1016/j.ccr.2013.01.013]

[107]   Muñoz-García, J.; Vázquez, L.; Cuerno, R.; Sánchez-García, J.A.; Castro, M.; Gago, R. *Self-organized*

surface nanopatterning by ion beam sputtering. InToward Functional Nanomaterials; Springer: New York, NY, **2009**, pp. 323-398.

[108]   Nie, M.; Sun, K.; Meng, D.D. Formation of metal nanoparticles by short-distance sputter deposition in a reactive ion etching chamber. *J. Appl. Phys., * **2009**, *106*(5), 054314.
[http://dx.doi.org/10.1063/1.3211326]

[109]   www.AngstromSciences.com

[110]   Jones, A.C.; Hitchman, M.L. Overview of chemical vapour deposition. Chemical Vapour Deposition: Precursors. *Processes and Applications., * **2009**, *1*, 1-36.

[111]   https://www.silcotek.com/semi-coating-blog/chemical-vapor-deposition-explained.-its-benefits-and-drawbacks

[112]   Ago, H. *CVD growth of high-quality single-layer graphene. InFrontiers of Graphene and Carbon Nanotubes*; Springer: Tokyo, **2015**, pp. 3-20.

[113]   Wu, Q.; Wongwiriyapan, W.; Park, J.H.; Park, S.; Jung, S.J.; Jeong, T.; Lee, S.; Lee, Y.H.; Song, Y.J. In situ chemical vapor deposition of graphene and hexagonal boron nitride heterostructures. *Curr. Appl. Phys., * **2016**, *16*(9), 1175-1191.
[http://dx.doi.org/10.1016/j.cap.2016.04.024]

[114]   Machac, P.; Cichon, S.; Lapcak, L.; Fekete, L. Graphene prepared by chemical vapour deposition process. *Graphene Technol., * **2020**, *5*(1-2), 9-17.
[http://dx.doi.org/10.1007/s41127-019-00029-6]

[115]   https://ninithi.wordpress.com/topdown_methods/

[116]   Choy, KL Vapor Processing of nanostructured materials. *In Handbook of nanostructured materials and nanotechnology,* Academic Press, **2000**, 533-577.

[117]   Arthur, J.R., Jr Interaction of Ga and As2 molecular beams with GaAs surfaces. *J. Appl. Phys., * **1968**, *39*(8), 4032-4034.
[http://dx.doi.org/10.1063/1.1656901]

[118]   Cho, A.Y.; Cheng, K.Y. Growth of extremely uniform layers by rotating substrate holder with molecular beam epitaxy for applications to electro-optic and microwave devices. *Appl. Phys. Lett., * **1981**, *38*(5), 360-362.
[http://dx.doi.org/10.1063/1.92377]

CHAPTER 2

Photocatalytic and Sensing Applications of Semiconductor Nanostructures

Seema Maheshwari[1], Shikha Bhogal[2], Kuldeep Kaur[1,*] and **Ashok Kumar Malik[2]**

[1] *Department of Chemistry, Mata Gujri College, Fatehgarh Sahib, Punjab, India*

[2] *Department of Chemistry, Punjabi University, Patiala, Punjab, India*

Abstract: Semiconductor Nanostructures (SCNSs) are of great interest due to their excellent optical and electronic properties. As a result of their unique properties, semiconductor nanostructures have found applications in several fields, including optoelectronics, solar energy conversion, photocatalysis, and sensing. SCNSs show promising prospects in photocatalytic and sensing applications. Photocatalytic application of SCNSs provides potential solutions for environmental remediation and energy generation. Several strategies have been developed to achieve high efficiency for photocatalytic processes using semiconductor nanostructures. Efforts have also been made to achieve high sensitivities in sensing applications using SCNSs. In the present chapter, the photocatalysis activity of semiconductor nanostructures has been discussed along with the photocatalytic mechanism and strategies for enhancing photocatalytic efficiency. Several applications of semiconductor photocatalysis in wastewater treatment, hydrogen production, and air purification are cited in recent literature. The sensing applications of semiconductor nanostructures have also been discussed, including their use as chemical sensors, gas sensors, and biosensors.

Keywords: Biosensors, Fluorescence sensors, Gas sensors, Photocatalysts, Photocatalytic efficiency, Semiconductor nanostructures.

INTRODUCTION

The semiconducting nanostructures (SCNSs) with sizes between 1 and 100 nm are particularly attractive in material science due to the quantum confinement effects, which attribute unique optical and electronic properties to these materials [1]. The unique properties attract researchers to investigate these materials for their broader applications in several fields, including optoelectronics, solar energy

---
* **Corresponding author Kuldeep Kaur**: Mata Gujri College, Fatehgarh Sahib, Punjab, India;
E-mail: shergillkk@gmail.com

Karamjit Singh Dhaliwal (Ed.)

conversion, photocatalysis, and sensing [2, 3]. Semiconductors possess band structures that can be conveniently adjusted to improve their properties [4]. Photocatalytic application of SCNSs provides potential solutions for various energy and environment-related issues [5]. The increasing discharge of contaminants into freshwater systems from a wide variety of industrial, municipal, and agricultural sources has seriously affected water quality. These contaminants, including pharmaceuticals, pesticides, dyes, personal care products, and synthetic hormones, present adverse health effects [6]. Several conventional physical and chemical processes have been adopted for the removal of organic pollutants. However, these conventional techniques are not very effective and have some limitations. Many contaminants are recalcitrant and cannot be degraded by conventional methods resulting in their presence in treated wastewater. Moreover, many conventional processes can convert organic pollutants from one phase to another, which requires further treatment to avoid secondary pollution. Advanced oxidation processes (AOP) have been proposed for the efficient removal of toxic pollutants. The AOP processes involve the generation of hydroxyl or sulphate radicals in sufficient quantity to remove organic and inorganic pollutants [7]. Photocatalysis using semiconductor materials is one of the most promising advanced oxidation processes (AOPs) for the treatment of contaminated wastewater. It has been demonstrated recently that the photocatalytic degradation of organic pollutants from the water by semiconductors is a viable alternative to conventional methods [8]. SCNSs possess high degradation and mineralization efficiency, low toxicity, and low cost over other AOPs, such as electrochemical oxidation, ozonation, H_2O_2/UV oxidation, and biodegradation [9]. Their use as catalysts under photolytic conditions can potentially solve the serious environmental and energy-related crisis the world is facing today [10]. Hydrogen is renewable energy of high calorific value and clean, pollution-free fuel. Hydrogen production with the use of semiconductor nanostructures as photocatalyst is the most promising technology [11]. Visible light-driven photocatalysis for the production of hydrogen from H_2O, H_2S, and organic waste has attracted attention due to the depletion of fossil fuel reserves [2]. Semiconductor nanostructures can also contribute to reducing the carbon dioxide concentration in the air. Semiconductor nanostructures of n- and p-type can potentially convert carbon dioxide into hydrocarbons, such as methane and ethane [12]. Human health, environmental protection and safety require regular sensing and monitoring of pollutants, toxic substances, and gases in real samples. Semiconductor nanomaterials have aroused great interest as sensors. This is due to the many desirable features offered by them, including unique optical and electrical properties, excellent light stability, narrow emission spectrum, and wide excitation spectrum [13]. SCNSs are capable of forming sensitive, selective, and stable sensing platforms, achieving great performance in sensing applications.

PHOTOCATALYTIC ACTIVITY

General Photocatalytic Mechanism

The characteristics of a bulk semiconductor depend largely on the bandgap energy (Eg). The Eg is the minimum energy required to excite an electron from the ground state valence band (VB) into the vacant conduction band (CB). The type of light energy absorbed by the semiconductor material depends on the bandgap. The wide bandgap semiconductor materials (TiO_2, ZnO) that absorb in the UV region pose a serious limitation [14]. They can absorb photons with light wavelengths shorter than 400 nm. This accounts for the loss of solar energy as UV radiations comprise only 5% of the total solar spectrum, whereas visible radiations comprise 45% of the total solar spectrum. To make maximum utilization of solar energy, various strategies have been adopted to synthesize visible-light-driven nanostructures [15]. Bandgap engineering is an efficient approach to developing visible-light-responsive photocatalysts [16]. Tuning the bandgap makes it possible to harvest maximum light energy and produce sufficient charge carriers, which are responsible for photoreduction and photooxidation in photocatalysis. The process of photocatalysis can be divided into three steps; (a) light absorption, (b) charge separation and transfer, and (c) surface reactions, which occur in sequence [17].

a. Upon sunlight irradiation, the photons with energy equal to or more than the optical band gap (*E*g) of the photocatalyst are absorbed and utilized for the generation of charge carriers. An electron in the valence band (VB) gets excited to the conduction band (CB) on the absorption of a photon. The negative electron and positive hole are mobilized with an electric field to generate the current. The electrostatically bonded electron-hole pair is named an exciton.

b. The generated charge carrier electrons (e^-) and holes (h^+) react with the surrounding oxygen and water molecules to form OH free radical ($OH^{\cdot}$) with powerful oxidation strength. The photogenerated holes (h^+) in the valence band react with either H_2O or OH^- to produce the $OH^{\cdot}$. The extremely reactive oxygen species thus formed transfer across the interface.

c. The transferred oxygen species are capable of destructive removal of adsorbed organic species on the semiconductor photocatalyst surface. The generated charge carriers can diffuse toward the surface of the photocatalyst particles. However, during this process, the electrons and holes tend to recombine either in bulk or on the surface of photocatalysts. The surviving photogenerated electrons and holes can reach the surface of particles and participate in the redox reactions on the surface with absorbed molecules. Adsorbed species can be reduced by CB electrons and can be oxidized by VB holes. In the presence of adsorbed water, electrons transfer from the water molecule to the positive holes to produce OH$^•$ radicals (Fig. **1**), which are powerful oxidants and react with organic and toxic compounds.

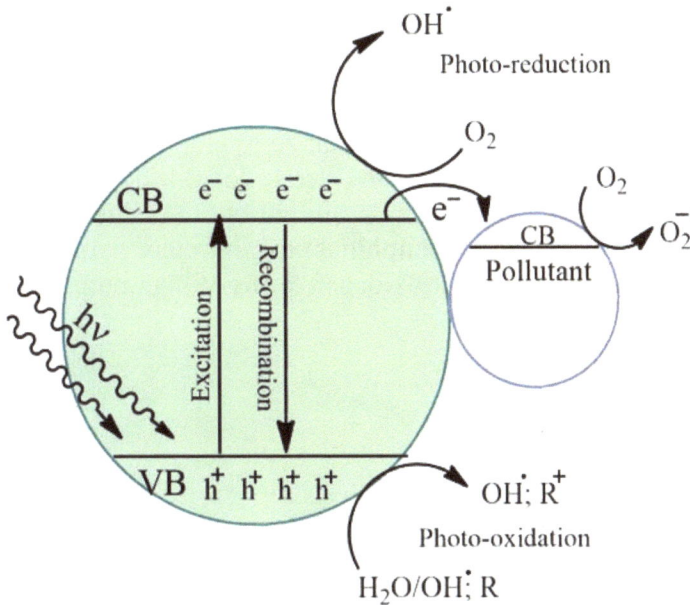

Fig. (1). Reaction with the absorbed molecule on the surface of SCNSs and formation of OH• radical, which initiate Advance Oxidative Processes (AOPs) [61].

Hydroxyl radicals play an important role in initiating oxidation reactions; a schematic illustration of the general mechanism of photocatalysis showing processes, such as the formation of electrons and holes with the absorption of energy suitable to the bandgap of SC NSs, as shown below:

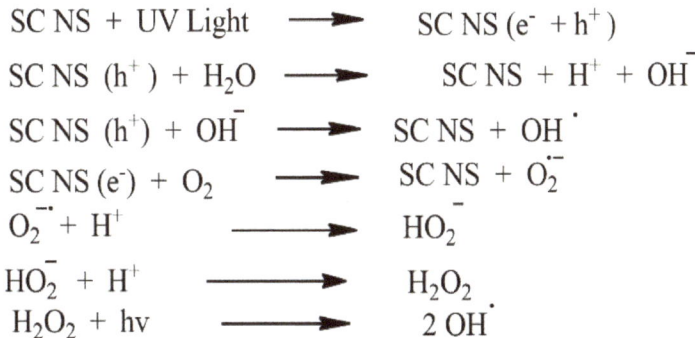

$$SC\ NS + UV\ Light \longrightarrow SC\ NS\ (e^- + h^+)$$

$$SC\ NS\ (h^+) + H_2O \longrightarrow SC\ NS + H^+ + OH^-$$

$$SC\ NS\ (h^+) + OH^- \longrightarrow SC\ NS + OH^•$$

$$SC\ NS\ (e^-) + O_2 \longrightarrow SC\ NS + O_2^{-•}$$

$$O_2^{-•} + H^+ \longrightarrow HO_2^-$$

$$HO_2^- + H^+ \longrightarrow H_2O_2$$

$$H_2O_2 + h\nu \longrightarrow 2\ OH^•$$

Classification of Photocatalysts

The semiconductor nanostructures can be classified into two categories based on chemical composition; pure and modified photocatalyst. Pure semiconductors can be further classified as oxides and non-oxides (Fig. **2**). Non-oxides mainly include chalcogenides (sulphides, selenides, and tellurides) and nitrides. Oxide semiconductors have been extensively studied for their transparency (wide material with a bandgap higher than 3eV), non-toxicity, and conductivity [18]. Chalcogenide semiconductors are semitransparent (Eg>2eV) or non-transparent (Eg< 2eV), and very few chalcogenides have a wide bandgap (ZnS: 3.6-3.8eV, CdS: 2.42eV) [19]. The bandgap of oxides is localized, and low valence state dispersion near the VB introduces a limitation to achieving high mobility. However, chalcogenides (mainly sulphides) exhibit more delocalized VB orbitals and hence offer an advantage over oxides in terms of high mobility [4].

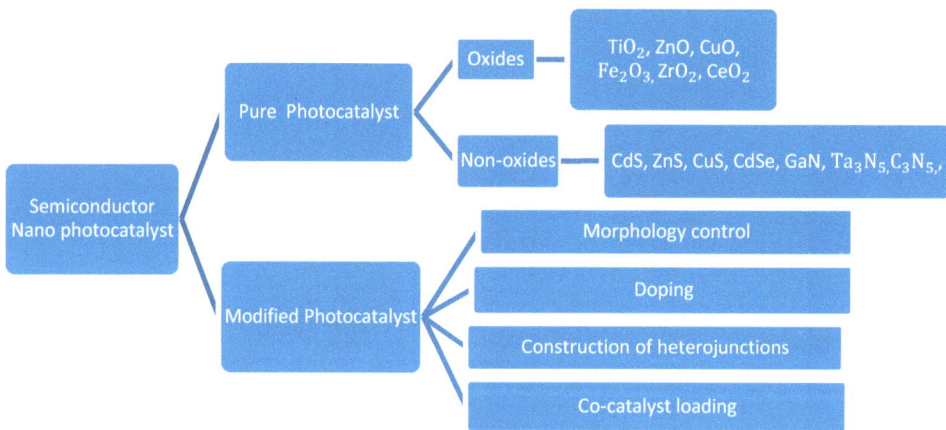

Fig. (2). Classification of semiconductor nano photocatalysts.

Oxide Photocatalyst

Among all the classes of materials, metal oxides play a significant role in photocatalysis [10]. Photocatalysis by semiconductor metal oxides can quickly oxidize a broad range of organic pollutants, pesticides, pharmaceutical by-products, and textile dyes. Pure metal oxide photocatalysts include mainly TiO_2, ZnO, CuO, Fe_2O_3, ZrO_2, and CeO_2. Among these materials, TiO_2 and ZnO have prompted a vast amount of research due to suitable bandgap energy and favourable band positions. TiO_2 crystallizes in three structure types: rutile, anatase, and brookite, which display different photocatalytic activities. Recently,

Advanced Oxidation Processes (AOPs) using TiO_2 have been of great interest in carrying out effective oxidation of a wide variety of organic pollutants, pesticides, and emerging contaminants [20].

ZnO has also been extensively studied because of its high photosensitivity, nontoxic nature, low cost, suitable bandgap (3.37 eV) and large excitation binding energy (60 meV) [21]. This large binding energy indicates that excitonic emission in ZnO can persist even at room temperature and higher. Moreover, the morphologies of ZnO photocatalysts can be easily controlled, which is an important factor influencing photocatalytic performances.

CuO is a promising material to be used in p-n junction-mediated photocatalytic degradation of organic pollutants because of its narrow bandgap (Eg=1.2-1.7 eV), p-type semiconducting properties, non-toxicity, and low production cost. Among all rare-earth oxides, CeO_2 has a wide bandgap of 3.2 eV. The less energy difference among lower (Ce^{3+}) and higher (Ce^{4+}) ions in ceria imparts superior oxygen mobility and storage capacity [22, 23]. Nano-ZrO_2 outperforms several binary nano photocatalysts (ZnO, CeO_2) due to its highly negative conduction band edge potential and wider optical band gaps for different crystalline phases, such as cubic (3.8 eV), tetragonal (4.11 eV), and monoclinic (4.51 eV).

Non-oxide Photocatalyst

In addition to metal oxides, several chalcogenides and nitrides have been investigated for their photocatalytic activity. The most widely investigated non-oxide semiconductor materials in recent years are chalcogenide semiconductors. These are inorganic compounds that consist of anion which belongs to the chalcogen family (sulphides, selenides, and tellurides). Although oxygen is also a member of the chalcogen group, oxides are typically distinguished from chalcogenides due to their different chemical behavior. Among them, metal sulphides (CdS, ZnS, CuS, Sb_2S_3, Bi_2S_3, MoS) are the most commonly studied semiconducting materials because of their narrow bandgap energy (1.3 – 2.40 eV), except for ZnS [4]. ZnS exists in two main crystalline forms, cubic (sphalerite) and hexagonal (wurtzite). The band gaps of cubic and hexagonal ZnS are 3.72 eV and 3.77 eV, respectively [24]. ZnS is responsive only to UV light absorption (λ<340 nm), which can generate electron-hole separation in ZnS [25]. Still, the use of ZnS as a photocatalyst offers advantages in terms of its non-toxic nature and good photocatalytic activity [26].

In the case of transition metal chalcogenides, the valence band and conduction band are composed of hybridized d-orbitals of the metal atom. When an electron is excited from the valence band d_z^2 orbital to the conduction band d_{xy}, d_{x2-y2} orbitals, the interatomic bonding does not get affected, and the compounds are generally resistant to the photo corrosion (photo-induced decomposition) process (an advantage over oxides) [21]. The Cd chalcogenides exhibit different band gaps E_g, decreasing with the increasing atomic number of the chalcogenide (CdS: 2.42 eV, CdSe: 1.84 eV, CdTe: 1.61 eV). Cadmium selenide and telluride can reduce water for hydrogen production because of the negative conduction band position [27]. The valence bands of sulphide semiconductors usually consist of 3p orbitals, which are in a more negative position than the 2p orbital of oxygen. CuS is a good candidate for visible light photocatalysis as it has a narrow bandgap (2.2 eV) and is non-toxic. Metal sulphides also undergo the process of photo corrosion but to less extent as compared to oxides. Despite this, they have attracted attention as photocatalysts with a visible light response. The photo corrosion can be considerably suppressed in the presence of sacrificial reagents such as methanol, triethanolamine, and sodium sulphide/sodium sulphite for oxide and sulphide photocatalysts. The representative chalcogenide photocatalyst is cadmium sulphide (CdS), which is active for H_2 evolution under visible light irradiation; but this material is known to be toxic in aqueous media as it undergoes photoanodic corrosion, releasing dangerous metal ions (Cd^{2+}) into the solution thus limiting CdS practical application for degradation of aqueous pollutants [28]. Moreover, these chalcogenides are unable to oxidize water because the valence band redox potential is lower than 1.23 V (as the redox potential of the reactions compared to SHE is $E(H^+/H_2) = 0$ V and $E(H_2O/O_2) = 1.23$V); therefore CdS, CdSe and CdTe exhibit poor photocatalytic activity for the splitting of water [15].

Semiconductor nitrides have also been explored as semiconductor photocatalysts. Graphite-like carbon nitride (g-C_3N_4) has a bandgap of 2.7 eV and can harvest visible light efficiently [29]. The g-C_3N_4 was able to degrade tylosin (TYL) (antibiotics) within 30 min under simulated sunlight irradiation. Kim *et al.* used urea-modified Ta_3N_5 for the degradation of methylene blue dye under visible light [30].

Methods for Improving Photocatalytic Efficiency

Many factors limit the efficiency of photocatalysis by semiconductor nanostructures. These include a mismatch between the semiconductor bandgap and the solar spectrum, inefficient charge separation, and uncontrolled photo corrosion. The strategies which can help to overcome these limitations and enhance the light absorption capacity in the visible region of the solar spectrum can help in enhancing photocatalytic efficiency. A wide range of structural modifications and crystal growth strategies have been proposed to enhance photocatalytic performance. Various such strategies include morphology control, doping, construction of heterostructures, co-catalyst loading and core-shell nanostructures.

Morphology Control

The efficiency of the photocatalyst also depends on the morphology of the crystal structure. It is widely believed that a larger surface area can provide a greater number of active sites, thus enhancing photocatalytic performance. Much work has been devoted to the controlled morphological synthesis of semiconductor nanostructures. John *et al.* synthesized ZnO nanorods, nanoflakes and quantum dots and studied their photocatalytic activity by taking natural dyes as analytes [31]. The ZnO quantum dots were found to exhibit higher photocatalytic activities as compared to other morphologies. This was due to a larger active surface area obtained in the case of ZnO quantum dots, leading to higher adsorption rates, larger charge separation, and slower recombination rates. Among the three morphologies, quantum dots showed maximum adsorption of dye with kinetic rate constant, k of $0.0093 min.t^{-1}$, which was three times greater than rods and flakes. Abbas *et al.* also synthesized ZnO nanostructures and achieved two different morphologies at different pH values [32]. ZnO nanoparticles were found to have higher efficiency (97.4%) than nanoplates/nanoflowers (86%) towards photodegradation of methylene blue dye. Khan *et al.* synthesized different morphologies of TiO_2 such as nanorod, nanohelics, and nanozigzags [33]. Studies on the morphological influence of TiO_2 nanostructures on photocatalytic degradation of various dyes present in industrial wastewater were also carried out. The nanozigzag films were found to show better performance than nanohelices and nanorods. The effect was attributed to the presence of higher porosity on the surface, which helped in the dispersion of active sites at different scales of pores, large surface area for reaction and presence of oxygen vacancies.

Doping

Doping includes the incorporation of a foreign metal (transition or rare earth metal) or non-metal ion (*e.g.*, nitrogen, tungsten, carbon, fluorine, sulfur, *etc.*) in SCNSs. Metal doping replaces the metal ion from the crystal lattice, whereas doping with non-metal ions replaces the electronegative atom from the crystal lattice. Elements with lower electronegativity than oxygen and a similar size to that of lattice O atoms, such as carbon (C), nitrogen (N), and sulfur (S), can effectively dope metal oxides. Metal doping does not change the position of the conduction and valence band of the host, rather creates new intra-band energy levels, which are useful in the generation of high oxidant holes [34]. Non-metal doping causes bandgap reduction by extending the valence band of the host. Metal doping also protects wide-bandgap photocatalysts against photo-corrosion [34]. Doping changes the bandgap; mostly, the narrow energy bandgap in the electronic structure shifts the absorption range from the UV region to visible light. Thus, the light absorption becomes more efficient, which allows more production of photocarriers and the consequent formation of active species, such as OH˙ [35].

Dopant ions also act as electron scavengers, reducing the recombination rate of electron/hole pairs and consequently increasing the lifetime of charge carriers and also improving the interface and surface characteristics [36]. The energy level of the dopant also affects the trapping efficiency. The trapping of electrons makes it easier for holes to transfer onto the surface of the semiconductor and react with OH^- in an aqueous solution and form active oxygen species to carry out photocatalysis. A large number of studies have been devoted to the synthesis of metal-doped (*e.g.*, Mn, Cu, Pb, and Au) semiconductor nanostructures for their use as photocatalysts [37]. The doped nanostructures enable the utilization of the visible-light spectrum of sunlight and enhance photocatalytic efficiency. The doping of TiO_2 nanoflower thin films with Cu^{2+} decreased the bandgap of TiO_2 and enabled the harvesting of visible light by TiO_2 [38]. The Cu^{2+} ions served as active sites of electron traps and suppressed the electron-hole recombination. The charge transfer property and LSPR (Localized Surface Plasmon Resonance) effect of Cu NPs caused Cu/TiO_2 films to exhibit better catalytic activity toward the reduction of CO_2 to methanol. Yttrium-doped CeO_2 nanomaterial exhibiting high visible light photocatalytic activity towards the degradation of Rhodamine B was reported [39]. The yttrium-doped CeO_2 exhibited a lower bandgap and higher surface area as compared to undoped CeO_2, leading to higher visible light photocatalytic activity. Doping of Au in the crystal lattice of ZnO has shown remarkable results for the degradation of Rhodamine B (RhB) dye [17]. Significant improvement in the visible light photocatalytic activity of carbon-doped ZnO was observed as carbon doping reduced the bandgap of ZnO from ~3.19 eV to ~2.72 eV and improved the separation efficiency of electron-hole pairs [14]. The oxidation state of the dopant ion also affects photocatalytic activity. Nunez *et al.* observed higher photoactivity for Zn:Ga (III) and Zn:Ga (IV) photocatalysts, while Zn:Ga (V) showed lower rates for H_2 production [40].

Construction of Heterostructures

To enhance the photocatalytic performance of semiconductor nanostructures, heterostructures can be prepared from more than one kind of semiconductor material. Semiconductors with narrow bandgaps like CdS, PbS, In_2S_3, and Bi_2S_3 attached to the wide bandgap semiconductors (ZnO, TiO_2, ZnS, *etc.*) surface can make suitable heterojunctions. The formation of heterostructures also leads to changes in the bandgap, which can be engineered to meet specific applications. The heterostructures with a suitable bandgap can function as efficient photocatalysts. Heterostructures are classified as Type-I, Type-II, and Type-III based on the exchange energy between semiconductors (SCs) and the gap between the valence band (VB) and conduction band (CB) [41] (Fig. **3**). In the case of Type-I structure, the bandgap of one component is embedded within the other component. On excitation, both excited electron and hole carriers occupy the lowest electronic states owing to their applications in photoemission devices. In type-II structures, the band gaps of the two components are staggered to each other, which enables the separation of the excited carriers in different regions of the two compositions by excitation. Type-III has a large difference in their band gaps [42].

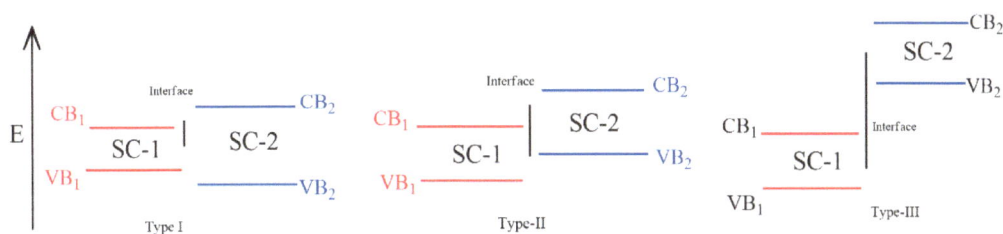

Fig. (3). Classification of heterojunctions of two semiconductors (SCs) as Type-I, Type-II, and Type-III.

Amongst these, type-II heterostructures are most efficient as photocatalysts because they possess efficient charge separation, which reduces the chances of the recombination of electrons and holes. Heterostructures of this type possess one semiconductor active in the UV region (ZnO, TiO_2), while another material is active in visible light (CdS, CdSe, *etc.*). If one excites it under UV radiation, both the band groups get excited, migrating electrons from CB_2 to CB_1, and holes remain with VB_2, resulting in efficient charge separation.

The efficiency of a photocatalyst also depends on the competition of different interface transfer processes. In some cases, lattice mismatch defects arise at the interface. These defects can trap the photogenerated electron or holes and decrease their deactivation by recombination. Further, charge separation arises due to the induction of an electric field at the interface because of lattice mismatch, which leads to higher photocatalytic activity [41]. Xiong *et al.* prepared a heterostructure, Pt-Cu_2O/TiO_2, where Pt and Cu_2O were co-deposited on TiO_2 crystals [43]. Pt captured photogenerated electrons, and co-deposited Cu_2O enhanced the CO_2 chemisorption resulting in enhancement of CO_2 reduction to CH_4. Dutta *et al.* synthesized heterostructures of Au-coupled semiconductor nanomaterials such as Au-SnS, Au−Bi_2S_3, and Au-ZnSe. These were found to have better efficiency than uncoupled SnS, Bi_2S_3, and ZnSe for organic dye degradation/reduction and/or photocatalytic water splitting for the generation of hydrogen [44]. To increase the stability and photocatalytic efficiency of CuO nanowires, n-type semiconductors magnetite (Fe_3O_4) and zinc oxide (ZnO) nanostructures were decorated over CuO nanowires to develop p-n heterojunction [45]. It was observed that 52% of the dye was degraded in 5h of treatment under UV light, while 90% of the dye was degraded within 5 h in the presence of visible light. This enhanced rate of dye degradation in visible light was attributed to the combination of photocatalysis and photosensitization phenomenon. Decorating CuS nanostructure on the surface of ZnO nanotubes led to the formation of a type-II semiconductor with p-n heterojunction [45]. The p-n type junction composites have increased light absorption and effective charge carrier separation and exhibit higher photocatalytic ability. At the equilibrium condition, the p-type becomes negatively charged, and the n-type becomes positively charged. Upon excitation, the negative pole (p-type) attracts holes towards it, and electrons move towards the positive pole (n-type): for example: Cu_2O/TiO_2, NiO/TiO_2, p-ZnO/-TiO_2, TiO_2/ZnO, TiO_2/ZnS, p-CuO/n-ZnO@ Fe_3O_4, MnO_x/TiO_2, CdS/ZnS [37, 46]. Multi-hetero nanostructures like CeO_2/Bi_2MoO_6 have been studied for photocatalytic degradation of rhodamine B dye, methyl orange dye, and tetracycline antibiotic under visible light irradiation [47]. The CeO_2/Bi_2MoO_6 heterojunction exhibited the highest photocatalytic activity with the Rhodamine B dye degradation efficiency of 100% in 75 min, which was considerably higher than those of pristine CeO_2 (26.8%) and Bi_2MoO_6 (80.3%).

The semiconductor heterojunction can also be realized in the form of core-shell geometry to enhance photocatalytic efficiency. In core-shell structure, the charge carriers (electrons and holes) produced under photolytic conditions are confined in either core or shell of the nanostructure [48]. This enables charge separation and reduces the recombination rate of photo-induced charge carriers, thus enhancing photocatalytic efficiency. Kanchandani *et al.* synthesized TiO_2/CuS core/shell nanostructures and compared the activity with designed TiO_2/CuS composite nanostructures under similar conditions [49]. The core/shell nanostructures were found to show 90% photocatalytic efficiency, whereas composite nanostructures exhibited only 58% efficiency. The control over the shell thickness of TiO_2 with the core of Cu_2S directly influenced the optical and surface-interface properties, resulting in enhanced H_2 production with the absorption of UV-visible light as compared to pristine materials [20]. The successful photodegradation and photo-mineralization of paracetamol with TiO_2/Fe_2O_3 core-shell nanostructures have also been reported [9].

Loading of Co-catalyst

The addition of electron and/or hole-transporting co-catalysts is a promising strategy to reduce the recombination of charge carriers and enhance photocatalytic efficiency. The co-loading of both electron and hole-transporting co-catalysts

(C@CoS$_2$ and TFA) on semiconductor hetero-phase homojunction CdS was reported to enhance the photocatalytic efficiency of the material nanostructure water splitting [50]. The co-catalyst Trifluoroacetic acid (TFA) was reported to accelerate the photogenerated hole transfer through a coupled reversible reaction (TFA/TFA$^-$) and act as an electron scavenger, which decreased photogenerated electron-hole recombination and accelerated the evolution of H$_2$. Pure GaN was reported to have very little photocatalytic activity, but the use of cocatalyst (Pt or Rh) greatly enhanced photocatalytic activity for H$_2$ evolution and CO$_2$ reduction compared to powdered GaN [51].

Applications

Hydrogen Production

The use of solar energy for hydrogen production using SCNSs is one of the most promising renewable energy technologies. Photocatalytic production of hydrogen from substrates like H$_2$O, H$_2$S, and organic waste using SCNSs is one of the most promising strategies for converting light energy into chemical energy. Hydrogen, a clean and renewable energy source, has become even more attractive with the depletion of fossil fuel reserves and the deterioration of the global environment [52]. The development of visible-light-driven photocatalysts for solar hydrogen production from water has been researched extensively [53]. For water splitting, the maximum of the VB (VB$_{max}$) should be lower than the oxidation potential of a donor, while the minimum of the CB (CB$_{min}$) is generally higher than the reductive potential of an acceptor. The bandgap energy of semiconductors should be larger than 1.23 eV, and the band edges should straddle the H$^+$/H$_2$ and H$_2$O/O$_2$ potentials for simultaneous H$_2$ and O$_2$ evolution [15].

The SCNSs utilize sunlight and produce the electrons and holes provided the energy state of the photoexcited electrons of the catalytic materials is above the redox potential of transformation of H$^+$ ions to molecular H$_2$. The semiconducting nanostructures having an ideal bandgap (Eg>1.23) can be used as photocatalysts for promoting this hydrogen generation process [53]. In the case of heterophase junction semiconductors, the photogenerated electrons could follow one of the two pathways to reduce the recombination rate and improve the H$^+$/H$_2$ reduction capability. First, the exciting photoelectrons can be transferred from one phase to another phase (cubic phase to hexagonal phase in case of heterophase CdS or ZnS) of nanostructure and effectively participate in hydrogen evolution. Second, the photoexcited electrons can shuttle from one semiconductor to another semiconductor (*e.g.*, CdS to C@CoS$_2$) nanostructure, which separates the photogenerated charge carriers followed by efficient H$^+$/H$_2$ reduction. For instance, considering the redox couple of H$^+$/H$_2$ and H$_2$O/O$_2$ for water splitting, the bandgap energy of semiconductors should be larger than 1.23 eV for simultaneous H_2 and O_2 evolution in the 2:1 ratio [15]. This reaction is named overall water splitting. In these systems, hole or electron scavengers such as CH_3OH or Ag^+ are added to reduce the recombination rate. Liao *et al.* published a comprehensive review on semiconductor-based photocatalytic hydrogen generation [54].

Wastewater Treatment

Water contamination is a global concern as it has adverse effects on the environment and human health along with impacts on economic growth and social perspectives. There is an urgent need to develop efficient and low-cost technologies for wastewater treatment [55]. Wastewater often contains pollutants, such as pesticides, organic dyes, heavy metals, pharmaceuticals, synthetic hormones, *etc.* As compared to traditional treatment methods, which are either less effective or cause secondary pollution, photocatalysis using semiconductor nanostructures is an effective technology for the removal of pollutants from water [56]. The majority of organic wastewater pollutants can be treated by semiconductor nanostructures (TiO_2 and ZnO are widely used) through photochemical transformation on the surface of catalysts. Inorganic pollutants like heavy metals are non-biodegradable and tend to accumulate in living systems. Semiconductor photocatalytic treatment processes can convert ionic species into their metallic solid forms and deposit them over the semiconductor surface or transform them into less toxic, soluble species [57]. The semiconductor material TiO_2 has shown excellent photocatalytic performance in the degradation of major organic pollutants, *e.g.*, pesticides and dyes present in water bodies [58]. Zinc oxide (ZnO) has also been widely used as a photocatalyst for the treatment of organic pollutants [59].

The pollutant species may undergo photoreduction or photooxidation on the surface of the catalyst. Photoreduction occurs if the CB of the semiconductor photocatalyst is more negative than the reduction potential of the chemical species, while photooxidation occurs when the valence band (VB) is more positive than the oxidation potential of the chemical species. The degradation of an organic species like a dye by the action of a semiconductor photocatalyst can take place in two ways; The first possibility is through a photosensitization process, where the radiation excites an electron from the dye and then it is injected into the conduction band of the semiconductor oxide. The second of which is by a true photocatalytic process, where radiation on the photocatalyst promotes an electron from its valence band to the conduction band, and then the electron-hole pair is formed. In both processes, a series of consecutive reactions lead to the eventual mineralization of organic dyes to CO_2 and H_2O. Details of the reaction mechanisms of degradation of dye with semiconductor (SC) photocatalyst by the above-mentioned processes of photosensitization and direct absorption of light are shown below:

a) Photosensitization:

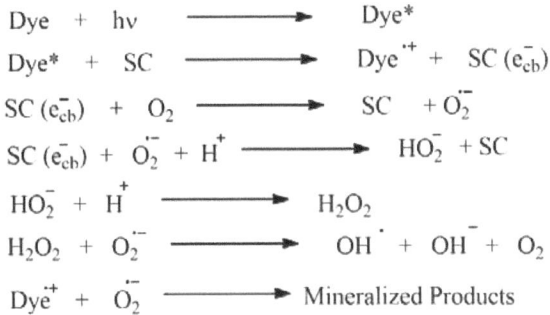

$$Dye + h\nu \longrightarrow Dye*$$

$$Dye* + SC \longrightarrow Dye^{\cdot+} + SC\ (e_{cb}^-)$$

$$SC\ (e_{cb}^-) + O_2 \longrightarrow SC + O_2^{\cdot-}$$

$$SC\ (e_{cb}^-) + O_2^{\cdot-} + H^+ \longrightarrow HO_2^- + SC$$

$$HO_2^- + H^+ \longrightarrow H_2O_2$$

$$H_2O_2 + O_2^{\cdot-} \longrightarrow OH^{\cdot} + OH^- + O_2$$

$$Dye^{\cdot+} + O_2^{\cdot-} \longrightarrow Mineralized\ Products$$

b) Direct activation of semiconductor (SC) by light irradiation:

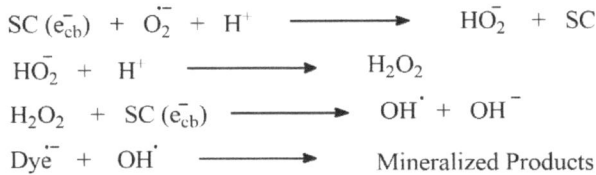

$$SC\ (e_{cb}^-) + O_2^{\cdot-} + H^+ \longrightarrow HO_2^- + SC$$

$$HO_2^- + H^+ \longrightarrow H_2O_2$$

$$H_2O_2 + SC\ (e_{cb}^-) \longrightarrow OH^{\cdot} + OH^-$$

$$Dye^{\cdot-} + OH^{\cdot} \longrightarrow Mineralized\ Products$$

The degradation of several contaminants present in wastewater has been achieved using photocatalysis by semiconductor nanostructures. The degradation of methyl orange (MO) dye was carried out using the heterojunction catalyst TiO_2/ZnO [10]. The two morphologies of the catalyst, hedgehogs and fan blade, were explored under UV light; 97% MO degradation was achieved within 30 mins using the hedgehog morphology. Baradaran *et al.* achieved a degradation efficiency of 95.2% for the degradation of methylene blue dye using Al-doped ZnO thin film [60]. A high catalytic efficiency of 97.2% was achieved for the degradation of common textile dye methylene blue using CdS/RGO composite material, which was higher than the efficiency achieved with pure CdS nanomaterial [61]. The degradation of pharmaceutical paracetamol was achieved using flower-like TiO_2/Fe_2O_3 core-shell photocatalysts [9]. The TiO_2/Fe_2O_3 catalyst with TiO_2 shell exhibited high catalytic efficiency and good reusability. Complete degradation of paracetamol was achieved in 90 min under light irradiation using TiO_2/Fe_2O_3 with 50% titania content [9]. The degradation of bisphenol A (BPA) was reported using $Ni-TiO_2$ catalysts. The complete degradation and 77% mineralization were achieved using a 1.0% $Ni–TiO_2$ catalyst at a time of 210 min [62]. Khan *et al.*

reported photocatalytic degradation of organophosphate pesticide chlorpyrifos using ZnO nanoparticles [63].

Carbon Dioxide Reduction

Semiconductor nanostructures have also contributed to reducing the environmental pollution. The CO_2 concentration has increased enormously in recent decades due to human activities. Semiconductor nanoparticles can be used in environmental remediation as they can potentially convert CO_2 into hydrocarbons like CH_4 and C_2H_4. Details of the reaction mechanism for the reduction of CO_2 using semiconductor nanostructures are shown below:

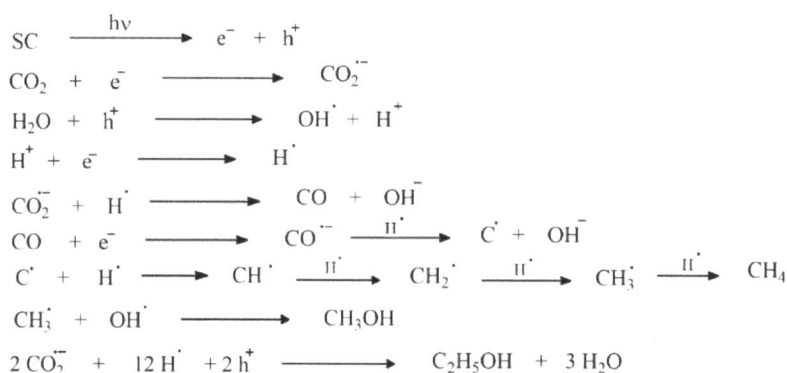

$$SC \xrightarrow{h\nu} e^- + h^+$$

$$CO_2 + e^- \longrightarrow CO_2^{\cdot -}$$

$$H_2O + h^+ \longrightarrow OH^{\cdot} + H^+$$

$$H^+ + e^- \longrightarrow H^{\cdot}$$

$$CO_2^{\cdot -} + H^{\cdot} \longrightarrow CO + OH^-$$

$$CO + e^- \longrightarrow CO^{\cdot -} \xrightarrow{H^{\cdot}} C^{\cdot} + OH^-$$

$$C^{\cdot} + H^{\cdot} \longrightarrow CH^{\cdot} \xrightarrow{H^{\cdot}} CH_2^{\cdot} \xrightarrow{H^{\cdot}} CH_3^{\cdot} \xrightarrow{H^{\cdot}} CH_4$$

$$CH_3^{\cdot} + OH^{\cdot} \longrightarrow CH_3OH$$

$$2\,CO_2^{\cdot -} + 12\,H^{\cdot} + 2\,h^+ \longrightarrow C_2H_5OH + 3\,H_2O$$

Photocatalytic reduction of CO_2 can also yield useful products like methanol and ethanol [64]. In the process of photocatalytic reduction of CO_2, firstly, the electron/hole pairs are generated on illumination with light of suitable energy. The electron-hole pairs are then transferred to the surface of the reactant, where redox reactions occur.

Subsequently, the electrons are transferred from the conduction band of the semiconductor nanostructure for reaction with CO_2 to form CO. Meanwhile, active valence band holes (h^+) react with absorbed H_2O molecules and oxidize water molecules on the surface of catalysts to form hydroxyl radicals ($OH^{\cdot}$), which can release O_2. Hydrogen ions (H^+) are oxidized with the excited electrons, leading to $H^{\cdot}$ radicals. In this case, the CO_2 anion radicals are further formed in water with a high dielectric constant and can be greatly stabilized by the solvent, which results in weakening the interaction of the radical with the catalyst surface. Then, the CO_2 radical anions react with $H^{\cdot}$ radicals to generate CO. At the same time, carbon radicals $C^{\cdot}$ develop by consecutive reactions, and then, $CH_3^{\cdot}$ radicals are formed. The $CH_3^{\cdot}$ radicals react with $OH^{\cdot}$ to produce methanol, and CO_2

radical anions are further oxidized to ethanol under the action of holes (h$^+$) [38]. The proposed mechanism of formation of CH_4, CH_3OH and C_2H_5OH on reduction of CO_2 by semiconductor nanostructures has already been discussed. High activity for photocatalytic reduction of CO_2 to CH_3OH was obtained using Cu/TiO_2 films. The methanol production rate reached 1.8 μmol cm^{-2} h^{-1} with an energy efficiency of 0.8% under UV and visible light irradiation [38]. Efficient photocatalytic reduction of CO_2 into CH_4 and CH_3OH was obtained using semiconductor $TiO_2/MnO_x/Pt$ semiconductor composite material [65].

SENSING APPLICATIONS

Semiconductor nanostructures have been extensively studied because of their outstanding size-dependent electronic, optical, and electrochemical properties. The increase in the bandgap with the decrease in the size of the particles is the most identified aspect of quantum confinement in semiconductors [66]. Luminescent properties of SCNSs have been examined extensively, which originate from electron-hole recombination. The electron relaxes to the lowest energy level in the conduction band before combining with the hole left behind in the valence band. The difference in the energy between the conduction and valence band is conserved in the form of an emitted photon; thus, the size of this gap dictates the wavelength of the light emitted from the SCNSs [67]. Semiconductor nanostructures have generated interest as fluorescent labels due to their excellent chemical and photo-stability, narrow bandwidths, exceptional brightness, and large surface area for surface functionalization [67, 68]. All these properties of SCNSs explain the several applications of these nanostructures when developing sensor probes. In the target analytes detection, the interaction between SCNSs and analytes can affect the fluorescence characteristic of sensors to output specific fluorescence signals, including fluorescence brightness (quenching or enhancement), characteristic wavelength (blue or redshift), anisotropy, and lifetime through different signal transduction mechanisms [69]. Fluorescence intensity changes are usually observed in SCNSs sensors upon analyte recognition, particularly the fluorescence quenching [68]. The signal transduction mechanism that happened upon the analyte binding with SCNSs can be explained by two processes, namely fluorescence resonance energy transfer (FRET) and photo-induced electron transfer (PET).

The semiconductor nanostructures (SNs) also show promise as electrochemical sensors. The SCNSs as sensing elements could be immobilized on working electrode surfaces by different methods, such as physical adsorption, chemical covalent bonding, or electro-polymerization [70]. The modified electrode could give a porous film with a high specific area, where the local environment of

SCNSs could be controlled by the cross-linking elements. The modified electrode may lead to specific and selective interactions with substrates. The catalytic properties of SCNSs could decrease the over-potential of reactions in the electrochemical analysis, which could further enhance the electrochemical responses greatly [70]. The constructed SNs-modified electrochemical sensors have special functions and could be readily applied to the sensing or biosensing of the target molecule.

Semiconductor nanostructures (SNs) have also been used as chemiluminescence sensors (CL) for the determination of analytes due to their high sensitivity, wide calibration range, low-cost analysis, and suitability for automation offered by such sensors. This section focuses on some important sensing applications of SCNSs as chemical sensors, gas sensors, and biosensors based on different transduction mechanisms (fluorescence, chemiluminescent, and electrochemical).

Chemical Sensors

A chemical sensor is a device that transforms chemical information ranging from the concentration of a specific sample component to the analysis of total composition into an analytically useful signal [71]. The desirable characteristics shown by SCNSs, *e.g.*, amazing optoelectronic properties, high emission quantum yield, size-tunable emission profiles, and narrow spectral bands, make them suitable for use as chemical sensors for a large number of organic and inorganic compounds.

Sensing of Organic Compounds

Organic compounds, including pesticides, drugs, explosives, and food additives, must be monitored regularly as they pose a serious threat to human health, public safety and the environment. Fluorescent SCNSs are suitable for rapid and sensitive detection of such analytes [72]. The herbicide paraquat was determined in real water samples using mercaptopropionic acid-modified CdSe/ZnS semiconductor quantum dots (SQD) based fluorescence probe. The mechanism of recognition was based on the fluorescence quenching of SQDs by paraquat herbicide, and a detection limit of 3 ng/L was achieved. Three carbamate pesticides (metolcarb, carbofuran, and carbaryl) were determined using a novel paper-based sensor based on CdTe SQDs and nanoZnTPyP "(nano zinc 5, 10, 15, 20-tetra (4-pyridyl)-21H 23H-porphine) [73]. A rapid method was developed for the detection of organic phosphorothioate pesticides based on the CdTe SQDs probe. The fluorescence of CdTe QDs was quenched by dithizone due to the FRET process [74]. Nitroaromatic compounds like trinitrotoluene (TNT) and trinitrophenol (TNP) are dangerous to humans and toxic to the environment. Fluorescent MoS_2 SQDs prepared through a simple and eco-friendly hydrothermal method have been used effectively for the detection of trinitrophenol (TNP) in aqueous media [75]. The fluorescence of MoS_2 QDs was selectively quenched by TNP *via* the radiative energy transfer (RET) process in the presence of other nitroaromatics (2,6-dinitrotoluene, 2-nitrotoluene, and 4-nitrotoluene). A novel, simple and inexpensive paper electrochemical platform modified with QDs was developed. The platform consisted of a carbon three-electrode setup integrated with the paper substrate [76]. The graphite working electrode was modified with CdSe/CdS QDs to improve the sensitivity achieving a detection limit of 96 nM.

The lack of selectivity of SCNSs-based probes was commonly reported, and the research on the development of novel and selective SCNSs-based sensors is a currently developing area. The specific recognition properties of SCNSs can be improved by their combination with molecularly imprinted polymers (MIPs). The high selectivity of MIPs is attributed to the recognition cavities generated after the analyte reaction with the functional monomer in the presence of a crosslinker [77]. These cavities are complementary to the template molecule in shape, size, and chemical functionality, ensuring selectivity.

A novel molecularly imprinted fluorescence sensor was constructed by developing mesoporous structured imprinting on the surface of CdTe QDs for the selective and sensitive detection of 2,4-dichlorophenoxyacetic acid *via* electron transfer-induced fluorescence quenching mechanism [78]. Hassanzadeh *et al.* developed a molecularly imprinted polymer (MIP) based chemiluminescence assay for its determination in environmental samples [79]. ZnO QDs were used as a support for the MIP layer, and synthesized ZnO-MIP composites showed a remarkable promoting effect on the weak chemiluminescence (CL) emission of alkaline permanganate rhodamine B ($KMnO_4$-RB). In the presence of TNT, its molecules were adsorbed on the specific sites of MIP and placed near the QDs, resulting in a sensible decrease in the CL intensity of $KMnO_4$-RB-MIP@QDs system with a detection limit of 6.8 pg/ml. Based on this interaction, a selective and sensitive CL assay was developed for TNT determination.

Sensing of Metal Ions

SCNSs have been frequently used in sensing metal ions, such as Hg, Pb, and Cu. The common mechanism involves the quenching of photoluminescence of SCNSs by the metal ion *via* an energy or charge transfer mechanism. The sensing of heavy metal ion Hg^{2+} was reported using mercaptopropionic acid (MPA) capped Mn-doped ZnSe/ZnS QDs. The system was able to detect Hg^{2+} ions at concentrations as low as 0.1 nM [80]. The monitoring of the Pb^{2+} in an aqueous medium has been reported using amine-functionalized MoS_2 QDs [81]. The PL intensity of MoS_2 QDs was quenched strongly in the presence of Pb^{2+} ions, which formed the basis of the development of a method for sensing Pb^{2+} with a detection limit of 50 μM. The detection of Cu^{2+} in water with high selectivity and sensitivity was achieved using hexadecyltrimethylammonium bromide (CTAB) modified ZnS/CdTe QD-based fluorescent probe [82].

Gas Sensors

Gas sensing technology is widely used in industrial and domestic fields, specifically in the automotive industry, indoor air quality control, and monitoring of the greenhouse effect [83]. The application of SCNSs, specifically oxides as gas sensors, is widespread. The most important parameters of gas sensing devices are sensitivity, operating temperature, selectivity, long-term stability, energy consumption, and finally, production cost [84]. Gas sensors based on ZnO, TiO_2, SnO/SnO_2, WO_3, CuO/Cu_2O, and V_2O_5 are commonly used to detect combustible, reducing and oxidizing gases. The sensing mechanism is mainly based on measurements of the change of resistance in response to the target gases [84]. The gas sensing mechanism can be different depending on the sensor materials and the gases to be detected. Metal oxide semiconductors, such as n-type SnO_2 and ZnO, share a common sensing mechanism involved in the surface reactions between analyte molecules and chemisorbed oxygen species [85]. Exposure to reductive gases, such as ethanol, will cause an increase in the conductivity for n-type semiconductors and a decrease for p-type materials, whereas the effect of oxidative gases is reversed [85]. The mechanism becomes more complex when pure metal oxide nanostructures are further functionalized with noble metal nanoparticles to improve sensor performances. Wang *et al.* reported that Pt-decorated In_2O_3 nanotubes were highly sensitive to H_2 and CO at room temperature [86]. The Pt nanoparticles supported on In_2O_3 enhanced the dissociation of oxygen molecules and hydrogen adsorption, which enabled the sensor to be sensitive at 25°C. The gas response to 1.5 vol% CO was ten times higher than that of H_2. The combination of different metal oxides to produce heterostructures can result in enhanced charge transduction and modulation potential barriers at the grain boundaries, which are advantageous for gas sensors. The p-n heterojunctions, such as p-type and n-type metal oxides, such as CuO-SnO_2, NiO-WO_3 and CuO-ZnO nanostructures, showed enhanced sensing properties [85]. A plate-like heterogenous NiO/WO_3 nanocomposites showed excellent sensitivity toward NO_2 and ultrafast response at room temperature due to their p-n heterogeneous characteristics [87].

Non-stoichiometric tungsten oxide ($W_{18}O_{49}$) can be used as a room-temperature gas sensor. This is because the large number of oxygen vacancies in $W_{18}O_{49}$ can serve as the adsorption sites and facilitate the chemisorption of oxygen at low temperatures. Zhao *et al.* reported that ultrathin $W_{18}O_{49}$ nanowire bundles exhibited room temperature ammonia sensing performance at sub-ppm level [88] compared to WO_3 nanowires, which exhibited sensitivity only at higher temperatures. WO_3 nanorods (WO_3 was a mixture of non-stoichiometric monoclinic $WO_{2.83}$ and $WO_{2.92}$ and tetragonal W_5O_{14} phases as analyzed by XRD) showed a p-type response behaviour to reducing gases (C_2H_5OH, CH_3OH, and C_3H_8O) at room temperature [89].

Biosensors

The unique properties of SCNSs make them one of the most exciting and versatile luminophores to be used as fluorescent labels in biosensing. Surface functionalization plays a key role in determining the biofunctionality of SCNSs. It provides hydrophilicity and stability to SCNSs in a range of biological media, including fluids, tissues, and within cells. Conjugation of SCNSs with bio-recognition elements is a critical step in enabling SCNSs to specifically recognize and bind with the biomolecular targets. Generally, there are two main strategies for the conjugation of SCNSs with bio-recognition elements, namely covalent and non-covalent [90]. Covalent or non-covalent conjugates of SCNSs with different bio-recognition elements (proteins, antibodies, aptamers, small molecules, nucleic acids, liposomes, and monosaccharides) can be used as fluorescence-encoded microbeads for multiple, sensitive, and fast sensing [91].

The use of SCNSs as donors in FRET has been well documented, with a wider approach to integrating SCNSs for bioanalysis. SCNSs FRET analysis was first reported by Mattoussi and co-workers, who used dye-labeled protein SCNSs conjugates to elucidate the fundamental spectroscopic properties of SCNSs as donors and acceptors [92]. Colistin-capped CdSe/ZnS QDs were used as a fluorescent probe to detect *E.coli,* and within 15 min, 28 *E.coli* cells/ml were detected [93]. *Staphylococcus aureus* secretes extracellular proteins, micrococcal nuclease (MNase) and *thermonuclease* (TNase), which are responsible for spoiling food and milk products. Water-soluble CdTe QDs were synthesized for their conjugation with antibodies through streptavidin coupling [94]. This was used for the development of a sandwich-type detection strip and FRET immunoassay protocol for the detection of TNase of *S. aureus* in contaminated food and water samples. The thrombin activity in human serum and whole blood was measured *via* FRET loss from CdSe/CdS/ZnS QDs donors to multiple A647 acceptor dye-labelled peptides. Recovery of SCNSs PL from the sample resulted in a loss of FRET due to thrombin activity [95]. A more recent approach for the use of SCNSs as energy acceptors in bioanalytical probes was established. SCNSs act much more effectively as FRET acceptors with lanthanide-based donors. FRET from Tb complexes to SCNSs (CdSe/CdTe/ZnS) with different core sizes has been reported using this approach, enabling a very sensitive fivefold multiplexed bioassay with time and colour-resolved detection of up to five biomarkers [96].

SCNSs-based chemiluminescence resonance energy transfer (CRET) modes have been used for bioanalysis. The DNAzyme generated CRET to CdSe/ZnS QDs is implemented to develop aptamer or DNA sensing platforms. This enables the chemiluminescent detection of Hg^{2+} [97]. SCNSs modified aptamer chemiluminescence probe was applied to detect carcinoembryonic antigen using capillary electrophoresis based on CRET from horseradish peroxidase and SNQDs [98].

Charge transfer reactions, such as PET between SCNSs and other redox-active species, provide a mechanism for the on/off switching of SCNSs photoluminescence. These reactions also led to the modulation of photocurrent intensity and therefore, can be used in photoelectrochemical (PEC) biosensing. PEC biosensors consist of SCNSs immobilized by a linking molecule to an electrode so that upon their photo-excitation, a PET-induced photocurrent is generated, which depends on the type and concentration of the bioanalyte providing the analytical signal [90]. SCNSs based PEC sensors provide an important branch of biosensing as they are superior to the other sensor types. A PEC glucose biosensor based on a dehydrogenase enzyme and the NAD (+)/ NADH with a glucose detection limit of 0.09 mM was reported [99].

CONCLUSION

The semiconductor nanostructures are highly promising materials offering solutions to many energy and environment-related issues the world is facing today. Semiconductor photocatalysis presents an effective and efficient technology for photocatalytic applications. The development of different strategies, such as morphology control, doping, and the formation of heterostructures loading of co-catalyst, has helped in increasing the efficiency of semiconductor nanostructures for photocatalytic applications. Semiconductor nanostructures have shown great potential in hydrogen production, wastewater treatment, and carbon dioxide reduction. The semiconductor nanostructures also form sensitive sensing platforms. The semiconductor nanostructures have been explored as chemical sensors for the detection of organic compounds and metal ions, gas sensors, and biosensors and demonstrate great potential for sensing applications.

ACKNOWLEDGMENTS

The authors (SM and KK) are thankful to Mata Gujri College, Fatehgarh Sahib, Punjab, for providing lab facilities. The authors (SB and AKM) are also thankful to UGC-SAP and Chemistry Department, Punjabi University, Patiala, Punjab, for providing lab and instrument facilities.

REFERENCES

[1]   Kilina, S.; Kilin, D.; Tretiak, S. Light-driven and phonon-assisted dynamics in organic and semiconductor nanostructures. *Chem. Rev.,* **2015**, *115*(12), 5929-5978.
[http://dx.doi.org/10.1021/acs.chemrev.5b00012] [PMID: 25993511]

[2]   Regulacio, M.D.; Han, M.Y. Multinary I-III-VI$_2$ and I$_2$-II-IV-VI$_4$ semiconductor nanostructures for photocatalytic applications. *Acc. Chem. Res.,* **2016**, *49*(3), 511-519.
[http://dx.doi.org/10.1021/acs.accounts.5b00535] [PMID: 26864703]

[3]   Liu, B.; Li, J.; Yang, W.; Zhang, X.; Jiang, X.; Bando, Y. Semiconductor solid-solution nanostructures: synthesis, property tailoring, and applications. *Small,* **2017**, *13*(45), 1701998.
[http://dx.doi.org/10.1002/smll.201701998] [PMID: 28961363]

[4]   Woods-Robinson, R.; Han, Y.; Zhang, H.; Ablekim, T.; Khan, I.; Persson, K.A.; Zakutayev, A. Wide band gap chalcogenide semiconductors. *Chem. Rev.,* **2020**, *120*(9), 4007-4055.
[http://dx.doi.org/10.1021/acs.chemrev.9b00600] [PMID: 32250103]

[5]   Xia, Y.; Wang, J.; Chen, R.; Zhou, D.; Xiang, L. A review on the fabrication of hierarchical ZnO nanostructures for photocatalysis application. *Crystals (Basel),* **2016**, *6*(11), 148.
[http://dx.doi.org/10.3390/cryst6110148]

[6]   Bashir, I; Lone, FA; Bhat, RA; Mir, SA; Dar, ZA; Dar, SA. Concerns and threats of contamination on aquatic ecosystems. *Biorem. Biotech.,* **2020**, 1-26.
[http://dx.doi.org/10.1007/978-3-030-35691-0_1]

[7]   Deng, Y.; Zhao, R. Advanced oxidation processes (AOPs) in wastewater treatment. *Curr. Pollut. Rep.,* **2015**, *1*(3), 167-176.
[http://dx.doi.org/10.1007/s40726-015-0015-z]

[8]   Dhandapani, C.; Narayanasamy, R.; Karthick, S.N.; Hemalatha, K.V.; Selvam, S.; Hemalatha, P.; kumar, M.S.; Kirupha, S.D.; Kim, H-J. Drastic photocatalytic degradation of methylene blue dye by neodymium doped zirconium oxide as photocatalyst under visible light irradiation. *Optik (Stuttg.),* **2016**, *127*(22), 10288-10296.
[http://dx.doi.org/10.1016/j.ijleo.2016.08.048]

[9]   Abdel-Wahab, A.M.; Al-Shirbini, A.S.; Mohamed, O.; Nasr, O. Photocatalytic degradation of paracetamol over magnetic flower-like TiO$_2$/Fe$_2$O$_3$ core-shell nanostructures. *J. Photochem. Photobiol. Chem.,* **2017**, *347*, 186-198.
[http://dx.doi.org/10.1016/j.jphotochem.2017.07.030]

[10] Zha, R.; Nadimicherla, R.; Guo, X. Ultraviolet photocatalytic degradation of methyl orange by nanostructured TiO$_2$/ZnO heterojunctions. *J. Mater. Chem. A Mater. Energy Sustain.,* **2015**, *3*(12), 6565-6574.
[http://dx.doi.org/10.1039/C5TA00764J]

[11] Ganguly, A.; Anjaneyulu, O.; Ojha, K.; Ganguli, A.K. Oxide-based nanostructures for photocatalytic and electrocatalytic applications. *Cryst. Eng. Comm.,* **2015**, *17*(47), 8978-9001.
[http://dx.doi.org/10.1039/C5CE01343G]

[12] Cheng, M.; Yang, S.; Chen, R.; Zhu, X.; Liao, Q.; Huang, Y. Copper-decorated TiO$_2$ nanorod thin films in optofluidic planar reactors for efficient photocatalytic reduction of CO$_2$. *Int. J. Hydrogen*

Energy, **2017**, *42*(15), 9722-9732.
[http://dx.doi.org/10.1016/j.ijhydene.2017.01.126]

[13] Xing, X.; Wang, D.; Chen, Z.; Zheng, B.; Li, B.; Wu, D. ZnTe quantum dots as fluorescence sensors for the detection of iron ions. *J. Mater. Sci. Mater. Electron.,* **2018**, *29*(16), 14192-14199.
[http://dx.doi.org/10.1007/s10854-018-9552-8]

[14] Zhang, X.; Qin, J.; Hao, R.; Wang, L.; Shen, X.; Yu, R.; Limpanart, S.; Ma, M.; Liu, R. Carbon-doped ZnO nanostructures: facile synthesis and visible light photocatalytic applications. *J. Phys. Chem. C,* **2015**, *119*(35), 20544-20554.
[http://dx.doi.org/10.1021/acs.jpcc.5b07116]

[15] Xiao, M.; Wang, Z.; Lyu, M.; Luo, B.; Wang, S.; Liu, G.; Cheng, H.M.; Wang, L. Hollow nanostructures for photocatalysis: advantages and challenges. *Adv. Mater.,* **2019**, *31*(38), 1801369.
[http://dx.doi.org/10.1002/adma.201801369] [PMID: 30125390]

[16] Yan, H.; Wang, X.; Yao, M.; Yao, X. Band structure design of semiconductors for enhanced photocatalytic activity: The case of TiO$_2$. *Prog. Nat. Sci.,* **2013**, *23*(4), 402-407.
[http://dx.doi.org/10.1016/j.pnsc.2013.06.002]

[17] Jiang, T.; Qin, X.; Sun, Y.; Yu, M. UV photocatalytic activity of Au@ZnO core–shell nanostructure with enhanced UV emission. *RSC Advances,* **2015**, *5*(80), 65595-65599.
[http://dx.doi.org/10.1039/C5RA11653H]

[18] Zhang, K.; Jin, L.; Yang, Y.; Guo, K.; Hu, F. Novel method of constructing CdS/ZnS heterojunction for high performance and stable photocatalytic activity. *J. Photochem. Photobiol. Chem.,* **2019**, *380*, 111859.
[http://dx.doi.org/10.1016/j.jphotochem.2019.111859]

[19] Avilés, M.A.; Córdoba, J.M.; Sayagués, M.J.; Gotor, F.J. Tailoring the band gap in the ZnS/ZnSe system: solid solutions by a mechanically induced self-sustaining reaction. *Inorg. Chem.,* **2019**, *58*(4), 2565-2575.
[http://dx.doi.org/10.1021/acs.inorgchem.8b03183] [PMID: 30694058]

[20] Navakoteswara Rao, V.; Lakshmana Reddy, N.; Mamatha Kumari, M.; Ravi, P.; Sathish, M.; Kuruvilla, K.M.; Preethi, V.; Reddy, K.R.; Shetti, N.P.; Aminabhavi, T.M.; Shankar, M.V. Photocatalytic recovery of H$_2$ from H$_2$S containing wastewater: Surface and interface control of photo-excitons in Cu$_2$S@TiO$_2$ core-shell nanostructures. *Appl. Catal. B,* **2019**, *254*, 174-185.
[http://dx.doi.org/10.1016/j.apcatb.2019.04.090]

[21] Yu, L.; Chen, W.; Li, D.; Wang, J.; Shao, Y.; He, M.; Wang, P.; Zheng, X. Inhibition of photocorrosion and photoactivity enhancement for ZnO *via* specific hollow ZnO core/ZnS shell structure. *Appl. Catal. B,* **2015**, *164*, 453-461.
[http://dx.doi.org/10.1016/j.apcatb.2014.09.055]

[22] Maria Magdalane, C.; Kaviyarasu, K.; Judith Vijaya, J.; Siddhardha, B.; Jeyaraj, B. Facile synthesis of heterostructured cerium oxide/yttrium oxide nanocomposite in UV light induced photocatalytic degradation and catalytic reduction: Synergistic effect of antimicrobial studies. *J. Photochem. Photobiol. B,* **2017**, *173*, 23-34.
[http://dx.doi.org/10.1016/j.jphotobiol.2017.05.024] [PMID: 28554073]

[23] Maria Magdalane, C.; Kaviyarasu, K.; Raja, A.; Arularasu, M.V.; Mola, G.T.; Isaev, A.B.; Al-Dhabi, N.A.; Arasu, M.V.; Jeyaraj, B.; Kennedy, J.; Maaza, M. Photocatalytic decomposition effect of erbium doped cerium oxide nanostructures driven by visible light irradiation: Investigation of cytotoxicity, antibacterial growth inhibition using catalyst. *J. Photochem. Photobiol. B,* **2018**, *185*, 275-282.
[http://dx.doi.org/10.1016/j.jphotobiol.2018.06.011] [PMID: 30012250]

[24] Lee, J.; Ham, S.; Choi, D.; Jang, D.J. Facile fabrication of porous ZnS nanostructures with a controlled amount of S vacancies for enhanced photocatalytic performances. *Nanoscale,* **2018**, *10*(29), 14254-14263.
[http://dx.doi.org/10.1039/C8NR02936A] [PMID: 30010687]

[25]   Lee, G.J.; Wu, J.J. Recent developments in ZnS photocatalysts from synthesis to photocatalytic applications — A review. *Powder Technol.,* **2017**, *318*, 8-22.
[http://dx.doi.org/10.1016/j.powtec.2017.05.022]

[26]   Kumar, G.A.; Naik, H.S.B.; Viswanath, R.; Gowda, I.K.S.; Santhosh, K.N. Tunable emission property of biotin capped Gd:ZnS nanoparticles and their antibacterial activity. *Mater. Sci. Semicond. Process.,* **2017**, *58*, 22-29.
[http://dx.doi.org/10.1016/j.mssp.2016.11.002]

[27]   Holmes, M.A.; Townsend, T.K.; Osterloh, F.E. Quantum confinement controlled photocatalytic water splitting by suspended CdSe nanocrystals. *Chem. Commun. (Camb.),* **2012**, *48*(3), 371-373.
[http://dx.doi.org/10.1039/C1CC16082F] [PMID: 22083249]

[28]   Cheng, L.; Xiang, Q.; Liao, Y.; Zhang, H. CdS-Based photocatalysts. *Energy Environ. Sci.,* **2018**, *11*(6), 1362-1391.
[http://dx.doi.org/10.1039/C7EE03640J]

[29]   Tian, N.; Huang, H.; Du, X.; Dong, F.; Zhang, Y. Rational nanostructure design of graphitic carbon nitride for photocatalytic applications. *J. Mater. Chem. A Mater. Energy Sustain.,* **2019**, *7*(19), 11584-11612.
[http://dx.doi.org/10.1039/C9TA01819K]

[30]   Kim, J.Y.; Lee, M.H.; Kim, J.H.; Kim, C.W.; Youn, D.H. Facile nanocrystalline Ta_3N_5 synthesis for photocatalytic dye degradation under visible light. *Chem. Phys. Lett.,* **2020**, *738*, 136900.
[http://dx.doi.org/10.1016/j.cplett.2019.136900]

[31]   John Peter, I.; Praveen, E.; Vignesh, G.; Nithiananthi, P. ZnO nanostructures with different morphology for enhanced photocatalytic activity. *Mater. Res. Express,* **2017**, *4*(12), 124003.
[http://dx.doi.org/10.1088/2053-1591/aa9d5d]

[32]   Abbas, K.N.; Bidin, N. Morphological driven photocatalytic activity of ZnO nanostructures. *Appl. Surf. Sci.,* **2017**, *394*, 498-508.
[http://dx.doi.org/10.1016/j.apsusc.2016.10.080]

[33]   Khan, S.B.; Hou, M.; Shuang, S.; Zhang, Z. Morphological influence of TiO 2 nanostructures (nanozigzag, nanohelics and nanorod) on photocatalytic degradation of organic dyes. *Appl. Surf. Sci.,* **2017**, *400*, 184-193.
[http://dx.doi.org/10.1016/j.apsusc.2016.12.172]

[34]   Samadi, M.; Zirak, M.; Naseri, A.; Khorashadizade, E.; Moshfegh, A.Z. Recent progress on doped ZnO nanostructures for visible-light photocatalysis. *Thin Solid Films,* **2016**, *605*, 2-19.
[http://dx.doi.org/10.1016/j.tsf.2015.12.064]

[35]   Sushma, C.; Girish Kumar, S. Advancements in the zinc oxide nanomaterials for efficient photocatalysis. *Chem. Pap.,* **2017**, *71*(10), 2023-2042.
[http://dx.doi.org/10.1007/s11696-017-0217-5]

[36]   Huang, F; Yan, A; Zhao, H Influences of doping on photocatalytic properties of TiO_2 photocatalyst. *Semiconductor Photocatalysis-Materials, Mechanisms and Applications.,* **2016**, 31-80.
[http://dx.doi.org/10.5772/63234]

[37]   Zhang, F.; Wang, X.; Liu, H.; Liu, C.; Wan, Y.; Long, Y.; Cai, Z. Recent advances and applications of semiconductor photocatalytic technology. *Appl. Sci. (Basel),* **2019**, *9*(12), 2489.
[http://dx.doi.org/10.3390/app9122489]

[38]   Liu, E.; Qi, L.; Bian, J.; Chen, Y.; Hu, X.; Fan, J.; Liu, H.; Zhu, C.; Wang, Q. A facile strategy to fabricate plasmonic Cu modified TiO_2 nano-flower films for photocatalytic reduction of CO_2 to methanol. *Mater. Res. Bull.,* **2015**, *68*, 203-209.
[http://dx.doi.org/10.1016/j.materresbull.2015.03.064]

[39]   Akbari-Fakhrabadi, A.; Saravanan, R.; Jamshidijam, M.; Mangalaraja, R.V.; Gracia, M.A. Preparation of nanosized yttrium doped CeO_2 catalyst used for photocatalytic application. *J. Saudi Chem. Soc.,*

2015, *19*(5), 505-510.
[http://dx.doi.org/10.1016/j.jscs.2015.06.003]

[40]   Núñez, J.; Fresno, F.; Platero-Prats, A.E.; Jana, P.; Fierro, J.L.G.; Coronado, J.M.; Serrano, D.P.; de la Peña O'Shea, V.A. Ga-promoted photocatalytic H_2 production over Pt/ZnO nanostructures. *ACS Appl. Mater. Interfaces*, **2016**, *8*(36), 23729-23738.
[http://dx.doi.org/10.1021/acsami.6b07599] [PMID: 27541830]

[41]   Xu, C.; Ravi Anusuyadevi, P.; Aymonier, C.; Luque, R.; Marre, S. Nanostructured materials for photocatalysis. *Chem. Soc. Rev.*, **2019**, *48*(14), 3868-3902.
[http://dx.doi.org/10.1039/C9CS00102F] [PMID: 31173018]

[42]   Prusty, G.; Guria, A.K.; Mondal, I.; Dutta, A.; Pal, U.; Pradhan, N. Modulated binary-ternary dual semiconductor heterostructures. *Angew. Chem. Int. Ed.*, **2016**, *55*(8), 2705-2708.
[http://dx.doi.org/10.1002/anie.201509701] [PMID: 26800297]

[43]   Xiong, Z.; Lei, Z.; Kuang, C.C.; Chen, X.; Gong, B.; Zhao, Y.; Zhang, J.; Zheng, C.; Wu, J.C.S. Selective photocatalytic reduction of CO_2 into CH_4 over Pt-Cu_2O TiO_2 nanocrystals: The interaction between Pt and Cu_2O cocatalysts. *Appl. Catal. B*, **2017**, *202*, 695-703.
[http://dx.doi.org/10.1016/j.apcatb.2016.10.001]

[44]   Dutta, S.K.; Mehetor, S.K.; Pradhan, N. Metal semiconductor heterostructures for photocatalytic conversion of light energy. *J. Phys. Chem. Lett.*, **2015**, *6*(6), 936-944.
[http://dx.doi.org/10.1021/acs.jpclett.5b00113] [PMID: 26262849]

[45]   Basu, M.; Garg, N.; Ganguli, A.K. A type-II semiconductor (ZnO/CuS heterostructure) for visible light photocatalysis. *J. Mater. Chem. A Mater. Energy Sustain.*, **2014**, *2*(20), 7517-7525.
[http://dx.doi.org/10.1039/C3TA15446G]

[46]   Li, Y.; Zhang, W.; Shen, X.; Peng, P.; Xiong, L.; Yu, Y. Octahedral Cu_2O-modified TiO_2 nanotube arrays for efficient photocatalytic reduction of CO_2. *Chin. J. Catal.*, **2015**, *36*(12), 2229-2236.
[http://dx.doi.org/10.1016/S1872-2067(15)60991-3]

[47]   Li, S.; Hu, S.; Jiang, W.; Liu, Y.; Zhou, Y.; Liu, J.; Wang, Z. Facile synthesis of cerium oxide nanoparticles decorated flower-like bismuth molybdate for enhanced photocatalytic activity toward organic pollutant degradation. *J. Colloid Interface Sci.*, **2018**, *530*, 171-178.
[http://dx.doi.org/10.1016/j.jcis.2018.06.084] [PMID: 29982008]

[48]   Lo, S.S.; Mirkovic, T.; Chuang, C.H.; Burda, C.; Scholes, G.D. Emergent properties resulting from type-II band alignment in semiconductor nanoheterostructures. *Adv. Mater.*, **2011**, *23*(2), 180-197.
[http://dx.doi.org/10.1002/adma.201002290] [PMID: 21069886]

[49]   Khanchandani, S.; Kumar, S.; Ganguli, A.K. Comparative study of TiO2/CuS core/shell and composite nanostructures for efficient visible light photocatalysis. *ACS Sustain. Chem.& Eng.*, **2016**, *4*(3), 1487-1499.
[http://dx.doi.org/10.1021/acssuschemeng.5b01460]

[50]   Reddy, D.A.; Kim, E.H.; Gopannagari, M.; Ma, R.; Bhavani, P.; Kumar, D.P.; Kim, T.K. Enhanced photocatalytic hydrogen evolution by integrating dual co-catalysts on heterophaseCdSnano-junctions. *ACS Sustain. Chem.& Eng.*, **2018**, *6*(10), 12835-12844.
[http://dx.doi.org/10.1021/acssuschemeng.8b02098]

[51]   Pang, H.; Liu, L.; Ouyang, S.; Xu, H.; Li, Y.; Wang, D. Structure, Optical Properties, and Photocatalytic Activity towards H_2 Generation and CO_2 Reduction of GaN Nanowires *via* Vapor-Liquid-Solid Process. *Int. J. Photoenergy*, **2014**, 894396.

[52]   Rosen, M.A.; Koohi-Fayegh, S. The prospects for hydrogen as an energy carrier: an overview of hydrogen energy and hydrogen energy systems. *Energy Ecol. Environ.*, **2016**, *1*(1), 10-29.
[http://dx.doi.org/10.1007/s40974-016-0005-z]

[53]   Shwetharani, R.; Sakar, M.; Fernando, C.A.N.; Binas, V.; Balakrishna, R.G. Recent advances and strategies to tailor the energy levels, active sites and electron mobility in titania and its

doped/composite analogues for hydrogen evolution in sunlight. *Catal. Sci. Technol.,* **2019**, *9*(1), 12-46.
[http://dx.doi.org/10.1039/C8CY01395K]

[54]   Liao, C.H.; Huang, C.W.; Wu, J.C.S. Hydrogen production from semiconductor-based photocatalysis
       via water splitting. *Catalysts,* **2012**, *2*(4), 490-516.
       [http://dx.doi.org/10.3390/catal2040490]

[55]   Yaqoob, A.A.; Parveen, T.; Umar, K.; Mohamad Ibrahim, M.N. Role of nanomaterials in the treatment
       of wastewater: A review. *Water,* **2020**, *12*(2), 495.
       [http://dx.doi.org/10.3390/w12020495]

[56]   Li, Y.; Chen, F.; He, R.; Wang, Y.; Tang, N. Semiconductor Photocatalysis for Water. In Nanoscale
       Materials in Water Purification., **2019**, pp. 689-705.
       [http://dx.doi.org/10.1016/B978-0-12-813926-4.00030-6]

[57]   Litter, M.I. Mechanisms of removal of heavy metals and arsenic from water by TiO_2 -heterogeneous
       photocatalysis. *Pure Appl. Chem.,* **2015**, *87*(6), 557-567.
       [http://dx.doi.org/10.1515/pac-2014-0710]

[58]   Kanan, S.; Moyet, M.A.; Arthur, R.B.; Patterson, H.H. Recent advances on TiO_2 -based photocatalysts
       toward the degradation of pesticides and major organic pollutants from water bodies. *Catal. Rev., Sci.
       Eng.,* **2020**, *62*(1), 1-65.
       [http://dx.doi.org/10.1080/01614940.2019.1613323]

[59]   Lee, K.M.; Lai, C.W.; Ngai, K.S.; Juan, J.C. Recent developments of zinc oxide based photocatalyst in
       water treatment technology: A review. *Water Res.,* **2016**, *88*, 428-448.
       [http://dx.doi.org/10.1016/j.watres.2015.09.045] [PMID: 26519627]

[60]   Baradaran, M.; Ghodsi, F.E.; Bittencourt, C.; Llobet, E. The role of Al concentration on improving the
       photocatalytic performance of nanostructured ZnO/ZnO:Al/ZnO multilayer thin films. *J. Alloys
       Compd.,* **2019**, *788*, 289-301.
       [http://dx.doi.org/10.1016/j.jallcom.2019.02.184]

[61]   Devendran, P.; Alagesan, T.; Nallamuthu, N.; Asath Bahadur, S.; Pandian, K. Single-precursor
       synthesis of sub-10 nm CdS nanoparticles embedded on graphene sheets nanocatalyst for active
       photodegradation under visible light. *Appl. Surf. Sci.,* **2020**, *534*, 147614.
       [http://dx.doi.org/10.1016/j.apsusc.2020.147614]

[62]   Blanco-Vega, M.P.; Guzmán-Mar, J.L.; Villanueva-Rodríguez, M.; Maya-Treviño, L.; Garza-Tovar,
       L.L.; Hernández-Ramírez, A.; Hinojosa-Reyes, L. Photocatalytic elimination of bisphenol A under
       visible light using Ni-doped TiO_2 synthesized by microwave assisted sol-gel method. *Mater. Sci.
       Semicond. Process.,* **2017**, *71*, 275-282.
       [http://dx.doi.org/10.1016/j.mssp.2017.08.013]

[63]   Heena Khan, S.; R, S.; Pathak, B.; Fulekar, M.H. Photocatalytic degradation of organophosphate
       pesticides (Chlorpyrifos) using synthesized zinc oxide nanoparticle by membrane filtration reactor
       under UV irradiation. *Fron. Nanosci. Nanotech.,* **2015**, *1*(1), 23-27.
       [http://dx.doi.org/10.15761/FNN.1000105]

[64]   Xie, S.; Zhang, Q.; Liu, G.; Wang, Y. Photocatalytic and photoelectrocatalytic reduction of CO_2 using
       heterogeneous catalysts with controlled nanostructures. *Chem. Commun. (Camb.),* **2016**, *52*(1), 35-59.
       [http://dx.doi.org/10.1039/C5CC07613G] [PMID: 26540265]

[65]   Meng, A.; Zhang, L.; Cheng, B.; Yu, J. TiO_2–MnO x–Pt hybrid multiheterojunction film photocatalyst
       with enhanced photocatalytic CO_2-reduction activity. *ACS Appl. Mater. Interfaces,* **2019**, *11*(6), 5581-
       5589.
       [http://dx.doi.org/10.1021/acsami.8b02552] [PMID: 29718652]

[66]   Li, S.; Li, X.; Zhang, Y.; Huang, F.; Wang, F.; Wei, X. Enhanced chemiluminescence of the
       luminol–KIO_4 system by ZnS nanoparticles. *Mikrochim. Acta,* **2009**, *167*(1-2), 103-108.
       [http://dx.doi.org/10.1007/s00604-009-0224-5]

[67]   Chern, M.; Kays, J.C.; Bhuckory, S.; Dennis, A.M. Sensing with photoluminescent semiconductor quantum dots. *Methods Appl. Fluoresc.,* **2019**, *7*(1), 012005.
[http://dx.doi.org/10.1088/2050-6120/aaf6f8] [PMID: 30530939]

[68]   Bhogal, S.; Kaur, K.; Malik, A.K.; Sonne, C.; Lee, S.S.; Kim, K.H. Core-shell structured molecularly imprinted materials for sensing applications. *Trends Analyt. Chem.,* **2020**, *133*, 116043.
[http://dx.doi.org/10.1016/j.trac.2020.116043]

[69]   Yang, Q.; Li, J.; Wang, X.; Peng, H.; Xiong, H.; Chen, L. Strategies of molecular imprinting-based fluorescence sensors for chemical and biological analysis. *Biosens. Bioelectron.,* **2018**, *112*, 54-71.
[http://dx.doi.org/10.1016/j.bios.2018.04.028] [PMID: 29698809]

[70]   Wang, F.; Hu, S. Electrochemical sensors based on metal and semiconductor nanoparticles. *Mikrochim. Acta,* **2009**, *165*(1-2), 1-22.
[http://dx.doi.org/10.1007/s00604-009-0136-4]

[71]   Wackerlig, J.; Lieberzeit, P.A. Molecularly imprinted polymer nanoparticles in chemical sensing – Synthesis, characterisation and application. *Sens. Actuators B Chem.,* **2015**, *207*, 144-157.
[http://dx.doi.org/10.1016/j.snb.2014.09.094]

[72]   Durán, G.M.; Contento, A.M.; Ríos, Á. Use of Cdse/ZnS quantum dots for sensitive detection and quantification of paraquat in water samples. *Anal. Chim. Acta,* **2013**, *801*, 84-90.
[http://dx.doi.org/10.1016/j.aca.2013.09.003] [PMID: 24139578]

[73]   Chen, H.; Hu, O.; Fan, Y.; Xu, L.; Zhang, L.; Lan, W.; Hu, Y.; Xie, X.; Ma, L.; She, Y.; Fu, H. Fluorescence paper-based sensor for visual detection of carbamate pesticides in food based on CdTe quantum dot and nano ZnTPyP. *Food Chem.,* **2020**, *327*, 127075.
[http://dx.doi.org/10.1016/j.foodchem.2020.127075] [PMID: 32446026]

[74]   Zhang, K.; Mei, Q.; Guan, G.; Liu, B.; Wang, S.; Zhang, Z. Ligand replacement-induced fluorescence switch of quantum dots for ultrasensitive detection of organophosphorothioate pesticides. *Anal. Chem.,* **2010**, *82*(22), 9579-9586.
[http://dx.doi.org/10.1021/ac102531z] [PMID: 20973515]

[75]   Sharma, P.; Mehata, M.S. Colloidal MoS_2 quantum dots based optical sensor for detection of 2,4,6-TNP explosive in an aqueous medium. *Opt. Mater.,* **2020**, *100*, 109646.
[http://dx.doi.org/10.1016/j.optmat.2019.109646]

[76]   de França, C.C.L.; Meneses, D.; Silva, A.C.A.; Dantas, N.O.; de Abreu, F.C.; Petroni, J.M.; Lucca, B.G. Development of novel paper-based electrochemical device modified with CdSe/CdS magic-sized quantum dots and application for the sensing of dopamine. *Electrochim. Acta,* **2021**, *367*, 137486.
[http://dx.doi.org/10.1016/j.electacta.2020.137486]

[77]   Chen, L.; Wang, X.; Lu, W.; Wu, X.; Li, J. Molecular imprinting: perspectives and applications. *Chem. Soc. Rev.,* **2016**, *45*(8), 2137-2211.
[http://dx.doi.org/10.1039/C6CS00061D] [PMID: 26936282]

[78]   Jia, M.; Zhang, Z.; Li, J.; Shao, H.; Chen, L.; Yang, X. A molecular imprinting fluorescence sensor based on quantum dots and a mesoporous structure for selective and sensitive detection of 2,4-dichlorophenoxyacetic acid. *Sens. Actuators B Chem.,* **2017**, *252*, 934-943.
[http://dx.doi.org/10.1016/j.snb.2017.06.090]

[79]   Hassanzadeh, J.; Khataee, A.; Mosaei Oskoei, Y.; Fattahi, H.; Bagheri, N. Selective chemiluminescence method for the determination of trinitrotoluene based on molecularly imprinted polymer-capped ZnO quantum dots. *New J. Chem.,* **2017**, *41*(19), 10659-10667.
[http://dx.doi.org/10.1039/C7NJ01802A]

[80]   Ke, J.; Li, X.; Zhao, Q.; Hou, Y.; Chen, J. Ultrasensitive quantum dot fluorescence quenching assay for selective detection of mercury ions in drinking water. *Sci. Rep.,* **2014**, *4*(1), 5624.
[http://dx.doi.org/10.1038/srep05624] [PMID: 25005836]

[81]   Sharma, P.; Mehata, M.S. Rapid sensing of lead metal ions in an aqueous medium by MoS_2 quantum

dots fluorescence turn-off. *Mater. Res. Bull.,* **2020**, *131*, 110978.
[http://dx.doi.org/10.1016/j.materresbull.2020.110978]

[82]   Jin, L.H.; Han, C.S. Ultrasensitive and selective fluorimetric detection of copper ions using thiosulfate-involved quantum dots. *Anal. Chem.,* **2014**, *86*(15), 7209-7213.
[http://dx.doi.org/10.1021/ac501515f] [PMID: 24981053]

[83]   Liu, X.; Cheng, S.; Liu, H.; Hu, S.; Zhang, D.; Ning, H. A survey on gas sensing technology. *Sensors (Basel),* **2012**, *12*(7), 9635-9665.
[http://dx.doi.org/10.3390/s120709635] [PMID: 23012563]

[84]   Nunes, D.; Pimentel, A.; Gonçalves, A.; Pereira, S.; Branquinho, R.; Barquinha, P.; Fortunato, E.; Martins, R. Metal oxide nanostructures for sensor applications. *Semicond. Sci. Technol.,* **2019**, *34*(4), 043001.
[http://dx.doi.org/10.1088/1361-6641/ab011e]

[85]   Zhang, J.; Liu, X.; Neri, G.; Pinna, N. Nanostructured materials for room-temperature gas sensors. *Adv. Mater.,* **2016**, *28*(5), 795-831.
[http://dx.doi.org/10.1002/adma.201503825] [PMID: 26662346]

[86]   Wang, Y.; Liu, B.; Cai, D.; Li, H.; Liu, Y.; Wang, D.; Wang, L.; Li, Q.; Wang, T. Room-temperature hydrogen sensor based on grain-boundary controlled Pt decorated In_2O_3 nanocubes. *Sens. Actuators B Chem.,* **2014**, *201*, 351-359.
[http://dx.doi.org/10.1016/j.snb.2014.05.013]

[87]   Bao, M.; Chen, Y.; Li, F.; Ma, J.; Lv, T.; Tang, Y.; Chen, L.; Xu, Z.; Wang, T. Plate-like p–n heterogeneous NiO/WO_3 nanocomposites for high performance room temperature NO_2 sensors. *Nanoscale,* **2014**, *6*(8), 4063-4066.
[http://dx.doi.org/10.1039/c3nr05268k] [PMID: 24603873]

[88]   Zhao, Y.M.; Zhu, Y.Q. Room temperature ammonia sensing properties of $W_{18}O_{49}$ nanowires. *Sens. Actuators B Chem.,* **2009**, *137*(1), 27-31.
[http://dx.doi.org/10.1016/j.snb.2009.01.004]

[89]   Wu, Y-Q.; Hu, M.; Wei, X-Y. A study of transition from n- to p-type based on hexagonal WO_3 nanorods sensor. *Chin. Phys. B,* **2014**, *23*(4), 040704.
[http://dx.doi.org/10.1088/1674-1056/23/4/040704]

[90]   Martynenko, I.V.; Litvin, A.P.; Purcell-Milton, F.; Baranov, A.V.; Fedorov, A.V.; Gun'ko, Y.K. Application of semiconductor quantum dots in bioimaging and biosensing. *J. Mater. Chem. B Mater. Biol. Med.,* **2017**, *5*(33), 6701-6727.
[http://dx.doi.org/10.1039/C7TB01425B] [PMID: 32264322]

[91]   Sukhanova, A.; Nabiev, I. Fluorescent nanocrystal-encoded microbeads for multiplexed cancer imaging and diagnosis. *Crit. Rev. Oncol. Hematol.,* **2008**, *68*(1), 39-59.
[http://dx.doi.org/10.1016/j.critrevonc.2008.05.006] [PMID: 18621543]

[92]   Medintz, I.L.; Clapp, A.R.; Brunel, F.M.; Tiefenbrunn, T.; Tetsuo Uyeda, H.; Chang, E.L.; Deschamps, J.R.; Dawson, P.E.; Mattoussi, H. Proteolytic activity monitored by fluorescence resonance energy transfer through quantum-dot–peptide conjugates. *Nat. Mater.,* **2006**, *5*(7), 581-589.
[http://dx.doi.org/10.1038/nmat1676] [PMID: 16799548]

[93]   Carrillo-Carrión, C.; Simonet, B.M.; Valcárcel, M. Colistin-functionalised CdSe/ZnS quantum dots as fluorescent probe for the rapid detection of Escherichia coli. *Biosens. Bioelectron.,* **2011**, *26*(11), 4368-4374.
[http://dx.doi.org/10.1016/j.bios.2011.04.050] [PMID: 21605965]

[94]   R, C.H.; Venkataramana, M.; Kurkuri, M.D.; Balakrishna R, G. Simple quantum dot bioprobe/label for sensitive detection of Staphylococcus aureus TNase. *Sens. Actuators B Chem.,* **2016**, *222*, 1201-1208.
[http://dx.doi.org/10.1016/j.snb.2015.07.121]

[95]   Petryayeva, E.; Algar, W.R. Single-step bioassays in serum and whole blood with a smartphone, quantum dots and paper-in-PDMS chips. *Analyst (Lond.),* **2015**, *140*(12), 4037-4045.
[http://dx.doi.org/10.1039/C5AN00475F] [PMID: 25924885]

[96]   Geißler, D.; Charbonnière, L.J.; Ziessel, R.F.; Butlin, N.G.; Löhmannsröben, H.G.; Hildebrandt, N. Quantum dot biosensors for ultrasensitive multiplexed diagnostics. *Angew. Chem. Int. Ed.,* **2010**, *49*(8), 1396-1401.
[http://dx.doi.org/10.1002/anie.200906399] [PMID: 20108296]

[97]   Freeman, R.; Liu, X.; Willner, I. Chemiluminescent and chemiluminescence resonance energy transfer (CRET) detection of DNA, metal ions, and aptamer-substrate complexes using hemin/G-quadruplexes and CdSe/ZnS quantum dots. *J. Am. Chem. Soc.,* **2011**, *133*(30), 11597-11604.
[http://dx.doi.org/10.1021/ja202639m] [PMID: 21678959]

[98]   Zhou, Z.M.; Feng, Z.; Zhou, J.; Fang, B.Y.; Ma, Z.Y.; Liu, B.; Zhao, Y.D.; Hu, X.B. Quantum dot-modified aptamer probe for chemiluminescence detection of carcino-embryonic antigen using capillary electrophoresis. *Sens. Actuators B Chem.,* **2015**, *210*, 158-164.
[http://dx.doi.org/10.1016/j.snb.2014.12.087]

[99]   Ertek, B.; Akgül, C.; Dilgin, Y. Photoelectrochemical glucose biosensor based on a dehydrogenase enzyme and NAD $^+$/NADH redox couple using a quantum dot modified pencil graphite electrode. *RSC Advances,* **2016**, *6*(24), 20058-20066.
[http://dx.doi.org/10.1039/C5RA25673A]

CHAPTER 3

Titania Nanoparticles: Electronic, Surface and Morphological Modifications for Photocatalytic Removal of Pesticides and Polycyclic Aromatic Hydrocarbons

Inderpreet Singh Grover[1] and **Rajeev Sharma**[2,*]

[1] *Department of Chemistry, Public College Samana, Patiala, Punjab, India*

[2] *Department of Chemistry, Multani Mal Modi College, Patiala, Punjab, India*

Abstract: Tailoring the electronic, surface and morphological properties alter the catalytic properties of the material(s), specifically at the nanoscale. In the past years, a plethora of research has been reported to find sustainable and eco-friendly catalysts for environmental pollution remediation. In this direction, titania nanoparticles have been intensively explored to check their potential for photocatalytic removal of various pollutants. In the current scenario, where the growing population needs to feed on an everyday basis, abundant pesticides indiscriminately are being used to increase crop yield, thus causing environmental pollution and ecological imbalance. In order to remove these environmental pollutants along with the polycyclic aromatic hydrocarbons (PAHs) that are formed by incomplete combustion of crop residue or any other organic matter have been studied, and the results reported for these two categories of pollutants are summarized in this chapter.

Keywords: Pesticides, Photocatalysis, Polycyclic aromatic hydrocarbons (PAHs), Titania nanoparticles.

PRINCIPLE OF TIO$_2$ PHOTOCATALYSIS

Nanocrystalline TiO$_2$, an n-type semiconductor of bandgap energy (Eg) ~3.0-3.2 eV [*vs.* NHE (Normal Hydrogen Electrode)], has been widely studied [1 - 10] as a photocatalyst for the decomposition of environmental pollutants. It mainly exists in three crystal phases, namely, anatase, rutile, and brookite. Generally, TiO$_2$ having anatase (~70%) and rutile (~30%) phases exhibit higher photocatalytic activity (for degradation of pollutants) than any of these phases alone and is found

---

* **Corresponding author Rajeev Sharma:** Department of Chemistry, Multanti Mal Modi College, Patiala, Punjab, India; E-mail: sharmarajeev0379@gmail.com

Karamjit Singh Dhaliwal (Ed.)

in commercially available P25-TiO$_2$ (P25). Upon photo-irradiation of TiO$_2$ (Fig. **1**) under ultraviolet (UV) and/or visible (Vis.) light from sunlight or an illuminated light source having energy ≥ its Eg, electrons (e$^-$) in its valence band (VB) get excited to the conduction band (CB) resulting into the formation of the positively charged holes (h$^+$) in the VB and diffuse [1 - 3] to the surface from the bulk of the catalyst.

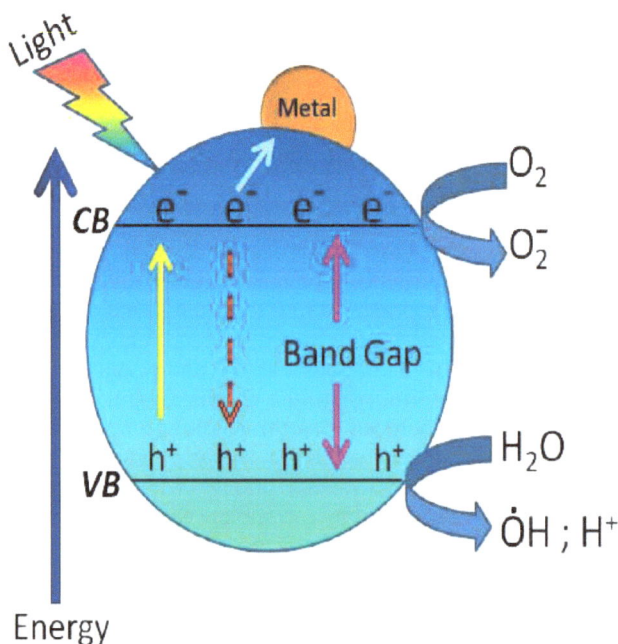

Fig. (1). Mechanism of photocatalysis.

The photogenerated h$^+$ having strong oxidizing power (+3.1 V *vs*. NHE) reacts with either adsorbed water molecules or other chemical species to form hydroxyl radicals or directly oxidizes [1 - 5] the molecules at the interface of TiO$_2$ and e$^-$ in the CB (-0.5 to -1.5 V *vs*. NHE) participates for reduction processes that convert the adsorbed molecular oxygen to oxidative superoxide radical anions (O$_2$$^{\cdot-}$) or reduces [1 - 10] the substrate molecule. In the presence of metal deposited onto its surface, there is an enhancement in the recombination time between the excited charge carriers that further improve their respective oxidation/reduction abilities. There are indeed other good reasons for favoring TiO$_2$, which include strong oxidizing power, photogenerated e$^-$ that reduces enough to produce superoxide from molecular oxygen, antibacterial activity, self-cleaning activity, cost-effective, chemically inert, and most importantly, its non-toxicity to living beings, favouring its use as a photocatalyst.

Influence of Metal Loading on Photocatalytic Performance of TiO_2

The photocatalytic activity of TiO_2 can be further improved by metal (M) loading and is frequently used for a variety of applications, *viz.*, in solar energy conversion, photocatalysis and electronic devices [1, 4, 5, 11]. These M-TiO_2 hybrid nanoparticles often exhibit noticeably different properties than the individual components due to separate electron and dielectric confinement in the TiO_2 and M parts, respectively [4, 5]. The photo-produced charge carriers in the TiO_2 can be transferred efficiently to the M part, resulting in the shifting of plasmon frequency and/or promotion of redox reactions by suppressing the recombination of excited charge species, as seen in Fig. (**1**). Another advantageous function of metal loading is comprehensible and incoherent interactions between M and TiO_2, leading to the broadening and extension of its absorbance spectra to the visible region [1, 2, 4, 5, 8]. Generally, noble metals (*e.g.*, Au, Ag, Ni, Cu, Pt, Rh, and Pd) with Fermi levels lower than TiO_2 should have been deposited on its surface for enhanced charge separation [1, 4, 5, 8], which is advantageous compared to other photosensitized systems because of its relatively more stability in an aerated atmosphere [1]. Therefore, these M-TiO_2 composites have been utilized for the photodegradation/decomposition of toxic organic pollutants, such as for the wastewater treatment and removals of other pollutants, such as dyes [1 - 6, 8 - 10], dehydrogenation of alcohols [12, 13], reduction of pharmaceutically important nitro compounds [14 - 16], degradation of polycyclic aromatic hydrocarbons (PAHs) [17 - 29] and other heteroatom-containing organic molecules, such as pesticides [30 - 55].

Role of Non-Metal Loading on Photocatalytic Activity of TiO_2

One of the major reasons for loading the TiO_2 nanoparticles with a non-metal dopant is to narrow down its E_g [56]. Initially, a non-metal dopant, such as nitrogen, that has an energy level slightly higher than that of oxygen, is introduced into the lattice of TiO_2, and the generation of a mid-gap energy level takes place [57, 58]. Upon irradiation with suitable light energy, the excitation of charge carriers takes from this newly created energy state rather than the VB of TiO_2. It means the E_g is now lower than before the loading with non-metal. This excited electron then shifts to the CB, creating a hole in the mid-way energy state. Afterward, the charge carriers thus generated migrate to the surface of nanoparticles, where they participate in the redox reactions. Apart from nitrogen, other non-metal dopants, such as boron, carbon, and sulphur, have also been studied and found [59] to impart a similar role except for the position of energy level created, which is different from that created by nitrogen.

Additionally, this type of modification also improves the charge density, creates oxygen vacancies [60 - 63], and develops a new orbital by mixing of original orbital of titania nanoparticles with added impurity [60, 62, 64, 65]; they are potential members to narrow down the E_g and shift the absorption towards the visible spectrum of light [56 - 58, 66 - 69]. In non-metal doped titania nanoparticles, excited electrons, upon reaching the surface, can be scavenged by the dissolved O_2 molecules to produce highly oxidative species, namely superoxides, followed by a series of reactive species, such as ·OH. On the other hand, holes reaching the surface of the photocatalyst are captured by hydroxyl ions to generate corresponding free radicals. These species generated then react with pollutants, such as pesticides or PAH, in multiple steps to ultimately yield water, carbon dioxide and oxides of heteroatom(s) present (if any) in the organic pollutant, such as in pesticide.

PESTICIDES

These are the organic compounds used [70, 71] to increase crop yield by reducing/inhibiting/mitigating/killing the target species. However, irregular and indiscriminate use of pesticides, acute toxicity at low concentrations, bio-reluctance nature, and inefficient transfer ($\sim 0.1\%$) to the target organism resulted in contamination of soil, water, and air, which can adversely affect [70, 72] the non-target organisms. Moreover, because of the longer half-life ($t_{1/2}$), *i.e.*, slow natural degradation, and overuse of pesticides led to their accumulation [70 - 73] in the environment. Once entered into the food chain, they undergo biomagnifications [74, 75], *i.e.*, their amount in tissue becomes much more fold than that found in their surroundings. This has been identified as the cause of disturbed ecological systems and a threat to humans too. For instance, the Malwa region of Punjab is found to be highly contaminated [76 - 78] by pesticides. Analysis of water and vegetable samples from this area showed the presence [77, 78] of pesticides, and most of them were unfit for human usage [79]. Along with this, studies reveal that majority of cases reported for deadly disease like cancer is from the malwa region of Punjab. Since pesticides are reported to be carcinogenic, the literature [75] evidences the association of cancer with pesticides. A number of pesticide formulations that are in use have been utilized in agricultural fields. Pesticides, depending upon their molecular structure, can be categorized into nine groups *viz.*, phenylurea (sulfosulfuron, amidosulfuron, imidacloprid, *etc.*), carbamates (methomyl, isolan, *etc.*), triazines (atrazine, simazine, *etc.*), organochlorines (endosulfan, aldrin, *etc.*), organophosphorus (methyl-, ethyl-parathion, propiconazole, *etc.*), phenoxy acids (propxour, MCPA *etc.*), quaternary ammonium salts (diquat, paraquat, *etc.*), thiocarbamate (butylate, eptam, *etc.*) and chlorophenols (pentachlorophenol, 2,3,4-trichlorphenol, *etc.*).

Among them, sulfosulfuron, imidacloprid (IMI), methylparathion (MP) and propiconazole have been widely used as herbicides/insecticides on a variety of crops over the past two decades [70, 71]. These pesticides possess variable $t_{1/2}$ (30-365 days), following the natural slow degradation process and transformed into some other heteroatom-containing intermediates that are sometimes reported to be more toxic than the parent compounds [70, 72]. As a result, these pesticides (Fig. 4 in chapter 1) have been considered for the present work and research works performed for their photodegradation in soil and water with/without the use of titania nanoparticles as a photocatalyst are being discussed in the upcoming sections.

Sulfosulfuron

It is an early-post emergent herbicide (Fig. 2) shows the molecular structure of sulfosulfuron), introduced in 1982 by Du Pont Crop Protection to control the isoproturon-resistant P. minor (unwanted plant in wheat). Its degradation in wheat straw and wheat grains was studied by Saha *et al.* [30], reporting 42% of its degradation under optimum conditions. Variously identified intermediates formed during its degradation were also isolated. Ramesh *et al.* [31] studied the dissipation of sulfosulfuron (1 and 2 ppm) in water and suggested its $t_{1/2} = 67-76$ days, and identified its metabolites (aminopyrimidine, demethylsulfosulfuron, guanidine, sulfonamide, ethyl sulfone and rearranged amines). The study related to the degradation of sulfosulfuron [32] under sunlight showed its photostability for up to 115 days and proceeded with the breaking of the sulfonylurea bridge, yielding different intermediates under alkaline and acidic conditions. Degradation of this pesticide in soil [33] revealed its varying half-life (11-28 days), and 14% of its residual amount was determined after 120 days of direct sunlight exposure. A very recent report [34] on the photocatalytic oxidation of sulfosulfuron (5 ppm) showed its 40% removal after 60 min of UV-light exposure in the presence of bare P25; however, in the presence of P25 and $Na_2S_2O_8$, its disappearance from solution became 80% after the same time interval and under similar reaction conditions. Arya *et al.* [33] studied its microbial degradation and reported 33.8-48.7% in 12h under their experimental conditions.

Imidacloprid

This pesticide (Fig. 3) was introduced in 1986 by Bayer Crop-Science and has been utilized in 120 nations worldwide with over 140 agricultural and horticultural cultivated plants [35]. Redlich *et al.* [36] studied the degradation of imidacloprid in an aqueous solution under-stimulated solar irradiation and found its $t_{1/2}$ to be 3 days, while some of the identified degraded intermediates of the same were more persistent ($t_{1/2}$ up to 660 days) than the imidacloprid itself toward their decomposition. However, Liu *et al.* [33] reported its $t_{1/2} = 30$ min when its aqueous solution was irradiated under UV light in the presence of P25.

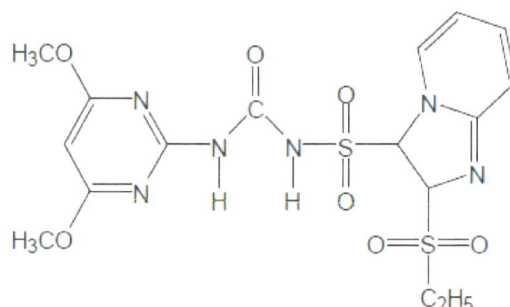

Fig. (2). Molecular Structure of Sulfosulfuron.

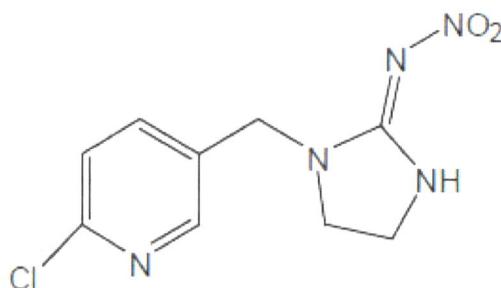

Fig. (3). Molecular Structure of Imidacloprid.

Kitsiou *et al.* [38] used P25 in relation to $P25/Fe^{+3}/H_2O_2$ system for its degradation and found an increase in mineralization efficiencies from 3.4 to 14.1%, respectively, after its photooxidation in UV-light. Various intermediates identified were 6-chloronicotic acid, methyl-6-chloronicotinate, N-ethylformaide, dimethyl foramide, amylnitrite and chlorine dioxide. Tang *et al.* [39] studied bare P25 in comparison to P25 supported on H-ZSM-15 for degradation of imidacloprid and reported a decrease in $t_{1/2}$ from 215 min to 10 min, respectively. The various intermediates formed during degradation were also reported. A report by Zabar *et al.* [41], which considered the degradation of three insecticides, including imidacloprid, showed 19% of actual mineralization despite its complete degradation in 2 h of photooxidation by using P25 as photocatalyst under UV-light. Recently, Changgen *et al.* [42] studied $H_3PW_{12}O_4/La$-TiO_2 for the oxidative removal of imidacloprid and found it to be 5-fold more active than bare TiO_2 for imidacloprid. Similarly, TiO_2 composite, namely $GO/Fe_3O_4/TiO_2$-NiO, degrades the same pesticide up to ~85% in 45 min under visible light irradiation [40].

Methyl Parathion

Pignatello *et al.* [43] reported the complete oxidation of methyl parathion (MP) (Fig. **4**) shows the molecular structure of MP) by a UV-assisted-fenton system and found to yield HNO_3, H_2SO_4, H_3PO_4, oxalic acid, 4-nitrophenol and dimethyl phosphoric acid as intermediate compounds. O'Shea *et al.* [44] demonstrated the degradation of MP and its other family members using TiO_2 as photocatalysts, revealing that their degradation can be accomplished in oxygenated aqueous solutions. The final products of this photodegradation were found to be CO_2 and phosphoric acid. Similar to this, Konstantinou *et al.* [45] also studied MP and its family members using an aqueous suspension of TiO_2 and identified their intermediates by mass spectroscopy.

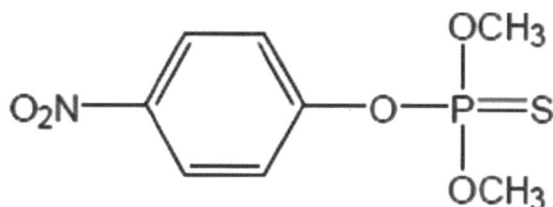

Fig. (4). Molecular Structure of Methyl Parathion.

Sanjuan *et al.* [46] described the use of 2,4,6-triphenylpyrylium ion encapsulated in Y-zeolite for the photocatalytic decomposition of MP. It was suggested that encapsulation stabilized the pyrylium ion, resulting in its comparable activity to that of P25 for the degradation of MP and was believed to work *via* the formation of radical intermediates. Evgenidou *et al.* [47] studied TiO_2 and ZnO as photocatalysts for the degradation and mineralization of MP and found that the activity of the former was 5 times higher than the latter and could be increased in the presence of $K_2S_2O_8$ oxidant. Various intermediates were also detected that were suggested to be formed by oxidation, hydroxylation, dealkylation and hydrolysis of its ester group. Moctezuma and his co-workers [48] investigated the degradation of MP using P25 as a photocatalyst and suggested a mechanism for its decomposition under acidic conditions, which was fast and more distinct than that found for alkaline conditions.

Propiconazole

Vialaton *et al.* [49] studied the influence of humic acid and water (natural and deionized) on propiconazole (Fig. **5**) shows the molecular structure of propiconazole) photostability under-stimulated solar light. It was found that degradation increases in the presence of humic acid, and its $t_{1/2}$ values were ca. to be 60±10 and 85±10 days in natural and pure water, respectively. It was suggested that its degradation proceeds *via*photocyclization and photohydrolysis routes. Pliego *et al*. [50] elucidated the use of high-temperature fenton treatment for the removal of 18 pesticides, including propiconazole, from the wastewater and optimized the conditions (pH and amount of fenton reagent) for its complete removal. This pesticide is also known to be degraded in soils [51] by hydroxylation of the n-propyl side chain and the dioxolane ring, as well as with the formation of 1,2,4-triazole with DT50 (loss of 50% of the parent) values of > 26 days under aerobic conditions at 25 °C. Another report [52] provided a range of its soil half-life of 40-51 days in aerobic soil under controlled conditions.

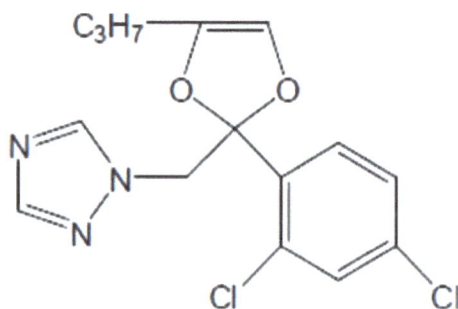

Fig. (5). Molecular Structure of Propiconazole.

In a laboratory study, the degradation of propiconazole and three other pesticides in four different types of soils showed it to be more persistent than the other pesticides, which was dependent upon the soil properties [53]. In laboratory studies on the formation and loss of bound residues of propiconazole in soils, Kim *et al*. [54] showed its $t_{1/2}$ ~ 315 days in sandy loam soil but was beyond the experimental time limit for silty clay loam soil. In a very recent report [55], the influence of ions on the degradation of propiconazole along with 15 more pesticides was studied, and it was found that the presence of nitrate ion accelerates its rate of degradation while bicarbonates retards the same by scavenging hydroxyl radicals.

POLYCYCLIC AROMATIC HYDROCARBONS

It is a class of environmental pollutants that are formed [70 - 75] largely by incomplete combustion and crop residue burning, which is one of its major sources in soil. These are found primarily in soil [80, 81], sediments [82], oil [83] and air [84]. Among a variety of PAHs, 16 are of special concern, and some of these are found to be carcinogenic [80 - 84]. Studies have shown that humans are exposed to these PAHs, and about 90% of PAHs intake comes from food [84], thereby making them another environmental pollutant that should be minimized. Indeed, there is a number of PAHs, but two PAHs, namely naphthalene (2 rings) and anthracene (3 rings), are initially formed during the combustion of agricultural waste [77, 84] and therefore are considered in the present study. The literature about naphthalene oxidation by photocatalysis showed its transformation into various compounds (naphthols, naphthoquinones, *etc.*) that finally mineralized to CO_2 and H_2O. Indeed, a number of authors [17 - 20] used cosolvents (acetonitrile, ethanol, *etc.*) which help in the dissolution of naphthalene but at the same time alter the rate of its photooxidation. The investigation on the role of non-ionic surfactants [21] and organic contaminants [22] during its photodegradation in the presence of P25 suggested that hydroxyl scavenging molecules reduced the rate of its degradation. Hykrdova *et al.* [19] studied the influence of quantum-sized TiO_2 nanoparticles on the degradation of naphthalene and proposed a mechanism for its degradation initiated by the attack of hydroxyl radicals. The role of photonic flux, temperature, solution pH, and amount of P25 on the degradation of naphthalene was studied by Liar *et al.* [23]. It was revealed that its degradation proceeds *via*hydroxylation and ring cleavage by the action of photoproduced hydroxyl and superoxide radicals. In a very recent report [24], the simultaneous degradation and hydrogen formation under inert conditions from an aqueous solution of naphthalene (5-20 ppm) were studied using P25 as a catalyst at pH = 3-8 under UV-light irradiation. The degradation process was found to follow pseudo-first-order kinetics. The conditions (airflow, catalyst dose and UV-light intensity) for photooxidation of naphthalene in a batch reactor using a slurry of P25 were optimized by Mahmoodi *et al.* [25], which were found to be in correlation with the theoretically predicted conditions. Ohno *et al.* [20] studied the influence of crystal phases of TiO_2 on the degradation of naphthalene. It was demonstrated that the synergetic effect between anatase and rutile always improved the photocatalytic activity (PCA) of TiO_2 more than any of these phases alone. The comparative photocatalytic (P25 as a catalyst) and electrochemical degradation of anthracene, along with four other PAHs in an aqueous medium, were studied by Cordeiro *et al.* [26]. It was proposed that photocatalytic degradation is initiated by hydroxyl radicals, while the direct electron transfer from adsorbed PAH to the electrode activates the electrochemical oxidation. The mechanism involved during the photodegradation of anthracene in the presence of P25 was proposed by Theurich *et al.* [27] and further verified by the work of Dass *et al.* [28]. Both of these reports indicated the attack of oxygenated radicals for the photooxidative transformation of anthracene into other intermediates. Woo *et al.* [29] demonstrated the degradation of naphthalene, anthracene and three other PAHs using P25. It was suggested that the presence of acetone as a cosolvent increased the solubility of these PAHs, but after adding ~16% of it, significant alteration in the mechanism for degradation of naphthalene and anthracene was observed. It was also shown that the attack site for reactive free radicals on the PAHs was dependent on the localization energies of the different positions of the compound. Thus, it is clear from the above studies that the PCA of TiO_2/P25 used for the degradation of the aforementioned pesticides and PAHs is either slow/requires a higher time for complete mineralization. Hence, there is a scope to increase the PCA of TiO_2, which can be achieved by its modifying size and shape since, at the nanoscale, the properties (optical, electronic *etc.*) become dependent on these parameters.

SIZE, SHAPE, AND SURFACE MODIFICATIONS OF TIO_2

It has been revealed from the studies that a decrease in the size of TiO_2 nanoparticles leads to the change in the position of CB and VB to higher and lower values, respectively, causing the widening of Eg (Fig. **6**), and is well-known as quantum confinement [1, 2, 4, 5, 8, 10]. As a result of this variation in positions of CB and VB, a considerable change in the recombination of the e^-/h^+ pair and hence in the PCA of TiO_2 is observed. Moreover, the S_{BET} and surface-to-volume ratio dramatically increases as the particle size of a material decreases [1, 2], as depicted in Fig. (**6**). For instance, Zhang *et al.* [89] studied the PCA of four sizes (~7, 15, 30, and 45 nm) of spherical TiO_2 nanoparticles on the decomposition of 2-propanol and found that 7 nm particles showed 1.6 times better PCA than commercially available P25. Anpo *et al.* [87] investigated the PCA of TiO_2 nanoparticles on hydrogenation reactions of CH_3OH with H_2O and observed an increase in PCA with a decrease in the diameter of the TiO_2 particles, especially below 10 nm.

However, alternation in the shape of TiO_2, such as by making anisotropic TiO_2 nanomaterials, such as nanorods (TNR), nanotubes or nanowires (TNW), *etc.*, have advantages with respect to the less recombination because of the short distance for the diffusion of charge carrier [1, 2, 88 - 94] and delocalization of photoexcited electrons along their length. For example, Wang *et al.* [97], with favor from other reports [95, 96], showed 10-70% higher PCA of rutile TNR for photooxidation of rhodamine-B. Testino *et al.* [97] showed superior activity of anatase TNR particles than rutile TNR and P25 particles for the degradation of phenol. They also found the comparatively higher activity of mixed-phased (anatase and rutile) TNR particles than TNR with any of the pure phases alone. Similarly, nanotubes, because of their enhanced surface area, layered structure and small wall thickness, have also been often studied and found to exhibit higher PCA for degradation of the dyes [98 - 104]. For example, Zhu *et al.* [43], Antonio *et al.* [44], and Kim *et al.* [45] showed that the nanotubes prepared by NaOH hydrothermal route exhibited 2-18 times higher PCA than P25 for the photooxidation of sulforhodamine, RB-69, rhodamine-B dyes, respectively. Similarly, TNW with its anatase and rutile phase is proven to show higher (~1.5-7.0 times) PCA than P25 for photooxidation of toluene [105], methyl orange [76] and for H_2 formation [77] by water splitting. The improved activity of TNW was ascribed to its higher surface area and delocalization of photoexcited electrons. A report by Liu *et al.* [88] demonstrated the formation of mixed-phase (brookite-anatase) TNWs having PCA double that of TNWs with any of these phases alone. They indicated the improved charge migration as a cause for their higher activity toward the degradation of methyl orange. The PCA of these anisotropic structures has also been found to improve [78, 79, 106 - 115] after loading of metals (Ag, Au, Cu, Pt, Rh, *etc.*) that were studied for the decomposition of dyes. For instance, loading of Au and Ag onto nanotubes [108] and TNWs [108 - 115] was reported to increase the rate of photooxidation of CO and methylene blue, respectively, and was ascribed to the decrease in the excited charge recombination process.

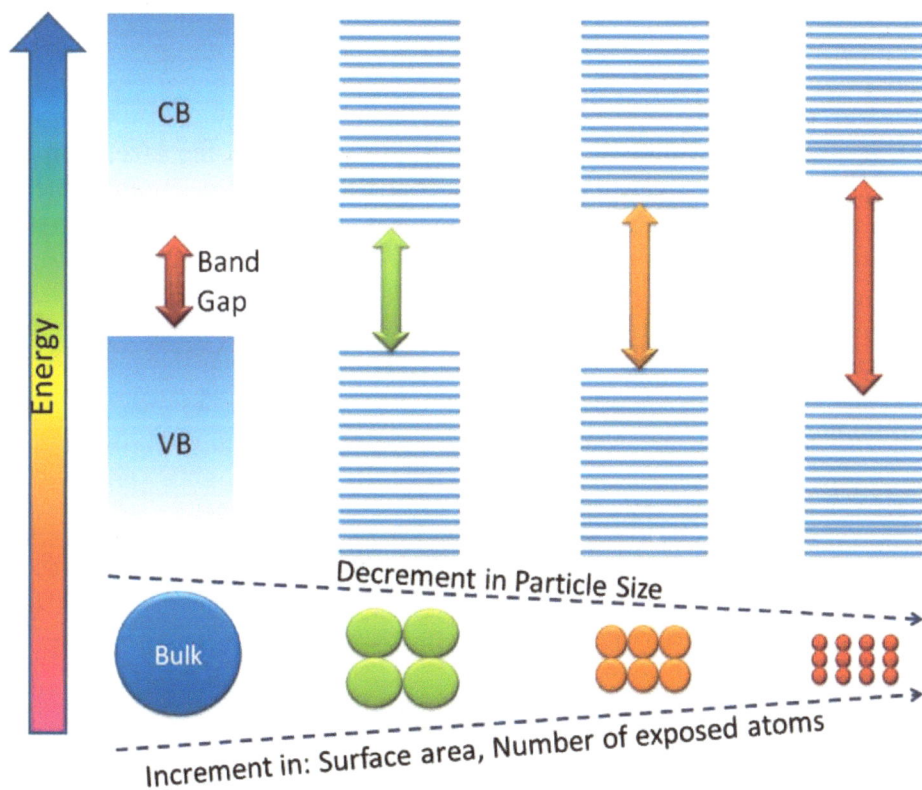

Fig. (6). Effect of particle size on electronic and surface properties of nanoparticles.

Additionally, an increase in the surface area of TiO_2 could also facilitate the improvement in its PCA by favoring the adsorption of the reacting molecule on its surface. In this connection, the surface area of TiO_2 nanoparticles has been found to generally increase by attaching it to inert silicon dioxide (SiO_2). The relevant examples consider the incorporation of P25 particles into mesoporous silica and the use of this composite material for the photooxidation of β-naphthol [116], methylene blue [117], rhodamine-6G [118], propyzamide [119], and propionaldehyde [120]. However, the thickness of the silica shell is a crucial parameter that needs to be controlled while optimizing the PCA of TiO_2 nanoparticles coated by SiO_2 since over-coating by SiO_2 reduces the light reaching the surface of TiO_2, ultimately decreasing the formation of e^-/h^+ pair and hence PCA. For example, in a recent report by Nussbaum *et al.* [121], it was suggested that the SiO_2 layer of thickness < 1 nm can enhance the activity of P25 for oxidation of salicylic acid than its bare analogy, while the same layer with a thickness of ~2-3 nm can deteriorate its PCA under similar experimental conditions.

CONCLUSION

It is clear from the aforementioned studies that changing the size, shape and loading of noble metal over titania and/or coating by SiO_2 of suitable thickness improves the PCA of titania nanoparticles that have been mainly studied for the degradation of dyes. They can also be used for the oxidative removal of other organic molecules, such as pesticides and PAHs. However, an efficient process leading to the complete removal of the parent molecule and the metabolites formed during the reaction with minimum catalyst under solar light irradiation is needed. The continuous research efforts that are being carried out in this area of research will soon bring the desired results.

ACKNOWLEDGEMENTS

Authors generously acknowledge the continuous support from their respective institutes and all the researchers for their continuous work in the concerned area that has been cited in the present work.

REFERENCES

[1]　Chen, X.; Mao, S.S. Titanium dioxide nanomaterials: synthesis, properties, modifications, and applications. *Chem. Rev.,* **2007**, *107*(7), 2891-2959.
[http://dx.doi.org/10.1021/cr0500535] [PMID: 17590053]

[2]　Nakata, K.; Fujishima, A. TiO_2 photocatalysis: Design and applications. *J. Photochem. Photobiol. Photochem. Rev.,* **2012**, *13*(3), 169-189.
[http://dx.doi.org/10.1016/j.jphotochemrev.2012.06.001]

[3]   Méndez-Arriaga, F.; Esplugas, S.; Giménez, J. Degradation of the emerging contaminant ibuprofen in water by photo-Fenton. *Water Res.*, **2010**, *44*(2), 589-595.
[http://dx.doi.org/10.1016/j.watres.2009.07.009] [PMID: 19656545]

[4]   Kansal, S.K.; Kundu, P.; Sood, S.; Lamba, R.; Umar, A.; Mehta, S.K. Photocatalytic degradation of the antibiotic levofloxacin using highly crystalline TiO_2 nanoparticles. *New J. Chem.*, **2014**, *38*(7), 3220-3226.
[http://dx.doi.org/10.1039/C3NJ01619F]

[5]   Litter, M. Heterogeneous photocatalysis Transition metal ions in photocatalytic systems. *Appl. Catal. B*, **1999**, *23*(2-3), 89-114.
[http://dx.doi.org/10.1016/S0926-3373(99)00069-7]

[6]   Sinha, A.K.; Jana, S.; Pande, S.; Sarkar, S.; Pradhan, M.; Basu, M.; Saha, S.; Pal, A.; Pal, T. New hydrothermal process for hierarchical TiO_2 nanostructures. *CrystEngComm*, **2009**, *11*(7), 1210-1212.
[http://dx.doi.org/10.1039/b906041n]

[7]   Sood, S.; Mehta, S.K.; Umar, A.; Kansal, S.K. The visible light-driven photocatalytic degradation of Alizarin red S using Bi-doped TiO_2 nanoparticles. *New J. Chem.*, **2014**, *38*(7), 3127-3136.
[http://dx.doi.org/10.1039/C4NJ00179F]

[8]   Naldoni, A.; D'Arienzo, M.; Altomare, M.; Marelli, M.; Scotti, R.; Morazzoni, F.; Selli, E.; Dal Santo, V. Pt and Au/TiO_2 photocatalysts for methanol reforming: Role of metal nanoparticles in tuning charge trapping properties and photoefficiency. *Appl. Catal. B*, **2013**, *130-131*, 239-248.
[http://dx.doi.org/10.1016/j.apcatb.2012.11.006]

[9]   Dozzi, M.V.; Selli, E. Effects of phase composition and surface area on the photocatalytic paths on fluorinated titania. *Catal. Today*, **2013**, *206*, 26-31.
[http://dx.doi.org/10.1016/j.cattod.2012.03.029]

[10]   Bernardini, C.; Cappelletti, G.; Dozzi, M.V.; Selli, E. Photocatalytic degradation of organic molecules in water: Photoactivity and reaction paths in relation to TiO_2 particles features. *J. Photochem. Photobiol. Chem.*, **2010**, *211*(2-3), 185-192.
[http://dx.doi.org/10.1016/j.jphotochem.2010.03.006]

[11]   Grover, I.S.; Singh, S.; Pal, B. The preparation, surface structure, zeta potential, surface charge density and photocatalytic activity of TiO_2 nanostructures of different shapes. *Appl. Surf. Sci.*, **2013**, *280*, 366-372.
[http://dx.doi.org/10.1016/j.apsusc.2013.04.163]

[12]   Lin, C.H.; Lee, C.H.; Chao, J.H.; Kuo, C.Y.; Cheng, Y.C.; Huang, W.N.; Chang, H.W.; Huang, Y.M.; Shih, M.K. Photocatalytic generation of H_2 gas from neat ethanol over Pt/TiO_2 nanotube catalysts. *Catal. Lett.*, **2004**, *98*(1), 61-66.
[http://dx.doi.org/10.1007/s10562-004-6450-x]

[13]   Kuo, H.L.; Kuo, C.Y.; Liu, C.H.; Chao, J.H.; Lin, C.H. A highly active bi-crystalline photocatalyst consisting of TiO_2 (B) nanotube and anatase particle for producing H_2 gas from neat ethanol. *Catal. Lett.*, **2007**, *113*(1-2), 7-12.
[http://dx.doi.org/10.1007/s10562-006-9009-1]

[14]   Grover, I.S.; Singh, S.; Pal, B. Enhanced photocatalytic activity of as-prepared sodium titanates for m-dinitrobenzene reduction and sulfosulfuron oxidation. *J. Nanosci. Nanotechnol.*, **2015**, *15*(2), 1490-1498.
[http://dx.doi.org/10.1166/jnn.2015.9072] [PMID: 26353678]

[15]   Cárdenas-Lizana, F.; Gómez-Quero, S.; Perret, N.; Keane, M.A. Gold catalysis at the gas–solid interface: role of the support in determining activity and selectivity in the hydrogenation of m-dinitrobenzene. *Catal. Sci. Technol.*, **2011**, *1*(4), 652-661.
[http://dx.doi.org/10.1039/c1cy00051a]

[16]   Cárdenas-Lizana, F.; Gómez-Quero, S.; Idriss, H.; Keane, M.A. Gold particle size effects in the gas-

phase hydrogenation of m-dinitrobenzene over Au/TiO$_2$. *J. Catal.,* **2009**, *268*(2), 223-234.
[http://dx.doi.org/10.1016/j.jcat.2009.09.020]

[17] Jianguang, J.; Teruhisa, O.; Yuji, M.; Michio, M. Dihydroxylation of naphthalene by molecular oxygen and water using TiO$_2$ photocatalysts. *Chem. Lett.,* **1999**, 963-964.

[18] Soana, F.; Sturini, M.; Cermenati, L.; Albini, A. Titanium dioxide photocatalyzed oxygenation of naphthalene and some of its derivatives. *J. Chem. Soc., Perkin Trans. 2,* **2000**, (4), 699-704.
[http://dx.doi.org/10.1039/a908945d]

[19] Hykrdová, L.; Jirkovský, J.; Mailhot, G.; Bolte, M. Fe(III) photoinduced and Q-TiO$_2$ photocatalysed degradation of naphthalene: comparison of kinetics and proposal of mechanism. *J. Photochem. Photobiol. Chem.,* **2002**, *151*(1-3), 181-193.
[http://dx.doi.org/10.1016/S1010-6030(02)00014-X]

[20] Ohno, T.; Tokieda, K.; Higashida, S.; Matsumura, M. Synergism between rutile and anatase TiO$_2$ particles in photocatalytic oxidation of naphthalene. *Appl. Catal. A Gen.,* **2003**, *244*(2), 383-391.
[http://dx.doi.org/10.1016/S0926-860X(02)00610-5]

[21] Pramauro, E.; Prevot, A.B.; Vincenti, M.; Gamberini, R. Photocatalytic degradation of naphthalene in aqueous TiO$_2$ dispersions: Effect of nonionic surfactants. *Chemosphere,* **1998**, *36*(7), 1523-1542.
[http://dx.doi.org/10.1016/S0045-6535(97)10051-0]

[22] Barrios, N.; Sivov, P.; D'andrea, D.; Núñez, O. Conditions for selective photocatalytic degradation of naphthalene in triton X-100 water solutions. *Int. J. Chem. Kinet.,* **2005**, *37*(7), 414-419.
[http://dx.doi.org/10.1002/kin.20094]

[23] Lair, A.; Ferronato, C.; Chovelon, J.M.; Herrmann, J.M. Naphthalene degradation in water by heterogeneous photocatalysis: An investigation of the influence of inorganic anions. *J. Photochem. Photobiol. Chem.,* **2008**, *193*(2-3), 193-203.
[http://dx.doi.org/10.1016/j.jphotochem.2007.06.025]

[24] Shaban, Y.A. Simultaneous hydrogen production with the degradation of naphthalene in seawater using solar light-responsive carbon-modified (CM)-n-TiO$_2$ photocatalyst. *Modern Research in Catalysis,* **2013**, *2*(3), 6-12.
[http://dx.doi.org/10.4236/mrc.2013.23A002]

[25] Mahmoodi, V.; Sargolzaei, J. Optimization of photocatalytic degradation of naphthalene using nano-TiO$_2$/UV system: statistical analysis by a response surface methodology. *Desalination Water Treat.,* **2014**, *52*(34-36), 6664-6672.
[http://dx.doi.org/10.1080/19443994.2013.861774]

[26] Cordeiro, D.S.; Corio, P. Electrochemical and photocatalytic reactions of polycyclic aromatic hydrocarbons investigated by raman spectroscopy. *J. Braz. Chem. Soc.,* **2009**, *20*(1), 80-87.
[http://dx.doi.org/10.1590/S0103-50532009000100014]

[27] Theurich, J.; Bahnemann, D.W.; Vogel, R.; Ehamed, F.E.; Alhakimi, G.; Rajab, I. Photocatalytic degradation of naphthalene and anthracene: GC-MS analysis of the degradation pathway. *Res. Chem. Intermed.,* **1997**, *23*(3), 247-274.
[http://dx.doi.org/10.1163/156856797X00457]

[28] Dass, S.; Muneer, M.; Gopidas, K.R. Photocatalytic degradation of wastewater pollutants. Titanium-dioxide-mediated oxidation of polynuclear aromatic hydrocarbons. *J. Photochem. Photobiol. Chem.,* **1994**, *77*(1), 83-88.
[http://dx.doi.org/10.1016/1010-6030(94)80011-1]

[29] Woo, O.T.; Chung, W.K.; Wong, K.H.; Chow, A.T.; Wong, P.K. Photocatalytic oxidation of polycyclic aromatic hydrocarbons: Intermediates identification and toxicity testing. *J. Hazard. Mater.,* **2009**, *168*(2-3), 1192-1199.
[http://dx.doi.org/10.1016/j.jhazmat.2009.02.170] [PMID: 19361920]

[30] Saha, S.; Singh, S.B.; Kulshrestha, G. High performance liquid chromatographic method for residue

determination of sulfosulfuron. *J. Environ. Sci. Health B,* **2003**, *38*(3), 337-347.
[http://dx.doi.org/10.1081/PFC-120019900] [PMID: 12716051]

[31]　Ramesh, A.; Sathiyanarayanan, S.; Chandran, L. Dissipation of sulfosulfuron in water – Bioaccumulation of residues in fish – LC-MS/MS-ESI identification and quantification of metabolites. *Chemosphere,* **2007**, *68*(3), 495-500.
[http://dx.doi.org/10.1016/j.chemosphere.2006.12.060] [PMID: 17289110]

[32]　Saha, S.; Kulshrestha, G. Degradation of sulfosulfuron, a sulfonylurea herbicide, as influenced by abiotic factors. *J. Agric. Food Chem.,* **2002**, *50*(16), 4572-4575.
[http://dx.doi.org/10.1021/jf0116653] [PMID: 12137477]

[33]　Arya, R.; Mishra, NK.; Sharma, AK. Brevibacillusborstelensis and Streptomyces albogriseolus have roles to play in degradation of herbicide, sulfosulfuron. Biotech. **2016**, *6*(2), 1-7.

[34]　Fenoll, J.; Sabater, P.; Navarro, G.; Vela, N.; Pérez-Lucas, G.; Navarro, S. Abatement kinetics of 30 Sulfonylurea herbicide residues in water by photocatalytic treatment with semiconductor materials. *J. Environ. Manage.,* **2013**, *130*, 361-368.
[http://dx.doi.org/10.1016/j.jenvman.2013.09.006] [PMID: 24121550]

[35]　Grover, I.S.; Singh, S.; Pal, B. Photodegradation of imidacloprid insecticide by Ag-deposited titanate nanotubes: a study of intermediates and their reaction pathways. *J. Agric. Food Chem.,* **2014**, *62*(52), 12497-12503.
[http://dx.doi.org/10.1021/jf5041614] [PMID: 25458204]

[36]　Redlich, D.; Shahin, N.; Ekici, P.; Friess, A.; Parlar, H. Kinetical study of the photoinduced degradation of imidacloprid in aquatic media. *Clean (Weinh.),* **2007**, *35*(5), 452-458.
[http://dx.doi.org/10.1002/clen.200720014]

[37]　Liu, W.; Zheng, W.; Ma, Y.; Liu, K. Sorption and degradation of imidacloprid in soil and water. *J. Environ. Sci. Health B,* **2006**, *41*(5), 623-634.
[http://dx.doi.org/10.1080/03601230600701775] [PMID: 16785171]

[38]　Kitsiou, V.; Filippidis, N.; Mantzavinos, D.; Poulios, I. Heterogeneous and homogeneous photocatalytic degradation of the insecticide imidacloprid in aqueous solutions. *Appl. Catal. B,* **2009**, *86*(1-2), 27-35.
[http://dx.doi.org/10.1016/j.apcatb.2008.07.018]

[39]　Tang, J.; Huang, X.; Huang, X.; Xiang, L.; Wang, Q. Photocatalytic degradation of imidacloprid in aqueous suspension of TiO_2 supported on H-ZSM-5. *Environ. Earth Sci.,* **2012**, *66*(2), 441-445.
[http://dx.doi.org/10.1007/s12665-011-1251-1]

[40]　Soltani-nezhad, F.; Saljooqi, A.; Shamspur, T.; Mostafavi, A. Photocatalytic degradation of imidacloprid using $GO/Fe_3O_4/TiO_2$-NiO under visible radiation: Optimization by response level method. *Polyhedron,* **2019**, *165*, 188-196.
[http://dx.doi.org/10.1016/j.poly.2019.02.012]

[41]　Žabar, R.; Komel, T.; Fabjan, J.; Kralj, M.B.; Trebše, P. Photocatalytic degradation with immobilised TiO_2 of three selected neonicotinoid insecticides: Imidacloprid, thiamethoxam and clothianidin. *Chemosphere,* **2012**, *89*(3), 293-301.
[http://dx.doi.org/10.1016/j.chemosphere.2012.04.039] [PMID: 22668598]

[42]　Feng, C.; Xu, G.; Liu, X. Photocatalytic degradation of imidacloprid by composite catalysts $H_3PW_{12}O_{40}$/La-TiO_2. *J. Rare Earths,* **2013**, *31*(1), 44-48.
[http://dx.doi.org/10.1016/S1002-0721(12)60232-4]

[43]　Pignatello, J.J.; Sun, Y. Complete oxidation of metolachlor and methyl parathion in water by the photoassisted Fenton reaction. *Water Res.,* **1995**, *29*(8), 1837-1844.
[http://dx.doi.org/10.1016/0043-1354(94)00352-8]

[44]　O'Shea, K.E.; Beightol, S.; Garcia, I.; Aguilar, M.; Kalen, D.V.; Cooper, W.J. Photocatalytic decomposition of organophosphonates in irradiated TiO_2 suspensions. *J. Photochem. Photobiol.*

Chem., **1997**, *107*(1-3), 221-226.
[http://dx.doi.org/10.1016/S1010-6030(96)04420-6]

[45] Konstantinou, I.K.; Sakellarides, T.M.; Sakkas, V.A.; Albanis, T.A. Photocatalytic degradation of selected s-triazine herbicides and organophosphorus insecticides over aqueous TiO_2 suspensions. *Environ. Sci. Technol.,* **2001**, *35*(2), 398-405.
[http://dx.doi.org/10.1021/es001271c] [PMID: 11347616]

[46] Sanjuán, A.; Aguirre, G.; Alvaro, M.; García, H. 2,4,6-Triphenylpyrylium ion encapsulated within Y zeolite as photocatalyst for the degradation of methyl parathion. *Water Res.,* **2000**, *34*(1), 320-326.
[http://dx.doi.org/10.1016/S0043-1354(99)00103-7]

[47] Evgenidou, E.; Konstantinou, I.; Fytianos, K.; Poulios, I.; Albanis, T. Photocatalytic oxidation of methyl parathion over TiO_2 and ZnO suspensions. *Catal. Today,* **2007**, *124*(3-4), 156-162.
[http://dx.doi.org/10.1016/j.cattod.2007.03.033]

[48] Moctezuma, E.; Leyva, E.; Palestino, G.; de Lasa, H. Photocatalytic degradation of methyl parathion: Reaction pathways and intermediate reaction products. *J. Photochem. Photobiol. Chem.,* **2007**, *186*(1), 71-84.
[http://dx.doi.org/10.1016/j.jphotochem.2006.07.014]

[49] Vialaton, D.; Pilichowski, J.F.; Baglio, D.; Paya-Perez, A.; Larsen, B.; Richard, C. Phototransformation of propiconazole in aqueous media. *J. Agric. Food Chem.,* **2001**, *49*(11), 5377-5382.
[http://dx.doi.org/10.1021/jf010253r] [PMID: 11714331]

[50] Pliego, G.; Zazo, J.A.; Blasco, S.; Casas, J.A.; Rodriguez, J.J. Treatment of highly polluted hazardous industrial wastewaters by combined coagulation–adsorption and high-temperature Fenton oxidation. *Ind. Eng. Chem. Res.,* **2012**, *51*(7), 2888-2896.
[http://dx.doi.org/10.1021/ie202587b]

[51] Katagi, T. Photodegradation of pesticides on plant and soil surfaces. *Rev. Environ. Contam. Toxicol.,* **2004**, *182*, 1-189.
[PMID: 15217019]

[52] Garrison, A.W.; Avants, J.K.; Miller, R.D. Loss of propiconazole and its four stereoisomers from the water phase of two soil-water slurries as measured by capillary electrophoresis. *Int. J. Environ. Res. Public Health,* **2011**, *8*(8), 3453-3467.
[http://dx.doi.org/10.3390/ijerph8083453] [PMID: 21909317]

[53] Thorstensen, C.W.; Lode, O. Laboratory degradation studies of bentazone, dichlorprop, MCPA, and propiconazole in Norwegian soils. *J. Environ. Qual.,* **2001**, *30*(3), 947-953.
[http://dx.doi.org/10.2134/jeq2001.303947x] [PMID: 11401285]

[54] Kim, I.S.; Shim, J.H.; Suh, Y.T. Laboratory studies on formation of bound residues and degradation of propiconazole in soils. *Pest Manag. Sci.,* **2003**, *59*(3), 324-330.
[http://dx.doi.org/10.1002/ps.642] [PMID: 12639050]

[55] Zeng, T.; Arnold, W.A. Pesticide photolysis in prairie potholes: probing photosensitized processes. *Environ. Sci. Technol.,* **2013**, *47*(13), 6735-6745.
[http://dx.doi.org/10.1021/es3030808] [PMID: 23116462]

[56] Zaleska, A. Doped-TiO_2: a review. *Recent Pat. Eng.,* **2008**, *2*(3), 157-164.
[http://dx.doi.org/10.2174/187221208786306289]

[57] Xu, H.; Ouyang, S.; Liu, L.; Reunchan, P.; Umezawa, N.; Ye, J. Recent advances in TiO_2 -based photocatalysis. *J. Mater. Chem. A Mater. Energy Sustain.,* **2014**, *2*(32), 12642-12661.
[http://dx.doi.org/10.1039/C4TA00941J]

[58] Ansari, S.A.; Khan, M.M.; Ansari, M.O.; Cho, M.H. Nitrogen-doped titanium dioxide (N-doped TiO_2) for visible light photocatalysis. *New J. Chem.,* **2016**, *40*(4), 3000-3009.
[http://dx.doi.org/10.1039/C5NJ03478G]

[59]   Shvadchina, Y.O.; Cherepivskaya, M.K.; Vakulenko, V.F.; Sova, A.N.; Stolyarova, I.V.; Prikhodko, R.V. The study of properties and catalytic activity of titanium dioxide doped with sulphure. *J. Water Chem. Technol.*, **2015**, *37*(6), 283-288.
[http://dx.doi.org/10.3103/S1063455X15060041]

[60]   Serpone, N. Is the band gap of pristine TiO_2 narrowed by anion- and cation-doping of titanium dioxide in second-generation photocatalysts? *J. Phys. Chem. B,* **2006**, *110*(48), 24287-24293.
[http://dx.doi.org/10.1021/jp065659r] [PMID: 17134177]

[61]   Grabowska, E.; Reszczyńska, J.; Zaleska, A. Retracted: Mechanism of phenol photodegradation in the presence of pure and modified-TiO_2: A review. *Water Res.*, **2012**, *46*(17), 5453-5471.
[http://dx.doi.org/10.1016/j.watres.2012.07.048] [PMID: 22921392]

[62]   Banerjee, S.; Dionysiou, D.D.; Pillai, S.C. Self-cleaning applications of TiO_2 by photo-induced hydrophilicity and photocatalysis. *Appl. Catal. B,* **2015**, *176-177*, 396-428.
[http://dx.doi.org/10.1016/j.apcatb.2015.03.058]

[63]   Lu, J.; Wang, Y.; Huang, J.; Fei, J.; Cao, L.; Li, C. In situ synthesis of mesoporous C-doped TiO_2 single crystal with oxygen vacancy and its enhanced sunlight photocatalytic properties. *Dyes Pigments,* **2017**, *144*, 203-211.
[http://dx.doi.org/10.1016/j.dyepig.2017.05.033]

[64]   Asahi, R.; Morikawa, T.; Ohwaki, T.; Aoki, K.; Taga, Y. Visible-light photocatalysis in nitrogen-doped titanium oxides. *Science,* **2001**, *293*(5528), 269-271.
[http://dx.doi.org/10.1126/science.1061051] [PMID: 11452117]

[65]   Di Valentin, C.; Pacchioni, G. Trends in non-metal doping of anatase TiO_2: B, C, N and F. *Catal. Today,* **2013**, *206*, 12-18.
[http://dx.doi.org/10.1016/j.cattod.2011.11.030]

[66]   Gao, H.; Liu, Y.; Ding, C.; Dai, D.; Liu, G. Synthesis, characterization, and theoretical study of N, S-codoped nano-TiO_2 with photocatalytic activities. *Int. J. Miner. Metall. Mater.*, **2011**, *18*(5), 606-614.
[http://dx.doi.org/10.1007/s12613-011-0485-y]

[67]   Shi, W.; Yang, W.; Li, Q.; Gao, S.; Shang, P.; Shang, J.K. The synthesis of nitrogen/sulfur co-doped TiO_2 nanocrystals with a high specific surface area and a high percentage of 001 facets and their enhanced visible-light photocatalytic performance. *Nanoscale Res. Lett.*, **2012**, *7*(1), 590.
[http://dx.doi.org/10.1186/1556-276X-7-590] [PMID: 23095371]

[68]   Niu, J.; Lu, P.; Kang, M.; Deng, K.; Yao, B.; Yu, X.; Zhang, Q. P-doped TiO_2 with superior visible-light activity prepared by rapid microwave hydrothermal method. *Appl. Surf. Sci.*, **2014**, *319*, 99-106.
[http://dx.doi.org/10.1016/j.apsusc.2014.07.048]

[69]   Samokhvalov, A. Hydrogen by photocatalysis with nitrogen codoped titanium dioxide. *Renew. Sustain. Energy Rev.,* **2017**, *72*, 981-1000.
[http://dx.doi.org/10.1016/j.rser.2017.01.024]

[70]   Arias-Estévez, M.; López-Periago, E.; Martínez-Carballo, E.; Simal-Gándara, J.; Mejuto, J.C.; García-Río, L. The mobility and degradation of pesticides in soils and the pollution of groundwater resources. *Agric. Ecosyst. Environ.,* **2008**, *123*(4), 247-260.
[http://dx.doi.org/10.1016/j.agee.2007.07.011]

[71]   Ahmed, S.; Rasul, M.G.; Brown, R.; Hashib, M.A. Influence of parameters on the heterogeneous photocatalytic degradation of pesticides and phenolic contaminants in wastewater: A short review. *J. Environ. Manage.,* **2011**, *92*(3), 311-330.
[http://dx.doi.org/10.1016/j.jenvman.2010.08.028] [PMID: 20950926]

[72]   Wilhelm, P.; Stephan, D. Photodegradation of rhodamine B in aqueous solution *via* SiO_2@TiO_2 nano-spheres. *J. Photochem. Photobiol. Chem.,* **2007**, *185*(1), 19-25.
[http://dx.doi.org/10.1016/j.jphotochem.2006.05.003]

[73]   Pang, S.; Lin, Z.; Zhang, Y.; Zhang, W.; Alansary, N.; Mishra, S.; Bhatt, P.; Chen, S. Insights into the

toxicity and degradation mechanisms of imidacloprid *via*physicochemical and microbial approaches. *Toxics,* **2020**, *8*(3), 65.
[http://dx.doi.org/10.3390/toxics8030065] [PMID: 32882955]

[74]   Thakur, J.; Rao, B.; Rajwanshi, A.; Parwana, H.; Kumar, R. Epidemiological study of high cancer among rural agricultural community of Punjab in Northern India. *Int. J. Environ. Res. Public Health,* **2008**, *5*(5), 399-407.
[http://dx.doi.org/10.3390/ijerph5050399] [PMID: 19151435]

[75]   Mittal, S.; Kaur, G.; Vishwakarma, G.S. Effects of environmental pesticides on the health of rural communities in the Malwa Region of Punjab, India: a review. *Hum. Ecol. Risk Assess.,* **2014**, *20*(2), 366-387.
[http://dx.doi.org/10.1080/10807039.2013.788972]

[76]   Parayil, S.K.; Kibombo, H.S.; Wu, C.M.; Peng, R.; Kindle, T.; Mishra, S.; Ahrenkiel, S.P.; Baltrusaitis, J.; Dimitrijevic, N.M.; Rajh, T.; Koodali, R.T. Synthesis-dependent oxidation state of platinum on TiO_2 and their influences on the solar simulated photocatalytic hydrogen production from water. *J. Phys. Chem. C,* **2013**, *117*(33), 16850-16862.
[http://dx.doi.org/10.1021/jp405727k]

[77]   Liu, B.; Khare, A.; Aydil, E.S. TiO_2-B/anatase core-shell heterojunction nanowires for photocatalysis. *ACS Appl. Mater. Interfaces,* **2011**, *3*(11), 4444-4450.
[http://dx.doi.org/10.1021/am201123u] [PMID: 22008419]

[78]   Song, X.; Yang, E.; Zheng, Y. Synthesis of $M_x H_y Ti_3O_7$ nanotubes by simple ion-exchanged process and their adsorption property. *Chin. Sci. Bull.,* **2007**, *52*(18), 2491-2495.
[http://dx.doi.org/10.1007/s11434-007-0337-3]

[79]   https://www.downtoearth.org.in/news/agriculture/80-groundwater-in-punjab-s-malwa-unfit-for-drinking-60951

[80]   Nadal, M.; Schuhmacher, M.; Domingo, J.L. Levels of PAHs in soil and vegetation samples from Tarragona County, Spain. *Environ. Pollut.,* **2004**, *132*(1), 1-11.
[http://dx.doi.org/10.1016/j.envpol.2004.04.003] [PMID: 15276268]

[81]   Nam, J.J.; Song, B.H.; Eom, K.C.; Lee, S.H.; Smith, A. Distribution of polycyclic aromatic hydrocarbons in agricultural soils in South Korea. *Chemosphere,* **2003**, *50*(10), 1281-1289.
[http://dx.doi.org/10.1016/S0045-6535(02)00764-6] [PMID: 12586160]

[82]   Cao, Z.; Liu, J.; Li, Y.; Ma, M. Distribution and source apportionment of polycyclic aromatic hydrocarbons (PAH) in water and sediments of the Luan River, China. *Toxicol. Environ. Chem.,* **2010**, *92*(4), 707-720.
[http://dx.doi.org/10.1080/02772240902984446]

[83]   Alomirah, H.; Al-Zenki, S.; Husain, A.; Sawaya, W.; Ahmed, N.; Gevao, B.; Kannan, K. Benzo[*a*]pyrene and total polycyclic aromatic hydrocarbons (PAHs) levels in vegetable oils and fats do not reflect the occurrence of the eight genotoxic PAHs. *Food Addit. Contam. Part A Chem. Anal. Control Expo. Risk Assess.,* **2010**, *27*(6), 869-878.
[http://dx.doi.org/10.1080/19440040903493793] [PMID: 20104381]

[84]   Ellickson, K.M.; McMahon, C.M.; Herbrandson, C.; Krause, M.J.; Schmitt, C.M.; Lippert, C.J.; Pratt, G.C. Analysis of polycyclic aromatic hydrocarbons (PAHs) in air using passive sampling calibrated with active measurements. *Environ. Pollut.,* **2017**, *231*(Pt 1), 487-496.
[http://dx.doi.org/10.1016/j.envpol.2017.08.049] [PMID: 28841501]

[85]   Jia, C.; Batterman, S. A critical review of naphthalene sources and exposures relevant to indoor and outdoor air. *Int. J. Environ. Res. Public Health,* **2010**, *7*(7), 2903-2939.
[http://dx.doi.org/10.3390/ijerph7072903] [PMID: 20717549]

[86]   Zhang, Z.; Wang, C.C.; Zakaria, R.; Ying, J.Y. Role of particle size in nanocrystalline TiO_2-based photocatalysts. *J. Phys. Chem. B,* **1998**, *102*(52), 10871-10878.
[http://dx.doi.org/10.1021/jp982948+]

[87]   Anpo, M.; Shima, T.; Kodama, S.; Kubokawa, Y. Photocatalytic hydrogenation of propyne with water on small-particle titania: size quantization effects and reaction intermediates. *J. Phys. Chem.,* **1987**, *91*(16), 4305-4310.
[http://dx.doi.org/10.1021/j100300a021]

[88]   Dong, F.; Zhao, W.; Wu, Z. Characterization and photocatalytic activities of C, N and S co-doped TiO_2 with 1D nanostructure prepared by the nano-confinement effect. *Nanotechnology,* **2008**, *19*(36), 365607.
[http://dx.doi.org/10.1088/0957-4484/19/36/365607] [PMID: 21828878]

[89]   Ge, L.; Liu, J. Preparation, characterization and photocatalytic performance of rutile TiO_2 nanowire arrays *via* a novel hydrothermal process. *J. Phys. Conf. Ser.,* **2009**, *188*(1), 012019.
[http://dx.doi.org/10.1088/1742-6596/188/1/012019]

[90]   Wu, M.C.; Liao, H.C.; Cho, Y.C.; Hsu, C.P.; Lin, T.H.; Su, W.F.; Sápi, A.; Kukovecz, Á.; Kónya, Z.; Shchukarev, A.; Sarkar, A.; Larsson, W.; Mikkola, J-P.; Mohl, M.; Tóth, G.; Jantunen, H.; Valtanen, A.; Huuhtanen, M.; Keiski, R.L.; Kordás, K. Photocatalytic activity of nitrogen-doped TiO_2-based nanowires: a photo-assisted Kelvin probe force microscopy study. *J. Nanopart. Res.,* **2014**, *16*(1), 2143.
[http://dx.doi.org/10.1007/s11051-013-2143-y]

[91]   Wang, C.; Yin, L.; Zhang, L.; Liu, N.; Lun, N.; Qi, Y. Platinum-nanoparticle-modified TiO_2 nanowires with enhanced photocatalytic property. *ACS Appl. Mater. Interfaces,* **2010**, *2*(11), 3373-3377.
[http://dx.doi.org/10.1021/am100834x] [PMID: 20961128]

[92]   Wen, B.; Liu, C.; Liu, Y. Bamboo-shaped Ag-doped TiO_2 nanowires with heterojunctions. *Inorg. Chem.,* **2005**, *44*(19), 6503-6505.
[http://dx.doi.org/10.1021/ic0505551] [PMID: 16156602]

[93]   Mandal, S.S.; Bhattacharyya, A.J. Titania nanowires as substrates for sensing and photocatalysis of common textile industry effluents. *Talanta,* **2010**, *82*(3), 876-884.
[http://dx.doi.org/10.1016/j.talanta.2010.04.021] [PMID: 20678640]

[94]   Wang, Y.; Zhang, L.; Deng, K.; Chen, X.; Zou, Z. Low temperature synthesis and photocatalytic activity of rutile TiO_2 nanorod superstructures. *J. Phys. Chem. C,* **2007**, *111*(6), 2709-2714.
[http://dx.doi.org/10.1021/jp066519k]

[95]   Long, Y.; Lu, Y.; Huang, Y.; Peng, Y.; Lu, Y.; Kang, S.Z.; Mu, J. Effect of C60 on the photocatalytic activity of TiO_2 nanorods. *J. Phys. Chem. C,* **2009**, *113*(31), 13899-13905.
[http://dx.doi.org/10.1021/jp902417j]

[96]   Wang, B.; Karthikeyan, R.; Lu, X.Y.; Xuan, J.; Leung, M.K.H. High photocatalytic activity of immobilized TiO_2 nanorods on carbonized cotton fibers. *J. Hazard. Mater.,* **2013**, *263*(Pt 2), 659-669.
[http://dx.doi.org/10.1016/j.jhazmat.2013.10.029] [PMID: 24220193]

[97]   Testino, A.; Bellobono, I.R.; Buscaglia, V.; Canevali, C.; D'Arienzo, M.; Polizzi, S.; Scotti, R.; Morazzoni, F. Optimizing the photocatalytic properties of hydrothermal TiO_2 by the control of phase composition and particle morphology. a systematic approach. *J. Am. Chem. Soc.,* **2007**, *129*(12), 3564-3575.
[http://dx.doi.org/10.1021/ja067050+] [PMID: 17341070]

[98]   Susumu, K.; Frail, P.R.; Angiolillo, P.J.; Therien, M.J. Conjugated chromophore arrays with unusually large hole polaron delocalization lengths. *J. Am. Chem. Soc.,* **2006**, *128*(26), 8380-8381.
[http://dx.doi.org/10.1021/ja0614823] [PMID: 16802786]

[99]   Toledo Antonio, J.A.; Cortes-Jacome, M.A.; Orozco-Cerros, S.L.; Montiel-Palacios, E.; Suarez-Parra, R.; Angeles-Chavez, C.; Navarete, J.; López-Salinas, E. Assessing optimal photoactivity on titania nanotubes using different annealing temperatures. *Appl. Catal. B,* **2010**, *100*(1-2), 47-54.
[http://dx.doi.org/10.1016/j.apcatb.2010.07.009]

[100] Kim, S.J.; Lee, Y.S.; Kim, B.H.; Seo, S.G.; Park, S.H.; Jung, S.C. Photocatalytic activity of titanate

nanotube powders in a hybrid pollution control system. *Int. J. Photoenergy,* **2012**, *2012*, 1-6.
[http://dx.doi.org/10.1155/2012/901907]

[101] Yu, J.; Yu, H.; Cheng, B.; Trapalis, C. Effects of calcination temperature on the microstructures and photocatalytic activity of titanate nanotubes. *J. Mol. Catal. Chem.,* **2006**, *249*(1-2), 135-142.
[http://dx.doi.org/10.1016/j.molcata.2006.01.003]

[102] Lee, C.K.; Wang, C.C.; Lyu, M.D.; Juang, L.C.; Liu, S.S.; Hung, S.H. Effects of sodium content and calcination temperature on the morphology, structure and photocatalytic activity of nanotubular titanates. *J. Colloid Interface Sci.,* **2007**, *316*(2), 562-569.
[http://dx.doi.org/10.1016/j.jcis.2007.08.008] [PMID: 17765912]

[103] Eduardo, P; Cristian, C; Francisco, H; Gerardo, C; Gisselle, A. Photocatalytic degradation of aqueous rhodamine 6G usingsupported TiO_2 catalysts. A model for the removal of organic contaminants from aqueous samples. Catalysis and Photocatalysis. *Front. Chem.,* **2020**, 8.
[http://dx.doi.org/10.3389/fchem.2020.00365]

[104] Akpan, U.G.; Hameed, B.H. Parameters affecting the photocatalytic degradation of dyes using TiO_2-based photocatalysts: A review. *J. Hazard. Mater.,* **2009**, *170*(2-3), 520-529.
[http://dx.doi.org/10.1016/j.jhazmat.2009.05.039] [PMID: 19505759]

[105] Yin, H.; Lin, T.; Yang, C.; Wang, Z.; Zhu, G.; Xu, T.; Xie, X.; Huang, F.; Jiang, M. Gray TiO_2 nanowires synthesized by aluminum-mediated reduction and their excellent photocatalytic activity for water cleaning. *Chemistry,* **2013**, *19*(40), 13313-13316.
[http://dx.doi.org/10.1002/chem.201302286] [PMID: 24014465]

[106] Si, X.; Li, F.; Sun, L.; Xu, F.; Liu, S.; Zhang, J.; Zhu, M.; Ouyang, L.Z.; Sun, D.; Liu, Y.L. Metals (Ni, Fe)-incorporated titanate nanotubes induced destabilization of $LiBH_4$. *J. Phys. Chem. C,* **2011**, *115*(19), 9780-9786.
[http://dx.doi.org/10.1021/jp111752w]

[107] Bavykin, D.V.; Lapkin, A.A.; Plucinski, P.K.; Torrente-Murciano, L.; Friedrich, J.M.; Walsh, F.C. Deposition of Pt, Pd, Ru and Au on the surfaces of titanate nanotubes. *Top. Catal.,* **2006**, *39*(3-4), 151-160.
[http://dx.doi.org/10.1007/s11244-006-0051-4]

[108] Tsai, J.Y.; Chao, J.H.; Lin, C.H. Low temperature carbon monoxide oxidation over gold nanoparticles supported on sodium titanate nanotubes. *J. Mol. Catal. Chem.,* **2009**, *298*(1-2), 115-124.
[http://dx.doi.org/10.1016/j.molcata.2008.10.019]

[109] Smith, W.; Mao, S.; Lu, G.; Catlett, A.; Chen, J.; Zhao, Y. The effect of Ag nanoparticle loading on the photocatalytic activity of TiO_2 nanorod arrays. *Chem. Phys. Lett.,* **2010**, *485*(1-3), 171-175.
[http://dx.doi.org/10.1016/j.cplett.2009.12.041]

[110] Kong, D.; Tan, J.Z.Y.; Yang, F.; Zeng, J.; Zhang, X. Electrodeposited Ag nanoparticles on TiO_2 nanorods for enhanced UV visible light photoreduction CO_2 to CH_4. *Appl. Surf. Sci.,* **2013**, *277*, 105-110.
[http://dx.doi.org/10.1016/j.apsusc.2013.04.010]

[111] Kerkez, Ö.; Boz, İ. Photo (electro) catalytic activity of Cu^{2+}-modified TiO_2 nanorod array thin films under visible light irradiation. *J. Phys. Chem. Solids,* **2014**, *75*(5), 611-618.
[http://dx.doi.org/10.1016/j.jpcs.2013.12.019]

[112] Zhao, Z.; Xu, J.; Liaw, P.K.; Wu, B.; Wang, Y. One-step formation and photocatalytic performance of spindle-like TiO_2 nanorods synthesized by dealloying amorphous $Cu_{50}Ti_{50}$ alloy. *Corros. Sci.,* **2014**, *84*, 66-73.
[http://dx.doi.org/10.1016/j.corsci.2014.03.014]

[113] Zhang, H.; Zhao, G.; Zhang, T.; Teng, F. Syntheses and photocatalytic performances of vertically grown Fe_2O_3 and TiO_2/Fe_2O_3 nanorods on coated glass substrates. *J. Alloys Compd.,* **2014**, *603*, 35-41.
[http://dx.doi.org/10.1016/j.jallcom.2014.03.042]

[114] Zhou, W.; He, Y. Ho/TiO$_2$ nanowires heterogeneous catalyst with enhanced photocatalytic properties by hydrothermal synthesis method. *Chem. Eng. J.,* **2012**, *179*, 412-416.
[http://dx.doi.org/10.1016/j.cej.2011.10.094]

[115] Jitputti, J.; Suzuki, Y.; Yoshikawa, S. Synthesis of TiO$_2$ nanowires and their photocatalytic activity for hydrogen evolution. *Catal. Commun.,* **2008**, *9*(6), 1265-1271.
[http://dx.doi.org/10.1016/j.catcom.2007.11.016]

[116] Qourzal, S.; Barka, N.; Tamimi, M.; Assabbane, A.; Nounah, A.; Ihlal, A.; Ait-Ichou, Y. Sol–gel synthesis of TiO$_2$–SiO$_2$ photocatalyst for β-naphthol photodegradation. *Mater. Sci. Eng. C,* **2009**, *29*(5), 1616-1620.
[http://dx.doi.org/10.1016/j.msec.2008.12.024]

[117] Mei, F.; Liu, C.; Zhang, L.; Ren, F.; Zhou, L.; Zhao, W.K.; Fang, Y.L. Microstructural study of binary TiO$_2$:SiO$_2$ nanocrystalline thin films. *J. Cryst. Growth,* **2006**, *292*(1), 87-91.
[http://dx.doi.org/10.1016/j.jcrysgro.2006.04.084]

[118] Anderson, C.; Bard, A.J. An improved photocatalyst of TiO$_2$/SiO$_2$ prepared by a sol-gel synthesis. *J. Phys. Chem.,* **1995**, *99*(24), 9882-9885.
[http://dx.doi.org/10.1021/j100024a033]

[119] Dibble, L.A.; Raupp, G.B. Fluidized-bed photocatalytic oxidation of trichloroethylene in contaminated air streams. *Environ. Sci. Technol.,* **1992**, *26*(3), 492-495.
[http://dx.doi.org/10.1021/es00027a006]

[120] Takeda, N.; Torimoto, T.; Sampath, S.; Kuwabata, S.; Yoneyama, H. Effect of inert supports for titanium dioxide loading on enhancement of photodecomposition rate of gaseous propionaldehyde. *J. Phys. Chem.,* **1995**, *99*(24), 9986-9991.
[http://dx.doi.org/10.1021/j100024a047]

[121] Nussbaum, M.; Paz, Y. Ultra-thin SiO$_2$ layers on TiO$_2$: improved photocatalysis by enhancing products' desorption. *Phys. Chem. Chem. Phys.,* **2012**, *14*(10), 3392-3399.
[http://dx.doi.org/10.1039/c2cp23202b] [PMID: 22298253]

CHAPTER 4

Antimicrobial Properties of Semiconductor Nanoparticles

Amanpreet Kaur[1] and **Saurabh Gupta**[1,*]

[1] *Department of Microbiology, Mata Gujri College, Fatehgarh Sahib-140 407, Punjab, India*

Abstract: Several metals have been reported to possess antimicrobial properties. Out of these metal nanoparticles, some semiconductor nanoparticles are expectant solutions to the pathogenic activities of microorganisms. Many studies have proved that these nano-sized particles are effective against several Gram-positive as well as Gram-negative pathogenic bacteria. Different types of nanoparticles are synthesized from different metals, and according to their size, they show effective responses against the target microorganisms. The exact mechanism of the antimicrobial effect has not been confirmed, but some supposed methods have been described. These particles also help to decrease antibiotic pollution as the hefty use of antibiotics can bring drastic changes in the environment and livings beings in the form of side effects.

Keywords: Antibiotic pollution, Gram-positive bacteria, Gram-negative pathogens, Nanoparticles, Semiconductor nanoparticles.

INTRODUCTION

Antibiotic pollution reforms the pathogenic microorganisms through which they alter their resistivity against antibiotics to cause incurable diseases. Therefore, it is a need of the hour to explore alternative treatment ways to deal with these bacteria to lessen the side effects. Nanoparticles have become a promising solution to this problem. Nano-technology is an advanced technology that has become the first choice of researchers. Many studies proved that semiconductor nanoparticles might be used as the best alternative to treat pathogens. Semiconductor nanoparticles are generally crystalline solids that lie between conductor and insulator due to their structure. Metals are well-known solids for their antimicrobial activities. In Indian culture, silver, zinc and copper have been used for centuries [1]. A drastic improvement in the optical and electrical properties of these nanostructured materials has been reported that depends on the size and

---

[1*] **Corresponding author Saurabh Gupta:** Department of Microbiology, Mata Gujri College, Fatehgarh Sahib-140 407, Punjab, India; Email: sau27282@gmail.com

Karamjit Singh Dhaliwal (Ed.)

shape of the nanoparticles. This is due to the confinement of the charge carriers in the narrow space of the nano-crystals [2]. Although the exact mechanisms for the antimicrobial properties of these metal nanoparticles are still under investigation, two popularly proposed mechanisms for the antimicrobial properties are: (a) free metal ion toxicity arising from the dissolution of the metals from the surface of the nanoparticles and (b) oxidative stress *via* the generation of reactive oxygen species (ROS) on surfaces of the nanoparticles. Besides the nano nature of these metal particles, the morphological and physicochemical properties of these nanoparticles play a significant role in enhancing their antimicrobial activities. Bactericidal properties of nanoparticles are associated with size in such a way that the smaller the nanoparticles are, the higher the bactericidal potential. Since metals have a strong tendency to donate the electron, hence are positively charged, which facilitates their binding to the negatively charged cell surface leading to an enhanced bactericidal effect.

A significant number of metals, such as silver, copper, gold, titanium, and zinc, have been accounted as bactericidal agents for centuries with wide properties and spectra of activity [3]. Different foods and daily use commodities, such as toothpaste, are now composed of powdered zinc citrate or acetate to combat dental plaque [4]. Titanium oxide has been used as a whiting agent in toothpaste for a long time [5]. Since zinc is an essential micronutrient required by the human body, zinc oxide (ZnO) has been treated as an important inorganic semiconductor material due to its GRAS (generally regarded as safe) status, high thermal and photostability, oxidation resistivity along with high electron mobility [6]. Zinc oxide nanoparticles are easy to synthesize, biologically safe, compatible with biological tissues as well as environment-friendly [7]. Hence, ZnO nanoparticles have been exploited for their wide application in different fields (Fig. **1**), such as biological sensing, gene delivery, drug delivery, wound dressing material, and antifungal and antibacterial activities [8, 9]. Along with all the above-mentioned properties of ZnO nanoparticles, these are quite cheap and readily available. ZnO nanoparticles exhibit morphological versatility, such as nanorods, nanoflowers, nanospheres, nanotubes, *etc.* and are more influential on rheological parameters [10]. The morphology and texture of these nanomaterials are significantly controlled by the organic components used during the synthesis process [11]. Besides the biological uses, ZnO semi-conductor nano-material also possesses photocatalytic properties that lead to use their great potential for decontamination of organic compounds in water along with associated environmentally friendly ease and complete mineralization of pollutants. Similarly, other semi-conductor photocatalysts have recently reported high photocatalytic activity for decontamination. Silver nanoparticles may also combine with semi-conductor material, leading to the separation of charges and producing more photo-generated charges. Heterogeneous hybrid systems have also been developed as compelling methodologies for the decontamination of aqueous solutions containing non-biodegradable compounds.

ANTIBIOTIC POLLUTION

Antibiotic pollution is a deadly threat to humans and the environment due to its hazardous consequences (Fig. **2**). The massive use of antibiotics and their improper disposal is a major cause of antibiotic pollution. This leads to an upsurge in antibiotic-resistant microorganisms. Incurable diseases are on the rise due to antibiotic waste, which leads to the development of antibiotic resistance among pathogenic microorganisms. Therefore, it is necessary to combat this pollution. Researchers find many different ways to tackle this pollution by targeting the pathogens which become resistant to the medicine. Many studies have been conducted to find an effective way to fight and degrade these microorganisms.

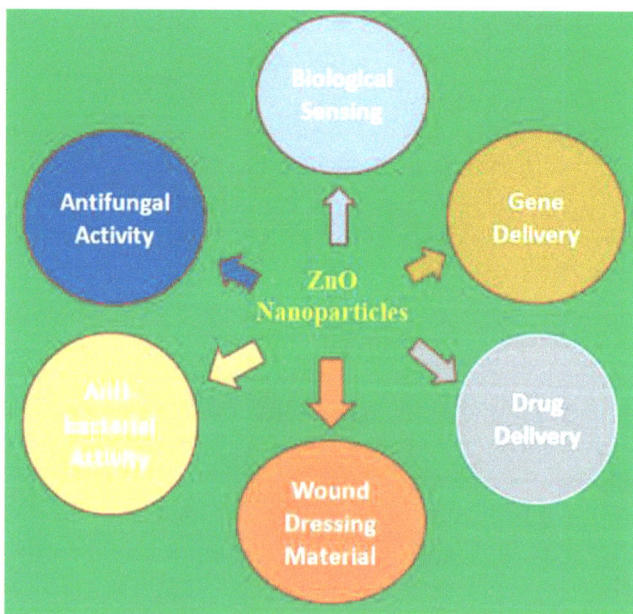

Fig. (1). Biological uses of ZnO nanoparticles.

NANOPARTICLES

Nanoparticles (NPs) are small materials with sizes ranging from 1 to 100 nm. Due to their high surface area to volume ratio and nano-scale size, they have unique chemical and physical properties compared to their bulk counterparts. Generally, the size, structure and shape of these particles have a great influence on their properties, such as reactivity, toughness, catalytic activity, and electronic and magnetic properties (Fig. **3**). These nano-materials may be explored for their wide applications in multidisciplinary fields, such as medical or environmental applications, imaging and energy-based research. Core-shell NPs are made of three layers; the first is the surface layer, which can be functionalized with various small molecules, polymers, metal ions and surfactants, the second one is the shell layer, which chemically differs from the core and the last one is the core, which is the centre of the NP [12].

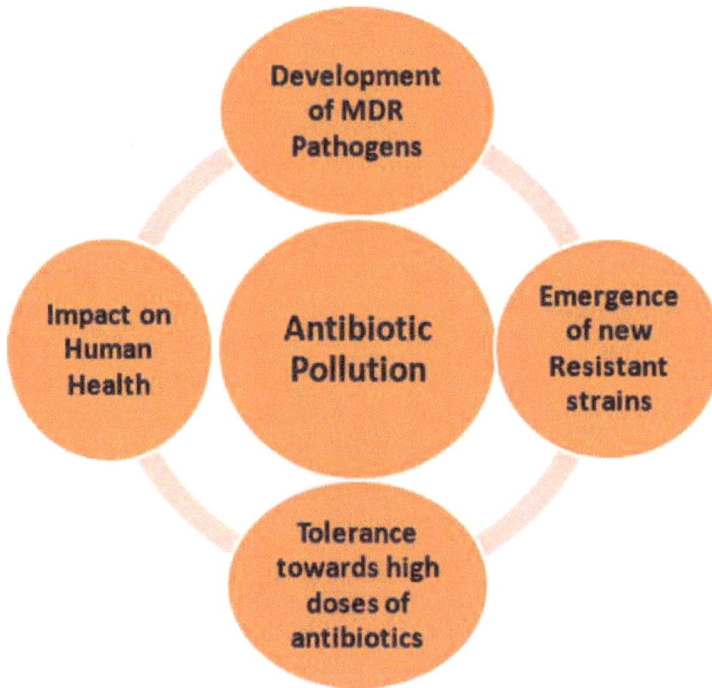

Fig. (2). Consequences of Antibiotic Pollution.

Classification of Nanoparticles

Nanoparticles have been classified in several ways based on their size, chemical properties, morphology and compositions. Broadly nanoparticles are divided into two categories, *i.e.*, inorganic and organic nanoparticles, including some subcategories. Inorganic nanoparticles include carbon-based NPs, metal NPs, ceramics NPs and semiconductor NPs. Carbon-based nanoparticles use carbon as a nano-filler. Carbon nanostructures (CB, CNT and graphene derivatives) are mainly used to improve the electrical conductive properties of polymer nano-composites, although their morphology influences the final properties [13].

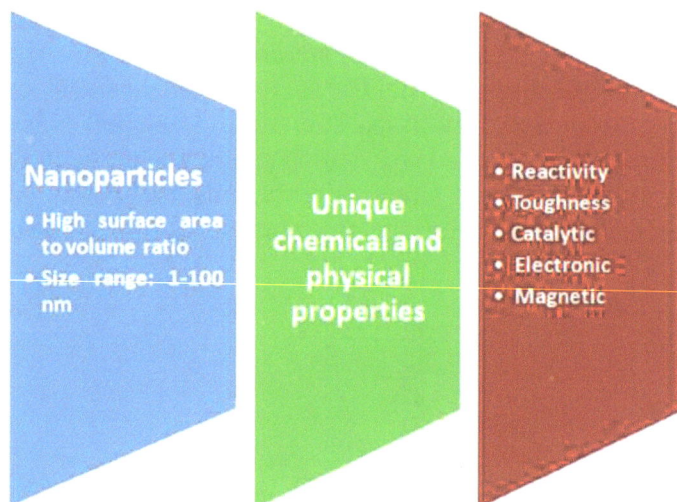

Fig. (3). Nanoparticles and associated characteristics.

Metal nanoparticles (synthesized from metals) are widely used in biomedical and engineering fields, such as biotechnology, magnetic separation, drug delivery, and diagnostic imaging. Moreover, magnetic nanoparticles, such as Fe_3O_4, silver and gold nanoparticles, nano-shells, and nano-cages, can be used in the diagnostic imaging and therapy of cancer [14]. Ceramics nanoparticles consist of oxides, carbides, phosphates and carbonates of metals and metalloids, such as silicon, titanium, and calcium, which have good heat resistance and chemical inactivity. These nanoparticles are used as carriers for genes, drugs, proteins and imaging agents. Additionally, these are used in drug delivery systems against bacterial infections, glaucoma, and cancer [15]. Semiconductor nanoparticles are small colloidal quantum dots (<10 nm), generally fluorescent semiconductor nanocrystals, having singular luminescent properties. These nanoparticles are made up of a semiconductor core, covered by semiconductive shells, such as cadmium selenide, cadmium telluride, zinc cadmium selenide, and zinc sulphide, to enhance their optical properties, and a cap to facilitate aqueous solubility for biological applications. The quantum dots can be used *in vitro* real-time bioimaging or monitoring of intracellular processes with time, diagnostic and imaging purposes because of their properties, such as narrow emission, great ultraviolet excitation or bright fluorescence. For example, cadmium selenium-zinc sulphide quantum dots can be utilized to visualize cerebral vasculature [16].

The organic category is composed of polymeric NPs and lipid-based NPs. Polymeric nanoparticles are colloidal particles (from 10 nm to 1 μm) made of biocompatible and biodegradable polymers of natural (such as alginate, albumin, chitosan) or synthetic (such as polylactide, polylactide– polyglycolide copolymers, polycaprolactones, and polyacrylates) origin [17]. On the other hand, lipid-based nanoparticle (LBNP) systems (for example, liposomes, solid lipid nanoparticles, and nanostructured lipid carriers) are used as colloidal carriers for bioactive organic molecules with many advantages, such as high temporal and thermal stability, excellent loading capacity, low fabrication costs, ease of fabrication, and ability to monitor the release of the drug and induce low or no toxicity. These can also transport hydrophobic and hydrophilic molecules [18].

Semiconductor Nanoparticles

A number of synthetic semiconductor nanoparticles, such as III-V semiconductors (GaN, GaP, InP, InAs), II-VI semiconductors (ZnO, ZnS, CdS, CdSe, CdTe) and group IV semiconductors (Si and Ge), have been synthesized using the advancement of technology [19, 20]. These different types of metal and metal oxide-based NPs, such as silver, copper, zinc oxide, titanium oxide, copper oxide, and nickel oxide NPs have demonstrated antimicrobial activity over the last years towards several pathogens [21, 22]. The antimicrobial potential of these nanoparticles depends on their composition, surface modification, intrinsic properties and the type of targeted microorganisms [23]. A special category of metallic NPs, *i.e.*, superparamagnetic iron-oxide nanoparticles (SPIONs) (*e.g.,* magnetite (Fe_3O_4) and maghemite (γ-Fe_2O_3) NPs) has been reported in the literature with an increase in antimicrobial activity upon the application of an external magnetic field [24]. An interesting strategy to increase the antimicrobial efficiency of metal-based nanoparticles is the use of silica and carbon compounds as delivery systems [25].

Antimicrobial Activity of Semiconductor Nanoparticles

Semiconductor nanoparticles have antimicrobial activities. Some of these agents act as cytotoxic agents against bacteria and have little impact on mammalian cells, thus making these NPs potential medicine for medical applications [26]. There are many studies to prove the effectiveness of these nanoparticles against pathogens. These particles have been used alone as well as in combination like ZnO-Ag [27] against Gram-positive (*Staphylococcus aureus* and *Candida albicans*). Another study shed light on the antimicrobial activity of ZnMgO against *Escherichia coli* and *Bacillus subtilis*. Nano-MgO has shown its bactericidal activity against both Gram-positive and Gram-negative bacteria and is highly dependent on particle size and concentration [28 - 31]. Nano-MgO could be used alone or in combination and has been proposed as a bactericide in food product treatment in order to improve microbiological food safety [32].

Zinc sulfide (ZnS) is one of the most significant compounds discovered as a semiconductor [33]. Cr-doped ZnS has been studied for its antimicrobial activities against MDR *S. aureus* with a significant zone of inhibition [34]. Jamzad and Bidkorpeh evaluated the antimicrobial properties of green synthesized iron oxide nanoparticles against Gram-positive bacterium, such as *listeria monocytogenes* and the fungi *Aspergillus flavus* and *Penicillium spinulosum* with significant antimicrobial properties [35]. Similarly, zinc-selenium nanoparticles (Zn-Se NPs) were explored by Mirzaei *et al.* and were used against four Gram-negative bacteria: *Escherichia coli* (ATCC 25922), *Pseudomonas aeruginosa* (ATCC 27853), *Klebsiella pneumoniae* (ATCC700603), *Salmonella typhi* (PTCC 1609) and two Gram-positive bacteria: *Staphylococcus aureus* (ATCC 12600) and *Bacillus subtilis* (ATCC 465) [36]. Furthermore, the gold nanoparticles-decorated hematite nanostructures (α-Fe$_2$O$_3$) showed a lethal effect on *E.coli* [37]. Jo, with his co-researchers, reported the antifungal activity of silver nanoparticles [38]. Shape-mediated antimicrobial activities of silver nanoparticles toward Gram-negative bacteria *E.coli* were founded by Pal in association with other researchers [39].

In dental diseases, dental caries and periodontal disease predominate and affect humans widely with the involvement of various microorganisms, leading to the development of biofilm on both natural and restored tooth surfaces [40]. The sulfide nanoparticles have been broadly studied for antibacterial, antifungal, and antiviral actions in association with the metals. On the other hand, inert silver nanoparticles have been used widely in biomedical applications, including the dental field [1]. Besides Ag NPs, zinc nanoparticles also inhibit the establishment and further progression of *Staphylococcus sobrinus* biofilm at appropriate concentrations [41]. Due to these inherited properties of metal nanoparticles on dental diseases, Malarkodi, with other co-authors, explored the antimicrobial activities of cadmium and zinc in their sulphide forms, such as cadmium sulphide (CdS) and zinc sulphide (ZnS) semiconductor nanoparticles against both bacterial and yeast oral pathogens. Their observation has confirmed an enhanced antimicrobial activity of sulphide-based metal nanoparticles [42].

Antimicrobial activity of different nanoparticles, such as CdSNPs, Ag NPs, and TiO$_2$ NPs, against several bacteria, *i.e.*, *Bacillus subtilis* and *Klebsiella planticola,* along with oral pathogenic species of *Streptococcus mutans,* has been documented by various researchers [5, 43]. Vazquez-Muñoz and others have investigated the antibacterial properties of silver nanoparticles against major urinary tract infections caused by the bacterium *Escherichia coli* [44]. Silver nanoparticles have also been reported for their potent antimicrobial activity against viruses, bacteria and fungi [45 - 47]. Moreover, AgNPs may exhibit synergistic effects with antibiotics and can be applied as sanitizers [48]. Ag NPs have also been investigated for their antimicrobial activity against different bacterial strains of birds origin isolated from live bird markets surrounding the Bangladesh Agricultural University (BAU) campus [49]. Silver NPs have a positively charged surface, ligand replacement ability and oxidative dissolution capabilities that facilitate their binding to the negatively charged surface of bacteria, resulting in an enhancement of the bactericidal effect [50]. Similarly, Ahamed *et al.* reported copper oxide (CuO) as another effective antimicrobial semiconductor nanomaterial with its remarkable inhibition potential on some of the human pathogenic bacteria, such as *Escherichia coli, Pseudomonas aeruginosa, Klebsiella pneumoniae, Enterococcus faecalis, Shigella flexneri, Salmonella typhimurium,* and *Proteusvulgaris* along with one gram-positive bacterium *Staphylococcus aureus* [51]. One more study on CuO NPs was conducted by Pagar with his co-authors using CDBE-mediated CuO NPs as an antimicrobial agent against three bacterial strains, including *Bacillus subtilis* MTCC 441, *Escherichia coli* MTCC 44, and *Staphylococcus aureus* MTCC 96 [52]. Five bacterial strains, including gram-negative bacteria (*E. coli, P. aeruginosa* and *S. typhi*), Gram-positive bacteria (*S. aureus and B. subtilis*) along with opportunistic pathogenic yeast (*Candida albicans*) are examined for the antimicrobial capability of zinc oxide nanoparticles by Kasahun in association with other researchers. Gram-positive bacteria were found to be more sensitive to ZnO NPs than gram-negative bacteria and fungi. Moreover, the differences in antimicrobial activities of differently synthesized (green synthesis, chemical synthesis and thermal synthesis) zinc oxide nanoparticles have also been studied. Environmental benign synthesis of ZnO NPs showed more effective antimicrobial activity than the nanoparticles synthesized using the chemical or the thermal route. Zinc oxide nanoparticles are considered one of the most important materials due to their utilization in various fields, including biomedical applications [53].

Researchers have also explored the antimicrobial property of ZnO NPs against *Pseudomonas aeruginosa, Proteus mirabilis, Bacillus subtilis, Staphylococcus epidermidis, Klebsiella pneumonia, Streptococcus pyogenes, Enterococcus faecalis, Proteus vulgaris, Salmonella typhimurium, Shigella flexneri, Pseudomonas alcaligenes,* and *Enterobacter aerogenes* [54, 55]. Bala *et al.* examined the antimicrobial properties of ZnO nanoparticles synthesized *via* green routes against Gram-negative *Escherichia coli* and Gram-positive *Staphylococcus aureus* bacterial strains. The minimum inhibitory concentration of nanoparticles for these bacteria was observed at the concentration of 50 and 100 μg ml^{-1}, respectively [56]. Besides the antibacterial activities of green synthesized ZnO nanoparticles against *S. aureus*, a significant reduction in the growth of *Streptococcus pyogenes* and *E. coli* has also been observed with the same concentration [57]. Similarly, Jayaseelan *et al.* studied the antifungal potential of ZnO nanoparticles synthesized by bacteria against *Aspergillus flavus, A. niger,* and *Candida albicans* and a considerable reduction in the mycelia growth of these fungi was recorded at 25 µg ml^{-1} concentration [58]. In corroboration with these observations, inhibition of the fungus *Erythricium salmonicolor*, a coffee fungus, has been reported by Arciniegas-Grijalba *et al.* when the antifungal activity of ZnO nanoparticles was tested against this fungus [59]. On the contrary, an elevated concentration of ZnO nanoparticles was found effective against the pathogenic fungus *Penicillium expansum* compared to the other studies on these nanoparticles and fungi [60]. These reports also suggested the mechanism of antifungal activity of ZnO nanoparticles, *i.e.*, through reactive oxygen species as hydroxyl radicals formed in water suspensions of ZnO, leading to the cytotoxic effect of ZnO nanoparticles [61]. Similarly, NiO nanoparticles also showed effective antimicrobial activity against Gram-negative (*E. coli*) and Gram-positive (*Bacillus subtilis* and *Streptococcus pneumonia*) bacteria along with fungi *Aspergillus niger* and *A. fumigatus* [62].

Remarkable results are observed due to the consistent morphology of these particles, rendering/making these nanoparticles effective. For instance, successful application of this type of copper nanoparticles (Cu NPs) has been reported by many researchers [63 - 67] against several Gram-positive bacteria: *Staphylococcus aureus* [68], *Bacillus subtilis* [69, 70], and *Bacillus cereus* [71], Gram-negative bacteria like *Pseudomonas aeruginosa*, *Escherichia coli*, *Salmonella typhi*, *Klebsiella pneumonia* [69], and *Enterobacter, Micrococcus* [66], rice pathogen *Xanthomonas oryzae pv, Oryzae* [72] or as an antifungal agent against *Trichoderma viride* [73], *Aspergillus Niger, Candida albicans,* and *Curvularia* [74]. Similarly, the antibacterial activities of CNT-Ag and GO-Ag were also investigated using the growth curve method and minimal inhibitory concentrations against different Gram-negative and Gram-positive bacteria [75]. A team of researchers also made a bandage of poly (ε-caprolactone)/cellulose nanofiber containing ZrO$_2$ nanoparticles (PCL/CNF/ZrO$_2$) with antimicrobial properties for wounds, which was tested for *S. aureus* and *E. coli* taken as model gram-positive and gram-negative bacteria. *Candida albicans* was also used, and results showed its good antimicrobial effect both on bacterial and fungal strains [76].

Mechanisms of Action of Nanoparticles against Phytopathogens

Although several mechanisms have been suggested and validated for the antimicrobial activity of different nanoparticles, oxidative stress induction, metal ion release, and non-oxidative damage to microbial cells have been well documented with possible explanations. Reactive oxygen species (ROS) are reactive intermediates with great oxidative potential and are eventually toxic to all microorganisms, irrespective of their short life span [77]. The most common examples of ROS are superoxide radicals (O_2-), hydroxyl radicals (•OH), hydrogen peroxide (H_2O_2), and singlet oxygen ($^1O_2$). Reactive oxygen species are produced by nanoparticles due to their photocatalytic activity. It has been observed that metal compounds, particularly nano-sized forms, receive enough energy from light irradiation. These light irradiations lead to the excitation and mobilization of an electron from the valence band to the conduction band, leaving a highly reactive gap (H^+). Finally, this zone emerges as a ROS source when it interacts with H_2O or OH^- that surrounds the nanoparticles [78]. Generation of insufficient reactive species results in a set of redox reactions that leads to cell death by the alteration of different cell structures and pathways, such as plasma membrane, DNA, proteins, and electron transport chain, respectively, along with the metabolic routes, which are responsible for maintaining the normal morphological and physiological cellular functions [79]. Oxidative stress induction also leads to the release of metal ions from the metal oxide nanoparticles. These metal ions then pass through the plasma membrane and find their entry into the cytoplasm and organelles. Finally, these metallic ions can interact with the functional groups of proteins and nucleic acids, such as thiol (–SH), amino (–NH), and carboxyl (–COOH) groups, leading to impairment of the enzymatic activities and several protein structures. Although the metal ions released are not the main source of damage caused by NPs but play a significant role in association with other factors [80].

Nanoparticles find their effectiveness even at the molecular level within the cell. Proteomic analysis has reported the impairment of protein regulation for the proteins involved in nitrogen metabolism, electron transfer, and substance transport in the presence of CuO NPs [81]. Similarly, silver ions interact with sulphur and phosphorus present in the protein moiety of the ribosome subunit and control its expression. These silver nanoparticles also affect the proteins of the cell wall and plasma membrane of bacteria [82]. In corroboration with the observations of silver nanoparticles, Cui *et al.* reported the inhibition of attachment of tRNA with ribosomal subunit to prevent protein synthesis. These NPs also inhibit the ATPase activity that leads to collapsing the membrane potential along with reducing the ATP levels and stimulation of the ROS generation, which adversely affects other structures [83]. Reactive oxygen species generated by TiO_2 NPs severely affect microbial regulatory functions, such as replications, transcription of metabolic pathways genes, along with cell division genes through the mutations in DNA. Genomic analysis has confirmed that these species target the sugar-phosphate moieties or nucleobases and cause saccharide fragmentation. These modifications finally lead to the fragmentation of DNA [84]. Experimental evidence of this phenomenon has been studied and reported in the pBR322 plasmid in the presence of Ag NPs *via* electrophoresis [85]. Guanine was found to be the most severely affected nucleobase due to its low redox potential. Oxidation of guanine produces a wide variety of modifications that ultimately affect DNA function [86]. These NPs affect not only bacteria but also other complex multi-cellular organisms through induced genetic damage [87].

Nanoparticles have also been reported to affect the cell wall of all microorganisms. Every group of organisms possesses wide variation in their cell wall composition. For instance, fungi and yeast are mainly composed of chitin and polysaccharides; Gram-positive bacteria contain many layers of peptidoglycan and teichoic acid (20–50 nm), while in Gram-negative bacteria, a few layers of peptidoglycan are surrounded by a second lipid membrane containing lipopolysaccharides and lipoproteins [88, 89]. A number of different mechanisms have been proposed to explain the process of rupturing the cell wall of pathogenic bacteria. Lipopolysaccharides present in the outer membrane of Gram-negative bacteria harbour a negative charge that attracts positively charged metal nanoparticles [90]. On the contrary, Gram-positive bacteria lack the outer membrane possessed by Gram-negative bacteria and have a higher permeability for the entry of nanoparticles since the capacity of a single membrane lacks the ability to stop the entry of foreign molecules. Besides the permeability factor, the cell wall of Gram-positive bacteria has a high negative charge due to the peptidoglycan and teichoic acid contents, which strongly attract the metal NPs, resulting in cell membrane damage and cell death [84, 91]. Besides these inherited features of the cell wall of bacteria, NPs of different metals are also attracted to the cell wall by electrostatic attraction [92], van der Waals forces [93], and hydrophobic interactions [94]. All these interactions induce changes in the shape, function and permeability of the cells.

CONCLUSION

A number of semiconductor nanoparticles have been reported with their wide antimicrobial potential for almost all classes of microbial pathogens. These nanoparticles, individually or in association with certain antibiotics, will tend to exert an enhanced impact along with controlling the development of multidrug resistance among the different pathogens. However, much work is in progress to establish the mechanisms of their antimicrobial potential and their wide utilization at the commercial level. Besides this, the green synthesis of these nanoparticles may further help to combat environmental pollution.

REFERENCES

[1]   Malarkodi, C.; Rajeshkumar, S.; Paulkumar, K.; Vanaja, M.; Gnanajobitha, G.; Annadurai, G. Biosynthesis and antimicrobial activity of semiconductor nanoparticles against oral pathogens. *Bioinorg. Chem. Appl.,* **2014**, *2014*, 1-10.

[http://dx.doi.org/10.1155/2014/347167] [PMID: 24860280]

[2]   Steigerwald, M.L.; Brus, L.E. Semiconductor crystallites: a class of large molecules. *Acc. Chem. Res.,* **1990**, *23*(6), 183-188.
[http://dx.doi.org/10.1021/ar00174a003]

[3]   Vanaja, M.; Gnanajobitha, G.; Paulkumar, K.; Rajeshkumar, S.; Malarkodi, C.; Annadurai, G. Phytosynthesis of silver nanoparticles by Cissus quadrangularis: influence of physicochemical factors. *J. Nanostructure Chem.,* **2013**, *3*(1), 17.
[http://dx.doi.org/10.1186/2193-8865-3-17]

[4]   Giertsen, E. Effects of mouthrinses with triclosan, zinc ions, copolymer, and sodium lauryl sulphate combined with fluoride on acid formation by dental plaque *in vivo. Caries Res.,* **2004**, *38*(5), 430-435.
[http://dx.doi.org/10.1159/000079623] [PMID: 15316186]

[5]   Malarkodi, C.; Chitra, K.; Rajeshkumar, S.; Gnanajobitha, G.; Paulkumar, K.; Vanaja, M.; Annadurai, G. Novel eco-friendly synthesis of titanium oxide nanoparticles by using *Planomicrobium* sp. and its antimicrobial evaluation. *Pharm. Sin.,* **2013**, *4*(3), 59-66.

[6]   Prasad, A.R.; Garvasis, J.; Oruvil, S.K.; Joseph, A. Bio-inspired green synthesis of zinc oxide nanoparticles using Abelmoschus esculentus mucilage and selective degradation of cationic dye pollutants. *J. Phys. Chem. Solids,* **2019**, *127*, 265-274.
[http://dx.doi.org/10.1016/j.jpcs.2019.01.003]

[7]   Vaseem, M; Umar, A; Hahn, YB ZnO nanoparticles: growth, properties, and applications. Metal oxide nanostructures and their applications. 2010;5(1).

[8]   Sharma, D.; Rajput, J.; Kaith, B.S.; Kaur, M.; Sharma, S. Synthesis of ZnO nanoparticles and study of their antibacterial and antifungal properties. *Thin Solid Films,* **2010**, *519*(3), 1224-1229.
[http://dx.doi.org/10.1016/j.tsf.2010.08.073]

[9]   Seil, J.T.; Webster, T.J. Antimicrobial applications of nanotechnology: methods and literature. *Int. J. Nanomedicine,* **2012**, *7*, 2767-2781.
[PMID: 22745541]

[10]   Ahmad, M.; Ahmed, E.; Zhang, Y.; Khalid, N.R.; Xu, J.; Ullah, M.; Hong, Z. Preparation of highly efficient Al-doped ZnO photocatalyst by combustion synthesis. *Curr. Appl. Phys.,* **2013**, *13*(4), 697-704.
[http://dx.doi.org/10.1016/j.cap.2012.11.008]

[11]   Fattahi, M.; Kazemeini, M.; Khorasheh, F.; Rashidi, A.M. Morphological investigations of nanostructured V_2O_5 over graphene used for the ODHP reaction: from synthesis to physiochemical evaluations. *Catal. Sci. Technol.,* **2015**, *5*(2), 910-924.
[http://dx.doi.org/10.1039/C4CY01108B]

[12]   Liguori, F. *Role of LL37-Au NPs in Antimicrobial Activity and Dental Application* (Doctoral dissertation, Politecnico di Torino).

[13]   Cassagnau, P. *Rheology of Carbon Nanoparticle Suspensions and Nanocomposites.,* **2015**.
[http://dx.doi.org/10.1016/B978-1-78548-036-2.50003-4]

[14]   Mody, V.; Siwale, R.; Singh, A.; Mody, H. Introduction to metallic nanoparticles. *J. Pharm. Bioallied Sci.,* **2010**, *2*(4), 282-289.
[http://dx.doi.org/10.4103/0975-7406.72127] [PMID: 21180459]

[15]   Thomas, S.; Harshita, B.S.P.; Mishra, P.; Talegaonkar, S. Kumar Mishra P, Talegaonkar S. Ceramic nanoparticles: fabrication methods and applications in drug delivery. *Curr. Pharm. Des.,* **2015**, *21*(42), 6165-6188.
[http://dx.doi.org/10.2174/1381612821666151027153246] [PMID: 26503144]

[16]   Bose, A; Wong, TW Nanotechnology-enabled drug delivery for cancer therapy. In Nanotechnology applications for tissue engineering. **2015**, 173-193.

[17]   Sharma, M. Transdermal and intravenous nano drug delivery systems: present and future. In Applications of targeted nano drugs and delivery systems. **2019**, 499-550.

[18]   García-Pinel, B.; Porras-Alcalá, C.; Ortega-Rodríguez, A.; Sarabia, F.; Prados, J.; Melguizo, C.; López-Romero, J.M. Lipid-based nanoparticles: application and recent advances in cancer treatment. *Nanomaterials (Basel),* **2019**, *9*(4), 638.
[http://dx.doi.org/10.3390/nano9040638] [PMID: 31010180]

[19]   Burda, C.; Chen, X.; Narayanan, R.; El-Sayed, M.A. Chemistry and properties of nanocrystals of different shapes. *Chem. Rev.,* **2005**, *105*(4), 1025-1102.
[http://dx.doi.org/10.1021/cr030063a] [PMID: 15826010]

[20]   Jun, Y.; Seo, J.; Oh, S.; Cheon, J. Recent advances in the shape control of inorganic nano-building blocks. *Coord. Chem. Rev.,* **2005**, *249*(17-18), 1766-1775.
[http://dx.doi.org/10.1016/j.ccr.2004.12.008]

[21]   Biswal, A.K.; Misra, P.K. Biosynthesis and characterization of silver nanoparticles for prospective application in food packaging and biomedical fields. *Mater. Chem. Phys.,* **2020**, *250*, 123014.
[http://dx.doi.org/10.1016/j.matchemphys.2020.123014]

[22]   Chaudhary, J.; Tailor, G.; Yadav, B.L.; Michael, O. Synthesis and biological function of Nickel and Copper nanoparticles. *Heliyon,* **2019**, *5*(6), e01878.
[http://dx.doi.org/10.1016/j.heliyon.2019.e01878] [PMID: 31198877]

[23]   Maleki Dizaj, S.; Mennati, A.; Jafari, S.; Khezri, K.; Adibkia, K. Antimicrobial activity of carbon-based nanoparticles. *Adv. Pharm. Bull.,* **2015**, *5*(1), 19-23.
[PMID: 25789215]

[24]   Seabra, AB; Pelegrino, MT; Haddad, PS Antimicrobial Applications of Superparamagnetic Iron Oxide Nanoparticles: Perspectives and Challenges. InNanostructures for Antimicrobial Therapy 2017 Jan 1 (pp. 531-550). Elsevier.

[25]   Camporotondi, DE; Foglia, ML; Alvarez, GS; Mebert, AM; Diaz, LE; Coradin, T; Desimone, MF. Antimicrobial properties of silica modified nanoparticles. Microbial Pathogens and Strategies for Combating Them: Science, Technology and Education. Méndez-Vilas, A., Ed. **2013**, 283-90.

[26]   Taylor, E.; Webster, T.J. Reducing infections through nanotechnology and nanoparticles. *Int. J. Nanomedicine,* **2011**, *6*, 1463-1473.
[PMID: 21796248]

[27]   Zaheer, Z.; Albukhari, S.M. Fabrication of zinc/silver binary nanoparticles, their enhanced microbial and adsorbing properties. *Arab. J. Chem.,* **2020**, *13*(11), 7921-7938.
[http://dx.doi.org/10.1016/j.arabjc.2020.09.023]

[28]   Huang, L.; Li, D.Q.; Lin, Y.J.; Wei, M.; Evans, D.G.; Duan, X. Controllable preparation of Nano-MgO and investigation of its bactericidal properties. *J. Inorg. Biochem.,* **2005**, *99*(5), 986-993.
[http://dx.doi.org/10.1016/j.jinorgbio.2004.12.022] [PMID: 15833320]

[29]   Makhluf, S.; Dror, R.; Nitzan, Y.; Abramovich, Y.; Jelinek, R.; Gedanken, A. Microwave assisted synthesis of nanocrystalline MgO and its use as a bacteriocide. *Adv. Funct. Mater.,* **2005**, *15*(10), 1708-1715.
[http://dx.doi.org/10.1002/adfm.200500029]

[30]   Krishnamoorthy, V.; Hiller, D.B.; Ripper, R.; Lin, B.; Vogel, S.M.; Feinstein, D.L.; Oswald, S.; Rothschild, L.; Hensel, P.; Rubinstein, I.; Minshall, R.; Weinberg, G.L. Epinephrine induces rapid deterioration in pulmonary oxygen exchange in intact, anesthetized rats: a flow and pulmonary capillary pressure-dependent phenomenon. *Anesthesiology,* **2012**, *117*(4), 745-754.
[http://dx.doi.org/10.1097/ALN.0b013e31826a7da7] [PMID: 22902967]

[31]   Koper, O.B.; Klabunde, J.S.; Marchin, G.L.; Klabunde, K.J.; Stoimenov, P.; Bohra, L. Nanoscale powders and formulations with biocidal activity toward spores and vegetative cells of bacillus species, viruses, and toxins. *Curr. Microbiol.,* **2002**, *44*(1), 49-55.

[http://dx.doi.org/10.1007/s00284-001-0073-x] [PMID: 11727041]

[32]   Jin, T.; He, Y. Antibacterial activities of magnesium oxide (MgO) nanoparticles against foodborne pathogens. *J. Nanopart. Res.,* **2011**, *13*(12), 6877-6885.
[http://dx.doi.org/10.1007/s11051-011-0595-5]

[33]   Bodke, M.R.; Purushotham, Y.; Dole, B.N. Crystallographic and optical studies on Cr doped ZnS nanocrystals. *Ceramica,* **2014**, *60*(355), 425-428.
[http://dx.doi.org/10.1590/S0366-69132014000300015]

[34]   Aqeel, M.; Ikram, M.; Asghar, A.; Haider, A.; Ul-Hamid, A.; Naz, M.; Imran, M.; Ali, S. Synthesis of capped Cr-doped ZnS nanoparticles with improved bactericidal and catalytic properties to treat polluted water. *Appl. Nanosci.,* **2020**, *10*(6), 2045-2055.
[http://dx.doi.org/10.1007/s13204-020-01268-3]

[35]   Jamzad, M.; Kamari Bidkorpeh, M. Green synthesis of iron oxide nanoparticles by the aqueous extract of Laurus nobilis L. leaves and evaluation of the antimicrobial activity. *J. Nanostructure Chem.,* **2020**, *10*(3), 193-201.
[http://dx.doi.org/10.1007/s40097-020-00341-1]

[36]   Mirzaei, S.Z.; Ahmadi Somaghian, S.; Lashgarian, H.E.; Karkhane, M.; Cheraghipour, K.; Marzban, A. Phyco-fabrication of bimetallic nanoparticles (zinc–selenium) using aqueous extract of Gracilaria corticata and its biological activity potentials. *Ceram. Int.,* **2021**, *47*(4), 5580-5586.
[http://dx.doi.org/10.1016/j.ceramint.2020.10.142]

[37]   Alp, E.; İmamoğlu, R.; Savacı, U.; Turan, S.; Kazmanlı, M.K.; Genç, A. Plasmon-enhanced photocatalytic and antibacterial activity of gold nanoparticles-decorated hematite nanostructures. *J. Alloys Compd.,* **2021**, *852*, 157021.
[http://dx.doi.org/10.1016/j.jallcom.2020.157021]

[38]   Jo, Y.K.; Kim, B.H.; Jung, G. Antifungal activity of silver ions and nanoparticles on phytopathogenic fungi. *Plant Dis.,* **2009**, *93*(10), 1037-1043.
[http://dx.doi.org/10.1094/PDIS-93-10-1037] [PMID: 30754381]

[39]   Pal, S.; Nisi, R.; Stoppa, M.; Licciulli, A. Silver-functionalized bacterial cellulose as antibacterial membrane for wound-healing applications. *ACS Omega,* **2017**, *2*(7), 3632-3639.
[http://dx.doi.org/10.1021/acsomega.7b00442] [PMID: 30023700]

[40]   Marsh, PD; Martin, MV; Lewis, MA; Williams, D Oral microbiology e-book. Elsevier Health Sciences. **2009**.

[41]   Rahiotis, C.; Vougiouklakis, G.; Eliades, G. Characterization of oral films formed in the presence of a CPP–ACP agent: An *in situ* study. *J. Dent.,* **2008**, *36*(4), 272-280.
[http://dx.doi.org/10.1016/j.jdent.2008.01.005] [PMID: 18291571]

[42]   Malarkodi, C.; Rajeshkumar, S.; Paulkumar, K.; Jobitha, G.G.; Vanaja, M.; Annadurai, G. Biosynthesis of semiconductor nanoparticles by using sulfur reducing bacteria Serratia nematodiphila. *Adv. Nano Res.,* **2013**, *1*(2), 83-91.
[http://dx.doi.org/10.12989/anr.2013.1.2.083]

[43]   Besinis, A.; De Peralta, T.; Handy, R.D. The antibacterial effects of silver, titanium dioxide and silica dioxide nanoparticles compared to the dental disinfectant chlorhexidine on *Streptococcus mutans* using a suite of bioassays. *Nanotoxicology,* **2014**, *8*(1), 1-16.
[http://dx.doi.org/10.3109/17435390.2012.742935] [PMID: 23092443]

[44]   Vazquez-Muñoz, R.; Bogdanchikova, N.; Huerta-Saquero, A. Beyond the nanomaterials approach: influence of culture conditions on the stability and antimicrobial activity of silver nanoparticles. *ACS Omega,* **2020**, *5*(44), 28441-28451.
[http://dx.doi.org/10.1021/acsomega.0c02007] [PMID: 33195894]

[45]   Bogdanchikova, N.; Muñoz, R.V.; Saquero, A.H.; Jasso, A.P.; Uzcanga, G.A.; Díaz, P.L.P.; Pestryakov, A.; Burmistrov, V.; Martynyuk, O.; Gómez, R.L.V.; Almanza, H. Silver nanoparticles

composition for treatment of distemper in dogs. *Int. J. Nanotechnol.*, **2016**, *13*(1/2/3), 227-237.
[http://dx.doi.org/10.1504/IJNT.2016.074536]

[46]   Vazquez-Muñoz, R.; Arellano-Jimenez, M.J.; Lopez, F.D.; Lopez-Ribot, J.L. Protocol optimization for a fast, simple and economical chemical reduction synthesis of antimicrobial silver nanoparticles in non-specialized facilities. *BMC Res. Notes*, **2019**, *12*(1), 773.
[http://dx.doi.org/10.1186/s13104-019-4813-z] [PMID: 31775864]

[47]   Lara, H.H.; Ixtepan-Turrent, L.; Jose Yacaman, M.; Lopez-Ribot, J. Inhibition of Candida auris biofilm formation on medical and environmental surfaces by silver nanoparticles. *ACS Appl. Mater. Interfaces*, **2020**, *12*(19), 21183-21191.
[http://dx.doi.org/10.1021/acsami.9b20708] [PMID: 31944650]

[48]   Fayaz, A.M.; Balaji, K.; Girilal, M.; Yadav, R.; Kalaichelvan, P.T.; Venketesan, R. Biogenic synthesis of silver nanoparticles and their synergistic effect with antibiotics: a study against gram-positive and gram-negative bacteria. *Nanomedicine*, **2010**, *6*(1), 103-109.
[http://dx.doi.org/10.1016/j.nano.2009.04.006] [PMID: 19447203]

[49]   Roy, KJ; Rahman, A; Hossain, KS; Rahman, MB; Kafi, MA Antibacterial Investigation of Silver Nanoparticle against Staphylococcus, E.. **2020**.

[50]   Le Ouay, B.; Stellacci, F. Antibacterial activity of silver nanoparticles: A surface science insight. *Nano Today*, **2015**, *10*(3), 339-354.
[http://dx.doi.org/10.1016/j.nantod.2015.04.002]

[51]   Ahamed, M; Alhadlaq, HA; Khan, MA; Karuppiah, P; Al-Dhabi, NA *Synthesis, characterization, and antimicrobial activity of copper oxide nanoparticles.*, **2014**.
[http://dx.doi.org/10.1155/2014/637858]

[52]   Pagar, T.; Ghotekar, S.; Pansambal, S.; Pagar, K.; Oza, R. Biomimetic Synthesis of CuO Nanoparticle using Capparis decidua and their Antibacterial Activity. *Advanced Journal of Science and Engineering.*, **2020**, *1*(4), 133-137.

[53]   Kasahun, M.; Yadate, A.; Belay, A.; Belay, Z.; Ramalingam, M. Antimicrobial activity of chemical, thermal and green route-derived zinc oxide nanoparticles: a comparative analysis. *Nano Biomed. Eng.*, **2020**, *12*(1), 47-56.
[http://dx.doi.org/10.5101/nbe.v12i1.p47-56]

[54]   Ibrahem, E.J.; Thalij, K.M.; Saleh, M.K.; Badawy, A.S. Biosynthesis of zinc oxide nanoparticles and assay of antibacterial activity. *Am. J. Biochem. Biotechnol.*, **2017**, *13*(2), 63-69.
[http://dx.doi.org/10.3844/ajbbsp.2017.63.69]

[55]   Mohammadi, F.M.; Ghasemi, N. Influence of temperature and concentration on biosynthesis and characterization of zinc oxide nanoparticles using cherry extract. *J. Nanostructure Chem.*, **2018**, *8*(1), 93-102.
[http://dx.doi.org/10.1007/s40097-018-0257-6]

[56]   Bala, N.; Saha, S.; Chakraborty, M.; Maiti, M.; Das, S.; Basu, R.; Nandy, P. Green synthesis of zinc oxide nanoparticles using Hibiscus subdariffa leaf extract: effect of temperature on synthesis, anti-bacterial activity and anti-diabetic activity. *RSC Advances*, **2015**, *5*(7), 4993-5003.
[http://dx.doi.org/10.1039/C4RA12784F]

[57]   Bhuyan, T.; Khanuja, M.; Sharma, R.; Patel, S.; Reddy, M.R.; Anand, S.; Varma, A. A comparative study of pure and copper (Cu)-doped ZnO nanorods for antibacterial and photocatalytic applications with their mechanism of action. *J. Nanopart. Res.*, **2015**, *17*(7), 288.
[http://dx.doi.org/10.1007/s11051-015-3093-3]

[58]   Jayaseelan, C.; Rahuman, A.A.; Kirthi, A.V.; Marimuthu, S.; Santhoshkumar, T.; Bagavan, A.; Gaurav, K.; Karthik, L.; Rao, K.V.B. Novel microbial route to synthesize ZnO nanoparticles using Aeromonas hydrophila and their activity against pathogenic bacteria and fungi. *Spectrochim. Acta A Mol. Biomol. Spectrosc.*, **2012**, *90*, 78-84.
[http://dx.doi.org/10.1016/j.saa.2012.01.006] [PMID: 22321514]

[59]   Arciniegas-Grijalba, P.A.; Patiño-Portela, M.C.; Mosquera-Sánchez, L.P.; Guerrero-Vargas, J.A.; Rodríguez-Páez, J.E. ZnO nanoparticles (ZnO-NPs) and their antifungal activity against coffee fungus Erythricium salmonicolor. *Appl. Nanosci.,* **2017**, *7*(5), 225-241.
[http://dx.doi.org/10.1007/s13204-017-0561-3]

[60]   Sardella, D.; Gatt, R.; Valdramidis, V.P. Assessing the efficacy of zinc oxide nanoparticles against *Penicillium expansum* by automated turbidimetric analysis. *Mycology,* **2018**, *9*(1), 43-48.
[http://dx.doi.org/10.1080/21501203.2017.1369187] [PMID: 30123660]

[61]   Lipovsky, A.; Nitzan, Y.; Gedanken, A.; Lubart, R. Antifungal activity of ZnO nanoparticles—the role of ROS mediated cell injury. *Nanotechnology,* **2011**, *22*(10), 105101.
[http://dx.doi.org/10.1088/0957-4484/22/10/105101] [PMID: 21289395]

[62]   Srihasam, S.; Thyagarajan, K.; Korivi, M.; Lebaka, V.R.; Mallem, S.P.R. Phytogenic generation of NiO nanoparticles using Ste*via* leaf extract and evaluation of their *in-vitro* antioxidant and antimicrobial properties. *Biomolecules,* **2020**, *10*(1), 89.
[http://dx.doi.org/10.3390/biom10010089] [PMID: 31935798]

[63]   Subhankari, I.; Nayak, P.L. Antimicrobial activity of copper nanoparticles synthesised by ginger (Zingiber officinale) extract. *World Journal of Nano Science & Technology.,* **2013**, *2*(1), 10-13.

[64]   Caroling, G.; Vinodhini, E.; Ranjitham, A.M.; Shanthi, P. Biosynthesis of copper nanoparticles using aqueous Phyllanthus embilica (Gooseberry) extract-characterisation and study of antimicrobial effects. *Int J Nano Chem.,* **2015**, *1*(2), 53-63.

[65]   Joseph, A.T.; Prakash, P.; Narvi, S.S. Phytofabrication and Characterization of copper nanoparticles using Allium sativum and its antibacterial activity. *Int. J. Sci. Eng. Technol.,* **2016**, *4*, 463-472.

[66]   Kaur, P; Thakur, R; Chaudhury, A Biogenesis of copper nanoparticles using peel extract of Punica granatum and their antimicrobial activity against opportunistic pathogens. green chemistry letters and reviews. 2016 Jan 2;9(1):33-8.

[67]   Hassanien, R.; Husein, D.Z.; Al-Hakkani, M.F. Biosynthesis of copper nanoparticles using aqueous *Tilia* extract: antimicrobial and anticancer activities. *Heliyon,* **2018**, *4*(12), e01077.
[http://dx.doi.org/10.1016/j.heliyon.2018.e01077] [PMID: 30603710]

[68]   Angrasan, J.K.; Subbaiya, R. Biosynthesis of copper nanoparticles by Vitis vinifera leaf aqueous extract and its antibacterial activity. *Int. J. Curr. Microbiol. Appl. Sci.,* **2014**, *3*(9), 768-774.

[69]   Gopinath, M.; Subbaiya, R.; Selvam, M.M.; Suresh, D. Synthesis of copper nanoparticles from Nerium oleander leaf aqueous extract and its antibacterial activity. *Int. J. Curr. Microbiol. Appl. Sci.,* **2014**, *3*(9), 814-818.

[70]   Subbaiya, R.; Selvam, M.M. Green synthesis of copper nanoparticles from Hibicusrosasinensis and their antimicrobial, antioxidant activities. *Res. J. Pharm. Biol. Chem. Sci.,* **2015**, *6*(2), 1183-1190.

[71]   Hariprasad, S.; Bai, G.S.; Santhoshkumar, J.; Madhu, C.H.; Sravani, D. Green synthesis of copper nanoparticles by Arevalanata leaves extract and their anti-microbial activites. *Int. J. Chemtech Res.,* **2016**, *9*, 98-105.

[72]   Kala, A.; Soosairaj, S.; Mathiyazhagan, S.; Raja, P. Green synthesis of copper bionanoparticles to control the bacterial leaf blight disease of rice. *Curr. Sci.,* **2016**, *110*(10), 2011-2014.
[http://dx.doi.org/10.18520/cs/v110/i10/2011-2014]

[73]   Ranjitham, A.M.; Ranjani, G.S.; Caroling, G. Biosynthesis, characterization, antimicrobial activity of copper nanoparticles using fresh aqueous Ananas comosus L.(Pineapple) extract. *Int. J. Pharm. Tech. Res.,* **2015**, *8*(4), 750-769.

[74]   Jayandran, M.; Haneefa, M.M.; Balasubramanian, V. Green synthesis of copper nanoparticles using natural reducer and stabilizer and an evaluation of antimicrobial activity. *J. Chem. Pharm. Res.,* **2015**, *7*(2), 251-259.

[75]   Yun, H.; Kim, J.D.; Choi, H.C.; Lee, C.W. Antibacterial activity of CNT-Ag and GO-Ag

nanocomposites against gram-negative and gram-positive bacteria. *Bull. Korean Chem. Soc.,* **2013**, *34*(11), 3261-3264.
[http://dx.doi.org/10.5012/bkcs.2013.34.11.3261]

[76] khanmohammadi, S.; Karimian, R.; Ghanbari Mehrabani, M.; Mehramuz, B.; Ganbarov, K.; Ejlali, L.; Tanomand, A.; Kamounah, F.S.; Ahangarzadeh Rezaee, M.; Yousefi, M.; Sheykhsaran, E.; Samadi Kafil, H. Poly (ε-Caprolactone)/cellulose nanofiber blend nanocomposites containing ZrO_2 nanoparticles: A new biocompatible wound dressing bandage with antimicrobial activity. *Adv. Pharm. Bull.,* **2020**, *10*(4), 577-585.
[http://dx.doi.org/10.34172/apb.2020.069] [PMID: 33072535]

[77] Dickinson, B.C.; Chang, C.J. Chemistry and biology of reactive oxygen species in signaling or stress responses. *Nat. Chem. Biol.,* **2011**, *7*(8), 504-511.
[http://dx.doi.org/10.1038/nchembio.607] [PMID: 21769097]

[78] Yu, J.; Zhang, W.; Li, Y.; Wang, G.; Yang, L.; Jin, J.; Chen, Q.; Huang, M. Synthesis, characterization, antimicrobial activity and mechanism of a novel hydroxyapatite whisker/nano zinc oxide biomaterial. *Biomed. Mater.,* **2014**, *10*(1), 015001.
[http://dx.doi.org/10.1088/1748-6041/10/1/015001] [PMID: 25534679]

[79] Kiwi, J.; Rtimi, S. Mechanisms of the antibacterial effects of TiO2–FeOx under solar or visible light: Schottky barriers *versus* surface plasmon resonance. *Coatings,* **2018**, *8*(11), 391.
[http://dx.doi.org/10.3390/coatings8110391]

[80] Hussein, M.Z.; Al Ali, S.; Geilich, B.; El Zowalaty, M.; Webster, T. Synthesis, characterization, and antimicrobial activity of an ampicillin-conjugated magnetic nanoantibiotic for medical applications. *Int. J. Nanomedicine,* **2014**, *9*, 3801-3814.
[http://dx.doi.org/10.2147/IJN.S61143] [PMID: 25143729]

[81] Di Pasqua, R.; Betts, G.; Hoskins, N.; Edwards, M.; Ercolini, D.; Mauriello, G. Membrane toxicity of antimicrobial compounds from essential oils. *J. Agric. Food Chem.,* **2007**, *55*(12), 4863-4870.
[http://dx.doi.org/10.1021/jf0636465] [PMID: 17497876]

[82] Yamanaka, M.; Hara, K.; Kudo, J. Bactericidal actions of a silver ion solution on Escherichia coli, studied by energy-filtering transmission electron microscopy and proteomic analysis. *Appl. Environ. Microbiol.,* **2005**, *71*(11), 7589-7593.
[http://dx.doi.org/10.1128/AEM.71.11.7589-7593.2005] [PMID: 16269810]

[83] Cui, Y.; Zhao, Y.; Tian, Y.; Zhang, W.; Lü, X.; Jiang, X. The molecular mechanism of action of bactericidal gold nanoparticles on Escherichia coli. *Biomaterials,* **2012**, *33*(7), 2327-2333.
[http://dx.doi.org/10.1016/j.biomaterials.2011.11.057] [PMID: 22182745]

[84] Liu, Z.; Zhang, M.; Han, X.; Xu, H.; Zhang, B.; Yu, Q.; Li, M. TiO2 nanoparticles cause cell damage independent of apoptosis and autophagy by impairing the ROS-scavenging system in Pichia pastoris. *Chem. Biol. Interact.,* **2016**, *252*, 9-18.
[http://dx.doi.org/10.1016/j.cbi.2016.03.029] [PMID: 27041071]

[85] Gulbagca, F.; Ozdemir, S.; Gulcan, M.; Sen, F. Synthesis and characterization of *Rosa canina*-mediated biogenic silver nanoparticles for anti-oxidant, antibacterial, antifungal, and DNA cleavage activities. *Heliyon,* **2019**, *5*(12), e02980.
[http://dx.doi.org/10.1016/j.heliyon.2019.e02980] [PMID: 31867461]

[86] Matter, B.; Seiler, C.L.; Murphy, K.; Ming, X.; Zhao, J.; Lindgren, B.; Jones, R.; Tretyakova, N. Mapping three guanine oxidation products along DNA following exposure to three types of reactive oxygen species. *Free Radic. Biol. Med.,* **2018**, *121*, 180-189.
[http://dx.doi.org/10.1016/j.freeradbiomed.2018.04.561] [PMID: 29702150]

[87] Trouiller, B.; Reliene, R.; Westbrook, A.; Solaimani, P.; Schiestl, R.H. Titanium dioxide nanoparticles induce DNA damage and genetic instability *in vivo* in mice. *Cancer Res.,* **2009**, *69*(22), 8784-8789.
[http://dx.doi.org/10.1158/0008-5472.CAN-09-2496] [PMID: 19887611]

[88] Gow, NA; Latge, JP; Munro, CA *The fungal cell wall: structure, biosynthesis, and function.,* **2017**.

[http://dx.doi.org/10.1128/9781555819583.ch12]

[89]   Guerrero Correa, M.; Martínez, F.B.; Vidal, C.P.; Streitt, C.; Escrig, J.; de Dicastillo, C.L. Antimicrobial metal-based nanoparticles: a review on their synthesis, types and antimicrobial action. *Beilstein J. Nanotechnol.,* **2020**, *11*(1), 1450-1469.
[http://dx.doi.org/10.3762/bjnano.11.129] [PMID: 33029474]

[90]   Bahrami, A.; Delshadi, R.; Jafari, S.M. Active delivery of antimicrobial nanoparticles into microbial cells through surface functionalization strategies. *Trends Food Sci. Technol.,* **2020**, *99*, 217-228.
[http://dx.doi.org/10.1016/j.tifs.2020.03.008]

[91]   Sarwar, A.; Katas, H.; Samsudin, S.N.; Zin, N.M. Regioselective sequential modification of chitosan *via* azide-alkyne click reaction: synthesis, characterization, and antimicrobial activity of chitosan derivatives and nanoparticles. *PLoS One,* **2015**, *10*(4), e0123084.
[http://dx.doi.org/10.1371/journal.pone.0123084] [PMID: 25928293]

[92]   Li, H.; Chen, Q.; Zhao, J.; Urmila, K. Enhancing the antimicrobial activity of natural extraction using the synthetic ultrasmall metal nanoparticles. *Sci. Rep.,* **2015**, *5*(1), 11033.
[http://dx.doi.org/10.1038/srep11033] [PMID: 26046938]

[93]   Armentano, I.; Arciola, C.R.; Fortunati, E.; Ferrari, D.; Mattioli, S.; Amoroso, C.F.; Rizzo, J.; Kenny, J.M.; Imbriani, M.; Visai, L. The interaction of bacteria with engineered nanostructured polymeric materials: a review. *ScientificWorldJournal,* **2014**, *2014*, 1-18.
[http://dx.doi.org/10.1155/2014/410423] [PMID: 25025086]

[94]   Luan, B.; Huynh, T.; Zhou, R. Complete wetting of graphene by biological lipids. *Nanoscale,* **2016**, *8*(10), 5750-5754.
[http://dx.doi.org/10.1039/C6NR00202A] [PMID: 26910517]

CHAPTER 5

Metal-Organic Frameworks: Synthesis, Characterization, and Applications

Shikha Bhogal[1], **Irshad Mohiuddin**[1], **Aman Grover**[1], **Sandeep Kumar**[1], **Kuldeep Kaur**[2,*] and **Ashok Kumar Malik**[1]

[1] *Department of Chemistry, Punjabi University, Patiala-147 002, Punjab, India*

[2] *Department of Chemistry, Mata Gujri College, Fatehgarh Sahib-140 406, Punjab, India*

Abstract: Metal-organic frameworks (MOFs) are structurally complex structures constructed from inorganic and organic components. MOFs are highly ordered porous structures with special characteristics, such as high thermal stability, tunable surface properties, and large surface area. The MOFs demonstrate a wider range of potential applications in adsorption, gas storage, catalysis, drug delivery and sensing. As a result, the research in the area of MOFs is experiencing rapid growth. Considering the promising prospects of MOFs, this chapter presents an overview of the general synthesis and characterization methods for MOFs. Besides, the applications of MOFs in adsorption, sensing, and catalysis are also highlighted.

Keywords: Adsorbent, Catalyst, Characterization, Electrochemical, Mechanochemical, Microwave, Solvothermal, Sonochemical, Sensors.

INTRODUCTION

Metal-organic frameworks (MOFs) are an evolving group of crystalline porous substances comprising 3D structures containing metal ions that are coordinated by organic linkers [1, 2]. Multidentate organic ligands containing O or N donors (such as carboxylates, azoles, and nitriles) are employed as linkers to provide linkage sites with metal clusters (Fig. **1**) [3]. The characteristic properties of the MOFs network are reflected by metal ions as well as linkers. Moreover, solvent, pH, ligand metal ratio and temperature can influence the structural characteristics of the synthesized framework [4]. They can offer tunable pores with a precise dimensional distribution, arbitrary characteristics, reduced density, and regulable chemical affinities for unique applications like magnetism and luminescence [2]. MOFs have earned much consideration owing to their excellent characteristics,

---

* **Corresponding author Kuldeep Kaur:** Department of Chemistry, Mata Gujri College, Fatehgarh Sahib-140 406, Punjab, India, E-mail:shergillkk@gmail.com

Karamjit Singh Dhaliwal (Ed.)

such as high thermal stability, tunable surface properties, and larger surface area than other classical porous materials (like silica and zeolites) [5, 6]. All these characteristics make them desirable materials for gas storage, separation, catalysis, adsorption, and drug delivery [7 - 10]. The elegance of chemical structures and the strength of both organic and inorganic chemistry are epitomized by MOFs; two disciplines sometimes considered contrasting.

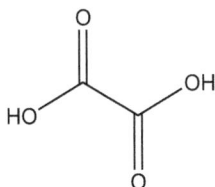

Ethane 1,2-dioic acid

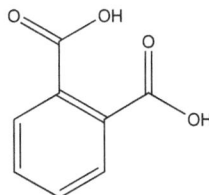

Benzene 1,2-dicarboxylic acid

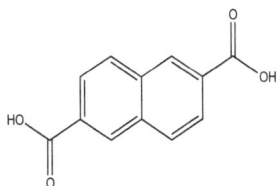

Napthalene-2,6-dicarboxylic acid

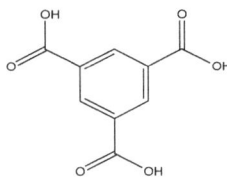

Benzene 1,3,5-tricarboxylic acid

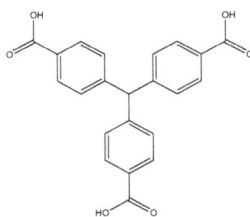

4,4',4"-Methanetriyltribenzoic acid

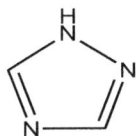

1H-1,2,3-triazole

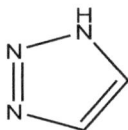

4H-imidazole

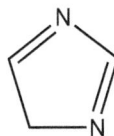

1,2,4-Triazole

(Fig. 1) contd.....

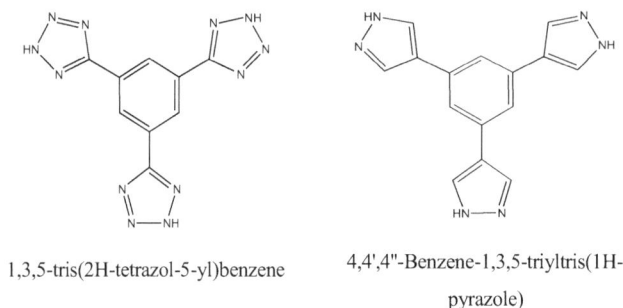

1,3,5-tris(2H-tetrazol-5-yl)benzene   4,4',4"-Benzene-1,3,5-triyltris(1H-pyrazole)

Fig. (1). Some common organic linkers used in MOF synthesis.

This field of chemistry has expanded almost incomparably since the 1990s, not only because of the large number of published research articles but also due to the growing scope of these investigations. Since the field of MOF is very broad, this chapter emphasizes the typical methods for synthesizing and characterizing these materials. The present chapter also covers the applications of MOFs in adsorption, sensing, separation, and catalysis.

SYNTHETIC METHODS FOR MOFS

A lot of attention has been paid to the synthetic methods used for MOFs because they have a wide range of interesting and variable structures that may also be of huge interest in a number of applications in several fields relevant to porous materials. The main objective of MOF synthesis is to create favourable synthetic conditions, leading to established inorganic building blocks without the deterioration of the linkers. At the same time, the kinetics of crystallization should be sufficient to permit nucleation and proper growth of the desired phase. Several synthesis methods have been reported to date, including but not limited to solvothermal, microwave, electrochemical, mechanochemical, and sonochemical (Fig. **2**) methods. Table **1** summarizes the representative MOFs synthesized by different synthetic methods.

Table 1. Different synthetic methods for MOFs.

Order	Name of MOF	Metal/Metal salt	Organic linker	Synthetic method	Refs
1	Amino-MIL-53(Al)	Aluminium chloride hydrate	Terephthalic acid	Solvothermal	[16]
2	Zr-MOF, UiO-66	ZrCl₄	Terephthalic acid	Solvothermal	[17]

(Table 1) cont.....

Order	Name of MOF	Metal/Metal salt	Organic linker	Synthetic method	Refs
3	Co-MOF-74	Cobalt nitrate hexahydrate	2,5-dihydroxybenzenedicarboxylic acid (H_4DHBDC)	Microwave	[21]
4	Co-based MOF-derived carbon (MOFDC)	Cobalt nitrate hexahydrate	4,4′-biphenyldicarboxylic acid (BPDC)	Microwave	[34]
5	UiO-66-NH_2	Zr	2-aminoterephthalic acid	Electrochemical	[35]
6	Cu_3(BTC)$_2$ MOF	Cu	1,3,5-benzene tricarboxylic acid	Electrochemical	[36]
7	Zn(II)-MOF	Zinc nitrate hexahydrate	4,4′-oxybis(benzoic acid) (H_2oba)	Sonochemical	[37]
8	TMU-8 and TMU-9	Cadmium acetate	1,4-bis(4-pyridyl)-2,3-diaza-1,3-butadiene (4-bpdb)	Sonochemical	[38]
9	MIL-78	Yttrium hydride	Trimesic acid	Mechanochemical	[39]
10	MOF-5	Zn(OAc)$_2$·$2H_2$O	Terephthalic acid (H2BDC)	Mechanochemical	[40]

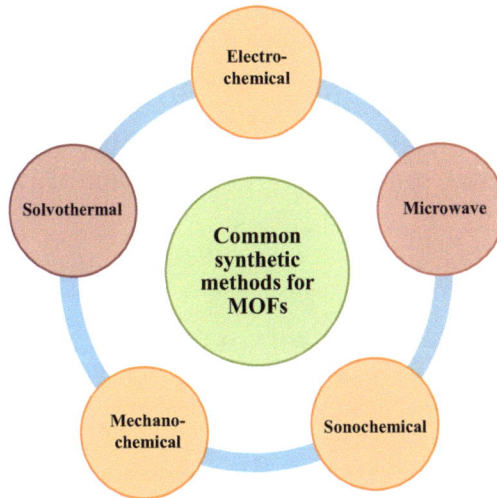

Fig. (2). General methods for MOFs synthesis.

Solvothermal Method

In special reactors close to or above solvent boiling temperatures, solvothermal synthesis is performed (Fig. **3**). Dialkylformamides, alcohols, and pyridines are the most common organic solvents used in this process. This approach contributes to the formation of morphologies at the nanoscale, high yields of the product, and better crystallinity. MOF-5, consisting of octahedral Zn_4O clusters bonded by benzene-1,4 dicarboxylic acid (BDC) having a formula of $Zn_4O(BDC)_3$, was prepared by a diffusion synthesis method but with a low yield [11]. It has now been replaced by a high-yield solvothermal method by heating $Zn(NO_3)_2.4H_2O$ and H_2BDC in an N,N-diethylformamide (DEF) at 105°C in a sealed vessel to give crystalline MOF-5 [12]. IRMOF-3 has an isoreticular structure, having Zn_4O clusters coordinated by BDC-NH_2. This was also synthesized by a similar solvothermal method under the given conditions containing the $Zn(NO_3)_2.6H_2O$ and $H_2BDC-NH_2$ suspended in DMF or DEF and heated in an oven at 90°C for 24 h [13]. While MOFs based on Zn(II) have mainly unfavourable characteristics, their poor stability against humidity and protic solvents has restricted their industrial use. Thus, more efforts are needed to build MOFs that are more stable and chemical-resistant. Porous chromium benzene dicarboxylate (MIL-101) is synthesized by hydrothermal method using a Cr salt and H_2BDC with a small amount of HF in an autoclave at 220°C for 8h [14]. Fe-MIL-100 is an iron (III) carboxylate prepared from the reaction mixture of Fe salt and H_3BTC at 150°C in an autoclave for 6 days [15]. Afterward, the synthesized orange crystals are filtered and washed with triply distilled water. Al-MIL-53 is synthesized in a mixed solvent of DEF and ethanol with the addition of $AlCl_3.H_2O$ and H_2BDC; the solution is heated at 110°C for 2 days [16]. In the same way, Al-MIL-53-NH_2 is synthesized by adding a mixture of $Al(NO_3)_3.9H_2O$ and $H_2BDC-NH_2$ in a DMF and is heated at 130°C for five days [16]. Zirconium is extremely corrosion-resistant and has a high affinity toward O-containing donor ligands. A Zr(IV)-based MOF, UiO-66, was synthesized from $ZrCl_4$ and H_2BDC in DMF at 120°C for 24h [17].

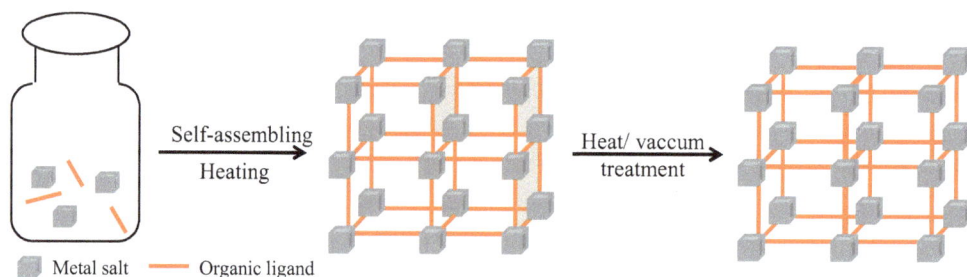

Metal salt   —— Organic ligand

Fig. (3). Solvothermal method of MOFs synthesis.

Microwave Method

Microwave (MW) aided synthesis depends on the electromagnetic interaction with mobile electrical charges. Microwave ovens suitable for material synthesis allow temperature and pressure throughout the reaction to be monitored and reaction conditions to be regulated more precisely. MW-assisted synthesis of

MOFs is frequently performed at temperatures above 100°C, with reaction times rarely exceeding 1h (Fig. **4**). Microwave-based MOFs have been made from different metal (III) carboxylate-based linkers (Fe, Al, Cr, V, Ce). Cr-MIL 100 was the first microwave-synthesized MOF [18]. It was prepared at 220°C under MW irradiation with a reaction time of 4h and 44% yield. A similar Fe MOF was synthesized in 30 min at 200°C [19]. Fe-MIL-101 nanocrystals with a size of 200 nm were prepared at 150°C (10 min) in DMF [20]. IRMOF-1 and HKUST-1 are the most studied MOFs synthesized by the MW method. The cubic microcrystals of IR-MOF-1 with a length of 2-4 μm were acquired in 9 min at 95°C with a 27% yield [21]. Compared to traditional ambient pressure dynamic electric synthesis, the MW-assisted synthesis of IRMOF-1 provided higher-quality crystals with excellent adsorption properties [22]. HKUST-1 was obtained using six different synthetic paths: conventional electric, ultrasonic, electrochemical, microwave, and mechanochemical methods [23]. Among the listed methods, the MW-assisted synthesis showed an upper edge for the synthesis of HKUST-1 in 30 min. In addition, MW-assisted synthesis yields high purity, high micropore volume, and large quantitative yields. A few studies mentioned the synthesis of azolate-based MOFs *via* MW radiation. One of the studies reported the synthesis of a pyrazolate-based compound $[Co_4O(BDPB)_3]$ at 155°C (2 min) *via* the MW method [24].

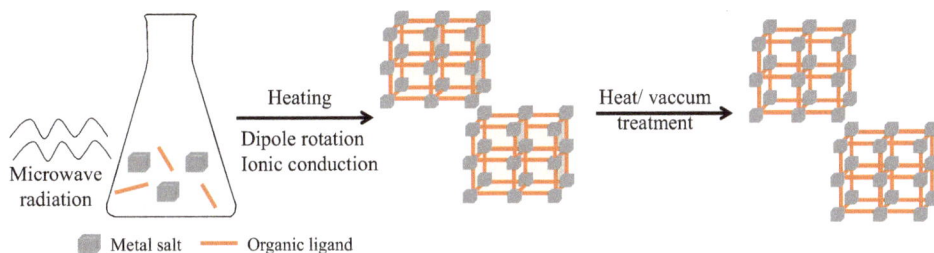

Fig. (4). Microwave method of MOFs synthesis.

Electrochemical Method

This method requires metal ions to be supplied continuously as a metal source by anodic dissolution (Fig. **5**). By using protic solvents, the deposition of metal on the cathode can be avoided [25]. HKUST was the first MOF to be synthesized by the electrochemical method in 2005 [26]. Bulk Cu plates act as an anode with methanol-dissolved H_3BTC and Cu as the cathode. The reaction was found to be complete within 150 min. HKUST-1, ZIF-8, Al-MIL-100, Al-MIL-53, and Al-MIL-53-NH_2 were the other MOFs prepared by the electrochemical method [27]. Many parameters which can affect the structural properties of MOFs need to be optimized, including electrolyte, solvent, temperature, and voltage-current density. The ethanol-water ratio during the synthesis of HKUST-1 affects the linker solubility as well as solution conductivity. Thus, it is concluded that deprotonation of the H_3BTC and ethanol content must be above 75% (vol%). It was claimed that the electrochemical method offers many benefits over traditional synthesis methods, which include (a) faster synthesis at lower temperatures, (b) no need for separation of anions from the synthesis solution before solvent recycling as metal salts are not required, (c) virtually exclusive use of the linker and high Faraday efficiencies.

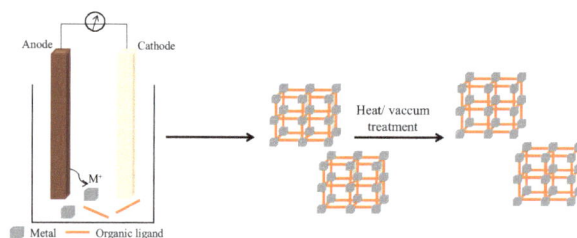

Fig. (5). Electrochemical method of MOFs synthesis.

Sonochemical Method

This method is rapid, energy-efficient, and environmentally friendly that can be easily carried out at room temperature. This technique is of particular interest for future applications of MOFs, as rapid reactions may enable MOFs to scale up (Fig. **6**). The MOFs which have been sonochemically synthesized include MOF-5 [28], MOF-177 [28], and HKUST-1 [23]. The first-ever sonochemical synthesized salt included Zn carboxylates [29]. $[Zn_3(BTC)_2]$ was constructed from a mixture of $Zn(CH_3COO)_2$ and H_3BTC in an ethanol-water solution at 25°C in an ultrasonic

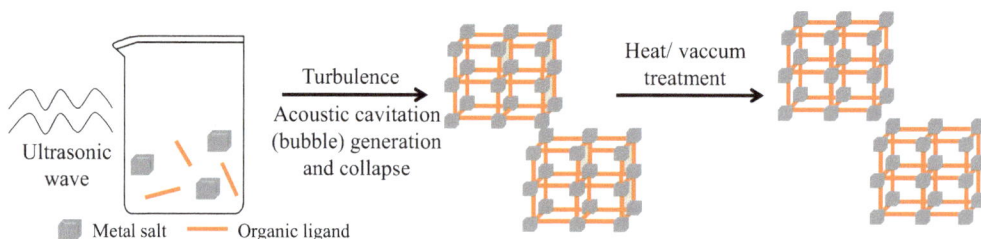

Fig. (6). Sonochemical method of MOFs synthesis.

bath. In contrast to solvothermal synthesis, sonochemical synthesis involves milder conditions and higher reaction rates. By varying the reaction time, HKUST-1 was synthesized using copper acetate and H_3BTC in an ultrasonic bath [30]. The nanocrystalline powder (10-40 nm) was finally collected after 5 min. By increasing the reaction time, larger size crystals (50-200 nm) and higher yields can be acquired. However, extended reaction time also results in partial degradation of the product.

Mechanochemical Method

This method involves the mechanical breakage of intramolecular bonds followed by chemical transformation (Fig. **7**). The first MOF synthesized by the given method was reported in 2006 [31]. Under solvent-free conditions, mechanochemical reactions can occur at ambient room temperature, particularly because the use of organic solvents can be avoided. In short reaction times (usually in the 10-60 min range), quantitative yields of small MOF particles can be obtained. HKUST-1 is successfully obtained in the absence of solvent [32]. HKUST-1 was formed by a solvent-less mechanochemical reaction between H_3BTC and copper acetate with a BET area of 1,084 m^2/g. Zeolite imidazolate frameworks (ZIF) were also synthesized *via*this method using ZnO and imidazole, 2-methylimidazole and 2-ethylimidazole as the starting material [33].

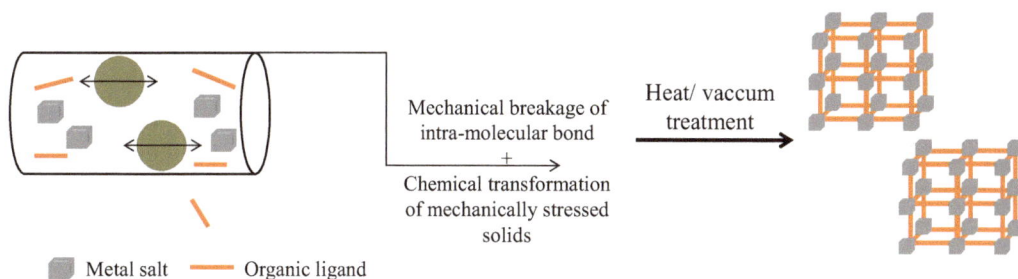

Mechanical breakage of intra-molecular bond
+
Chemical transformation of mechanically stressed solids

Heat/ vaccum treatment

▪ Metal salt ── Organic ligand

Fig. (7). Mechanochemical method of MOFs synthesis.

CHARACTERIZATION

Specific MOF characterization data requires powder X-ray diffraction (PXRD) data to assess the crystallinity and phase purity of the MOF and N_2 adsorption/desorption isotherm to validate porosity and a surface area. Some other characterization methods can include (1) scanning electron microscopy (SEM) for measurement of the morphological structure of MOF that can be combined with energy-dispersive X-ray spectroscopy (EDS) for knowledge of elemental composition, (2) thermogravimetric analysis (TGA) for determination of the thermal strength of MOF, (3) aqueous stability studies to determine MOF stability in the aqueous phase by variable pH levels, (4) NMR spectroscopic analysis for identification of sample MOFs, (5) inductively coupled plasma optical emission spectroscopy (ICP-OES) for determining the percentage purity/elemental ratios, and (6) diffuse reflectance infrared Fourier transform spectroscopy (DRIFTS) to identify functional groups in the framework.

Powder X-ray Diffraction

For evaluating the bulk crystallinity of MOF samples, PXRD patterns are used. Other details like unit cell size may be easily retrieved after confirmation of the crystalline nature of MOF. The loading of the pulverized sample onto a flat plate sample vessel (usually of plastic, glass, or aluminium) is a mostly applied method for preparing samples for PXRD analysis. Using oil or volatile solvents, samples can be mounted onto the sample holder.

Nitrogen Adsorption and Desorption Isotherms

Besides pore volumes and pore size distribution, adsorption isotherms for non-reactive gases may be utilized to assess surface areas for MOFs. The shape of an adsorption/desorption isotherm can provide valued data about the material [41]. Until an isotherm is obtained, the MOF should be properly enabled to obtain proper information. The selection of data analysis protocol is important for the isotherms to produce reliable results. Currently, the common practice is to use BET theory to measure apparent surface areas, as multi-layer gas adsorption is possible by the pores of most MOFs. To determine pore volume and pore size distribution for MOFs, density functional theory (DFT) and the Barret-Joyne--Halenda method are commonly applied. From X-ray crystal structures, expected pore size distributions may be conveniently obtained.

Thermogravimetric Analysis

In determining the thermal stability of MOF materials, TGA is widely used. The carrier gas species (N_2 or O_2) preferred for the experiment are significant when measuring the thermal stability of a MOF. From the TGA spectrum of a solvated sample, pore volume can be approximately calculated. The weight loss ascribed to the trapped solvent in the pores will, in this case, represent the pore volume of the MOF sample. Diffusion of the solvents within pores is ensured by soaking MOF in a high boiling point solvent for a reasonable duration. To extract any non-embedded solvent molecules in the pores, the MOF can then be filtered easily. The weight loss due to the release of the trapped solvent should correspond to the void space of MOF if done correctly.

Aqueous Stability Testing

MOFs are usually proposed as established supports for catalysis and chemical storage. The aqueous stability of MOFs is of considerable importance for some of the applications. An ample amount of water for every analysis should be utilized (*i.e.*, 20 mg of MOF suspended in 10 ml of water) as many functional groups of MOFs can be basic or acidic. The pH of the water has to be confirmed before adding any MOF sample. A precise weighted sum of the activated MOF has to be suspended in an aqueous solution of the required pH during solution preparation. At room temperature, the suspension has to be left uninterrupted for at least 12 h. The substance should be retrieved by careful filtration to gain access to the MOF stability after introduction to the aqueous solution. The quantity of material recovered should be calculated (mass $_{recovered}$/mass $_{initial}$ x 100 = yield %), ensuring that the solvent does not dissolve a part of the MOF.

Scanning Electron Microscopy

Scanning electron microscopy (SEM) is a suitable instrument for measuring a range of MOF properties, including the size of crystals, surface structure, and composition of components. Image artifacts, like charging effects, may also hamper the acquisition of quality SEM images due to the insulating nature of most MOFs. Coating the MOF with a conducting substance (*e.g.*, gold or osmium) to minimize the accumulation of charge from the electron gun is the most common way of mitigating these problems. Coupling SEM with EDS may provide data regarding metal dispersion/incorporation in sample MOFs. It is necessary, however, to choose the coating material because peaks either from the coating or from their pollutants can overlap with metals of interest.

Inductively Coupled Plasma Optical Emission Spectroscopy

ICP-OES or mass spectrometry (ICP-MS) is utilized for checking the percentage purity or elemental ratios in a MOF sample. To detect very low concentrations compared to ICP-OES, ICP-MS can be used for most components. For any analysis, before injection into the instrument, the sample must first be fully dissolved. Commonly used matrices for ICP-OES/MS samples include dilute (3-5%) acids, such as nitric or sulphuric acid. Different sample preparation procedures should be used, as many of the MOFs are not soluble in dilute acid at room temperature. Hydrofluoric acid can also be used before ICP analysis to dissolve MOF samples, although it is uncommon, and caution must be taken when handling HF.

NMR Spectroscopy

NMR spectroscopy is applied to determine the pureness of MOF samples, linker ratios, and leftover modulator, as well as the absence of solvent after activation. In traditional NMR solvents, many MOFs are not soluble; MOF solubility is typically necessary before a spectrum formation. The general procedure of dissolving MOF for NMR purposes involves the addition of 5-10 drops of D_2SO_4 to 1-2 mg of MOF and sonication of the mixture until the sample is well dispersed in the acid. For analysis, the sample should be completely dissolved in the solvent as in ICP-OES. For MOF characterization, solid-state NMR (SS-NMR) spectroscopy may be utilized, especially when testing the local chemical environment within the MOF. This method is applied for the identification of a particular functional group within a MOF [18]. SS-NMR can also be utilized to improve the proper understanding of the supramolecular interactions and kinetics of small molecule docking inside a MOF [42].

Diffuse Reflectance Infrared Fourier Transform Spectroscopy

The presence or absence of IR active functional groups in a MOF can be checked using diffuse reflectance infrared Fourier transform spectroscopy (DRIFTS). DRIFTS may be carried out on powdered MOF samples or dissolved MOF in a matrix, such as potassium bromide. The sample should be triggered before the DRIFTS analysis, or the sample may be heated in the spectrometer to ensure the complete evaporation of water.

APPLICATIONS

MOFs are built with highly flexible and adjustable structures for extensive applications, particularly gas storage and release, drug delivery, sensing, catalysis, chemical separations, light-harvesting, energy conversion, ion exchange, removal of toxic substances, and the degradation of chemical warfare agents. Here, we particularly focus on some of the common applications of MOFs, namely sensing, adsorption, and catalysis.

MOFs as Adsorbents

The excellent features of MOFs, including low framework density, designability in structure and properties, and high surface area, prove them as a unique sorbent in pre-concentration techniques. The adsorption of the target molecules on the

MOFs surface is facilitated owing to the porous structure and wide surface area. Fig. (**8**) represents the expansion of MOFs in diverse areas of analytical chemistry. MOFs have been used as an appropriate sorbent for the separation and detection of a range of analytes, such as volatile organic compounds, heavy metals, phthalates, and drugs. Different types of interactions between analytes and MOF (*e.g.*, π-π interactions, hydrogen bonding, and hydrophobic-hydrophilic interactions) play a major role in the adsorption of target analytes on MOF. Their use as SPE (solid phase extraction) material has been enormously simulated since the effective role of MOFs as adsorbent was realized. The potential of MOF as a novel medium for the extraction and enrichment of both organic and inorganic pollutants in a wide range of complex matrices has been discussed.

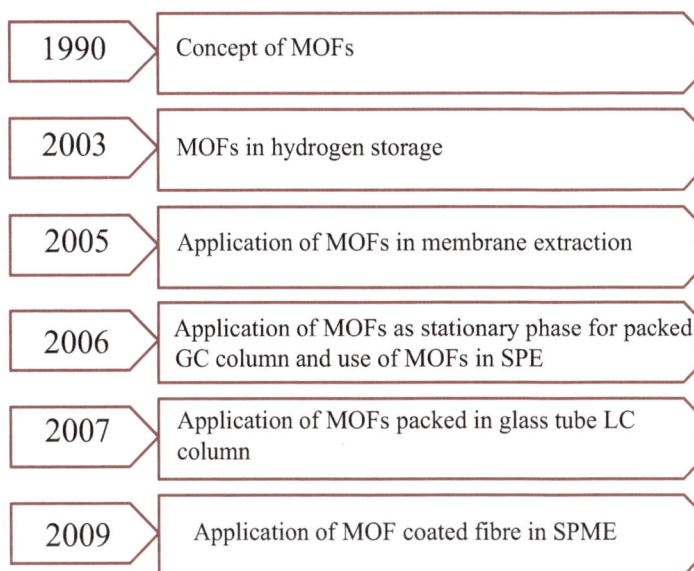

1990	Concept of MOFs
2003	MOFs in hydrogen storage
2005	Application of MOFs in membrane extraction
2006	Application of MOFs as stationary phase for packed GC column and use of MOFs in SPE
2007	Application of MOFs packed in glass tube LC column
2009	Application of MOF coated fibre in SPME

Fig. (8). Development of MOFs in analytical science.

Depending on their form, size, and hydrophilic and hydrophobic properties, selective sorting of the target compounds by MOFs is possible. In solid-phase extraction (SPE), many traditional sorbents have limits in selectiveness or poor ability to adsorb analytes. From this perspective, MOFs are quite flexible to address these limitations through modification of the pore surface for different target molecules. In 2006, Zhou and co-workers used MOFs for the first time as SPE sorbents [43]. They synthesized ($Cu(4-C_5H_4N-COO)_2(H_2O)_4$) using copper nitrate and isonicotinic acid as the precursors.

Furthermore, the synthesized MOF was applied for the flow injection SPE on line coupled with HPLC to trace the identification of polycyclic aromatic hydrocarbons (PAHs) in environmental samples (coal fly ash and local water). Magnetic MIL-101 was applied for the SPE of polycyclic aromatic hydrocarbons from lake water and wastewater samples [44]. Using these composites made it easier for target molecules to be removed from the solution without centrifugation or filtration. The pre-concentration of steroid hormones (methyltestosterone, testosterone, propionate, and nandrolone) from real samples (urine and water) was performed *via* an SPE method based on magnetic porous carbon composites of MOF [45]. Pure MOFs or hybrids can be efficiently used as sorbents for the extraction of inorganics (*e.g.*, mainly heavy metals) from real samples. Three MOFs, namely Zn(II)-MOFs (TMU-4, TMU-5, and TMU-6), were reported for the quantification of heavy metal ions, *e.g.*, Cu(II), Cd(II), Pb(II), Cr(II), and Co(II) from water samples [46]. Out of three MOFs, TMU-5 showed the highest extraction efficiency. This can be explained based on the higher basicity of the azine groups on the TMU-5 compared to the other synthesized MOFs. Zeolite imidazolate frameworks 8 (ZIF-8) were used as a sorbent for micro-solid phase extraction (μ-SPE) of six PAHs (naphthalene, flouineacenapthene, phenanthrene, anthracene, and fluoranthene) from environmental samples [47]. The remarkable properties of ZIF-8 facilitated the adsorption and extraction of trace-level target analytes. HKUST-1, MOF-5, and MIL-53 (Al) have been employed as pre-concentration material in vortex-assisted dispersive micro solid-phase extraction of parabens, followed by HPLC in different complex matrices (environmental waters, cosmetic creams, and human urine) [48]. The application of MOFs as sorbents for the enrichment of different analytes from a range of complex matrices is summarized in Table **2**.

Table 2. Application of MOFs for detection of organic pollutants from different matrices.

Order	MOF	Target Analytes	Sample Preparation Technique	Analytical Instrument	Matrices	LOD (ng/L)	Refs
1	$Fe_3O_4@SiO_2$-MIL-101(Cr)	PAHs	SPE	HPLC-PDA	Environmental water samples	2.8-27.2	[44]
2	MIL-53-C	Sex hormones	SPE	LC-UV	Water and human urine samples	5-300	[45]
3	MIL-101	Pyrazole/pyrolle pesticides	SPE	HPLC-DAD	Environmental water samples	0.3-1.5	[48]
4	MOF-5	PAHs and GAs	SPE	GC-MS	Soil, fish, and plant samples	0.91-80	[49]

(Table 2) cont.....

Order	MOF	Target Analytes	Sample Preparation Technique	Analytical Instrument	Matrices	LOD (ng/L)	Refs
5	$Fe_3O_4@SiO_2$-MOF-177	Phenols	SPE	GC-MS	River, lake, and wastewater	16.8-208.3	[50]
6	Zn(II)MOFs	Heavy metals	SPE	ICP-OES	Tap, mineral, and river water	10-1000	[46]
7	Yb-MOF	PAHs	SPME	GC-MS/MS	Environmental samples	0.07-1.67	[51]
8	MOF-199	Ethylene	SPME	GC-FID	Fruit samples	16	[52]

MOFs as Sensors

Owing to their phenomenal properties, such as tunable pore size, simple design/functionality and high thermal stability, MOFs have been particularly attractive as new sensing materials. Here, we have discussed popular optical and electrochemical sensors (Table **3**).

Table 3. Application of MOFs for sensing of target analytes

Order	MOF	Target Analyte	Type of sensor	Matrices	LOD (µM)	Ref
1	Tb-MOF	Folic acid	Fluorescence sensor	Tap water	0.28	[70]
2	Ln-MOFs MIL-91	Uric acid	Fluorescence sensor	Urine	1.6	[54]
3	Zr-MOFs	Human semen	Colorimetric sensor	Semen	-	[58]
4	Cu-MOFs	Alkaline phosphatase	Colorimetric sensor	Serum		[59]
5	Cu-BTC (HKUST-1)	Dopamine	Chemiluminescent sensor	Human plasma and urine	0.0023	[60]
6	Zn-hemin MOFs	Acute myocardial infarction related miRNAs	Chemiluminescent sensor	Human plasma	10^{-11}	[62]
7	MMPF-6 (Fe)	Hydroxylamine	Electrochemical sensor	Real pharmaceutical and water samples	0.004	[65]
8	$UiO-66-NH_2$	Nitrite	Electrochemical sensor	Sausage and pickle sample	0.01 Mm	[66]

Optical Sensors

Optical sensing has received widespread attention due to its simplicity and ease of operation. Optical sensors are categorized as colorimetric sensors, fluorescence sensors, and chemiluminescence sensors based on the mechanism of optical transduction.

Fluorescence Sensors

Fluorescence sensors captivate most of the attention owing to their high sensitivity, good selectivity, rapidity and subsequent analyte-specific optical response. The phenomena which describe the origin of luminescence in MOFs include mainly ligand-based emission, emission due to excimer or exciplex formation, guest-induced emission, etc [53]. The fluorescence sensing of various analyte species is usually performed *via* two mechanisms, namely "turn-on" (enhancement) and "turn-off" (quenching). In the case of "turn-on" sensors, an enhancement in the luminescence intensity of the signal occurs with the addition of the target analyte. On the other hand, a decrease in the fluorescence emission intensity with the addition of an analyte is termed "turn-off" sensing [53]. Table 3 summarizes the fluorescence sensing of various analytes achieved by MOFs in several complex matrices.

Lanthanide organic framework (Ln-MOF) offers the capability for sensing different metabolites. A phosphonate MOF, MIL-91 (Al: Eu), showing tremendous luminescence and remarkable stability, was synthesized for the identification of uric acid in urine samples [54]. In a comparison of carboxylate MOFs, MIL-91 constituting a piperazine derivative ligand N,N'-piperazinebismethylphosphonic acid was found to be more stable.

Ratiometric sensors with dual emission signals are generally superior to single-emitting signal sensors. The ratio of the two signals can provide a self-reference calibration to the intrinsic or extrinsic factors enhancing the sensitivity and accuracy [55]. Dual emission can be achieved from the metal centers and ligands of Ln-MOFs by the partial transfer of absorbed energy from the ligand to the lanthanide. Eu-based MOF, $[Eu_2(L)_3(H_2O)_2(DMF)_2].16\ H_2O$ (H_2L:1,4-bis(--carboxy-1H-benzimidazole-2-yl)benzene) showed dual emission, *i.e.*, from both ligand and lanthanide [56]. By using MOF, a self-calibrating ratiometric dual-functional luminescent sensor was developed for detecting temperature and humidity.

Colorimetric Sensors

Fe-MIL-88NH$_2$ MOF was prepared using 2-aminoterephthalic acid and ferric chloride as the starting materials in acetic acid [57]. MOF was combined with glucose oxidase to develop a colorimetric sensor for the detection of glucose. Moreover, it was known for the very first time that MOF could act as peroxidase and catalyze the oxidation of 3,3',5,5'-tetramethylbenzidine by H$_2$O$_2$. Inspired by this work, many researchers exploited the peroxidase-like activity of MOFs to construct colorimetric sensors. A colorimetric sensor was designed by coupling zirconium MOFs with single-stranded DNA (ss-DNA) decorated Au nanoparticles for the identification of human semen [58]. The differential interactions between semen and ss-DNA-Au nanoparticles and MOF generated unique colorimetric patterns. In contrast to traditional enzymes, enzyme mimics based on MOFs offer an excellent alternative as they possess superior catalytic activity and more sensitivity in colorimetric sensing. For the detection of alkaline phosphatase (ALP), a colorimetric test was performed with Cu^{2+}MOFs as a peroxidase mimic and pyrophosphate as a recognition element [59]. A good sensing performance was obtained with a detection limit of 0.19 U/l in serum samples.

Chemiluminescence Sensors

Apart from colorimetric and fluorescence sensors, chemiluminescence sensors have also been investigated in great detail. The chemiluminescence generated by the reaction of the luminol-H$_2$O$_2$ system catalyzed in the presence of HKUST-1 made the basis for the development of the chemiluminescence sensor for dopamine. Upon adding dopamine to the system, the chemiluminescence intensity was quenched [60]. Luo *et al.* embedded hemin into the HKUST-1 MOF and concluded that composites maintained the catalytic activity from hemin. They exploited the peroxidase-like activity of the material to develop a chemiluminescence sensor for H$_2$O$_2$ and glucose [61]. A double-amplified chemiluminescence sensor was developed for miRNA detection in real patients' blood [62].

Electrochemical Sensors

The use of MOFs for electrochemical sensing applications has considerably expanded. Amperometric sensors are the most widely used electrochemical sensing systems that use change in current for the targeted analyte as the sensing response. Modified electrodes comprising MOFs incorporated with active species are supposed to be good candidates for electrochemical sensing. However, the majority of the MOFs are less feasible to display electroanalytical performances in aqueous electrolytes without deteriorating the framework. One notable class of MOFs reported for electrochemical analysis is constructed from chromium-based nodes and carboxylate-based linkers. Cr-based MOFs designed from carboxylate linkers have shown remarkable chemical stability in aqueous solution. MIL-101 (Cr) has been reported to demonstrate reversible electrochemical activity and employed in electrochemical sensing for uric acid and dopamine [63]. A well-known strategy to construct carboxylate MOFs with excellent stability is to use higher valent metal cations as the nodes. MOFs synthesized from carboxylate-based linkers and group IV metal ions (Ti^{4+}, Zr^{4+}, and Hf^{4+}) have shown tremendous thermal and chemical stability [64]. Owing to their remarkable stability in neutral and acidic aqueous solutions, Zr-based MOFs have been considered potential candidates to be used as active thin films for electroanalytical purposes [65, 66]. MOFs constructed from N-containing linkers can obtain better chemical stability in water compared to carboxylate-based MOFs [67]. The most commonly reported MOFs used for electroanalytical purposes are ZIF-8 and ZIF-67, which are methylimidazole-based MOFs with Zn^{2+} and Co^{2+}, respectively [68, 69].

MOFs as Catalysts

The several features of MOFs, including unprecedented structural diversity, intrinsic hybrid inorganic-organic nature, the potential for rational design, and well-defined porosity, make them excellent candidates for catalysts. A summary of the different applications of MOFs for catalysis is provided in Table **4**.

Two CeO_2-CuO bimetallic metal oxides (CeO_2-CuO and Ce-Cu-Ox) obtained from a Ce-Cu-MOF were reported to have catalytic activities towards CO oxidation [71]. CeO_2-CuO was synthesized from the thermal treatment of a Ce-MOF, followed by CuO loading *via*the impregnation method, whereas Ce-Cu-Ox was synthesized by the direct calcination of a bimetallic Ce(Cu)-MOF at 600°C for 3 h. The CeO_2-CuO (80% CO conversion) material exhibited higher catalytic activity as compared to the Ce-Cu-Ox material (22% CO conversion). Cobalt nanoparticles supported on nanoporous N-doped carbon obtained from a Co/Zn ZIF-67 were explored as the catalyst for ammonia borane dehydrogenation [72]. ZIF-67/8, with a molar ratio of Co/Zn = 1, showed the highest catalytic activity.

Table 4. **Application of MOFs for catalysis.**

Order	Catalytic reactions	MOF	Refs
1	CO oxidation	Ce-Cu-MOF	[71]
2	Dehydrogenation of ammonia borane	Co/Zn ZIF-67	[72]

(Table 4) cont.....

Order	Catalytic reactions	MOF	Refs
3	Fischer-Tropsch	MIL-88	[73]
4	Styrene epoxidation	ZIF-67	[74]
5	Biofuel upgrade	ZIF-8	[75]
6	Alcohol esterification	ZIF-67	[76]
7	NO reduction	Ce-ZIF-67	[77]
8	Reverse water gas shift	Zn-Cu-BTC	[78]

CONCLUSION

MOFs are attracting much attention nowadays due to their extraordinary characteristics and feasibility of pre- and post-synthetic modifications. In this chapter, we summarized the common synthetic methods and characterization techniques used for MOFs. New synthetic methods (microwave, sonochemical, mechanochemical, and electrochemical) which provide fast crystallization, facile morphology control, and narrow particle size distribution are adopted for MOFs synthesis in comparison to the conventional (solvothermal) method. However, only well-known compounds have been prepared by these synthetic routes. The reproducibility of these alternative synthetic procedures needs to be further established.

The applications of MOFs are discussed with a focus on adsorption, sensing, and catalysis applications. MOFs have now emerged as outstanding sorbents for pre-treatment procedures in separation science. Different types of target pollutants, which include but are not limited to PAHs, phenols, drugs, and heavy metals from different matrices (water, food, and biological samples), have been extracted using MOFs as sorbents. The sensor performance, including sensitivity, selectivity, response time, and stability, can be readily enhanced by using MOFs as sensor probes. The presence of multiple emission centers in MOFs (ratiometric fluorescence) can bestow the sensors with a self-calibrating curve without being influenced by external stimuli and can also enhance the selectivity and sensitivity of MOFs. The applications of MOFs as catalysts have also been reviewed and found to show excellent catalytic properties. In the end, MOFs emerge as a novel and advanced porous material, which has the potential capability to be applied in different applications by considering some of the challenges, including the synthesis of MOFs with higher chemical stability, good selectivity, and better reusability.

ACKNOWLEDGEMENTS

The authors are thankful to the UGC-SAP and Chemistry Department, Punjabi University, Patiala, Punjab, for providing lab and instrument facilities. KK is also thankful to Mata Gujri College, Fatehgarh Sahib, Punjab, for providing lab facilities.

REFERENCES

[1]   Li, B.; Wen, H.M.; Cui, Y.; Zhou, W.; Qian, G.; Chen, B. Emerging multifunctional metal-organic framework materials. *Adv. Mater.,* **2016**, *28*(40), 8819-8860.
[http://dx.doi.org/10.1002/adma.201601133] [PMID: 27454668]

[2]   Joseph, L.; Jun, B.M.; Jang, M.; Park, C.M.; Muñoz-Senmache, J.C.; Hernández-Maldonado, A.J.; Heyden, A.; Yu, M.; Yoon, Y. Removal of contaminants of emerging concern by metal-organic framework nanoadsorbents: A review. *Chem. Eng. J.,* **2019**, *369*, 928-946.
[http://dx.doi.org/10.1016/j.cej.2019.03.173]

[3]   Furukawa, H.; Cordova, K.E.; O'Keeffe, M.; Yaghi, O.M. The chemistry and applications of metal-organic frameworks. *Science,* **2013**, *341*(6149), 1230444.
[http://dx.doi.org/10.1126/science.1230444] [PMID: 23990564]

[4]   Hashemi, B.; Zohrabi, P.; Raza, N.; Kim, K.H. Metal-organic frameworks as advanced sorbents for the extraction and determination of pollutants from environmental, biological, and food media. *Trends Analyt. Chem.,* **2017**, *97*, 65-82.
[http://dx.doi.org/10.1016/j.trac.2017.08.015]

[5]   Gangu, K.K.; Maddila, S.; Mukkamala, S.B.; Jonnalagadda, S.B. A review on contemporary Metal-Organic Framework materials. *Inorg. Chim. Acta,* **2016**, *446*, 61-74.
[http://dx.doi.org/10.1016/j.ica.2016.02.062]

[6]   Bagheri, N.; Khataee, A.; Habibi, B.; Hassanzadeh, J. Mimetic Ag nanoparticle/Zn-based MOF nanocomposite (AgNPs@ZnMOF) capped with molecularly imprinted polymer for the selective detection of patulin. *Talanta,* **2018**, *179*, 710-718.
[http://dx.doi.org/10.1016/j.talanta.2017.12.009] [PMID: 29310298]

[7]   Alezi, D.; Belmabkhout, Y.; Suyetin, M.; Bhatt, P.M.; Weseliński, Ł.J.; Solovyeva, V.; Adil, K.; Spanopoulos, I.; Trikalitis, P.N.; Emwas, A.H.; Eddaoudi, M. MOF crystal chemistry paving the way to gas storage needs: aluminum-based soc-MOF for CH_4, O_2, and CO_2 storage. *J. Am. Chem. Soc.,* **2015**, *137*(41), 13308-13318.
[http://dx.doi.org/10.1021/jacs.5b07053] [PMID: 26364990]

[8]   Zheng, H.; Zhang, Y.; Liu, L.; Wan, W.; Guo, P.; Nyström, A.M.; Zou, X. One-pot synthesis of metal-organic frameworks with encapsulated target molecules and their applications for controlled drug delivery. *J. Am. Chem. Soc.,* **2016**, *138*(3), 962-968.
[http://dx.doi.org/10.1021/jacs.5b11720] [PMID: 26710234]

[9]   Liu, C.; Yu, L.Q.; Zhao, Y.T.; Lv, Y.K. Recent advances in metal-organic frameworks for adsorption of common aromatic pollutants. *Mikrochim. Acta,* **2018**, *185*(7), 342.
[http://dx.doi.org/10.1007/s00604-018-2879-2] [PMID: 29951844]

[10]   Dhakshinamoorthy, A.; Li, Z.; Garcia, H. Catalysis and photocatalysis by metal organic frameworks. *Chem. Soc. Rev.,* **2018**, *47*(22), 8134-8172.
[http://dx.doi.org/10.1039/C8CS00256H] [PMID: 30003212]

[11]   Li, H.; Eddaoudi, M.; O'Keeffe, M.; Yaghi, OM. Design and synthesis of an exceptionally stable and highly porous metal-organic framework. *Nature,* **1999**, *402*(6759), 276-9.

[12]   Eddaoudi, M.; Kim, J.; Rosi, N.; Vodak, D.; Wachter, J.; O'Keeffe, M.; Yaghi, O.M. Systematic

design of pore size and functionality in isoreticular MOFs and their application in methane storage. *Science,* **2002**, *295*(5554), 469-472.
[http://dx.doi.org/10.1126/science.1067208] [PMID: 11799235]

[13]   Gascon, J.; Aktay, U.; Hernandezalonso, M.; Vanklink, G.; Kapteijn, F. Amino-based metal-organic frameworks as stable, highly active basic catalysts. *J. Catal.,* **2009**, *261*(1), 75-87.
[http://dx.doi.org/10.1016/j.jcat.2008.11.010]

[14]   Férey, G.; Mellot-Draznieks, C.; Serre, C.; Millange, F.; Dutour, J.; Surblé, S.; Margiolaki, I. A chromium terephthalate-based solid with unusually large pore volumes and surface area. *Science,* **2005**, *309*(5743), 2040-2042.
[http://dx.doi.org/10.1126/science.1116275] [PMID: 16179475]

[15]   Horcajada, P.; Surblé, S.; Serre, C.; Hong, D.Y.; Seo, Y.K.; Chang, J.S.; Grenèche, J.M.; Margiolaki, I.; Férey, G. Synthesis and catalytic properties of MIL-100(Fe), an iron(III) carboxylate with large pores. *Chem. Commun. (Camb.),* **2007**, (27), 2820-2822.
[http://dx.doi.org/10.1039/B704325B] [PMID: 17609787]

[16]   Kim, J.; Kim, W.Y.; Ahn, W.S. Amine-functionalized MIL-53(Al) for CO_2/N_2 separation: Effect of textural properties. *Fuel,* **2012**, *102*, 574-579.
[http://dx.doi.org/10.1016/j.fuel.2012.06.016]

[17]   Abid, H.R.; Tian, H.; Ang, H.M.; Tade, M.O.; Buckley, C.E.; Wang, S. Nanosize Zr-metal organic framework (UiO-66) for hydrogen and carbon dioxide storage. *Chem. Eng. J.,* **2012**, *187*, 415-420.
[http://dx.doi.org/10.1016/j.cej.2012.01.104]

[18]   Jhung, S.H.; Lee, J.H.; Chang, J.S. Microwave synthesis of a nanoporous hybrid material, chromium trimesate. *Bull. Korean Chem. Soc.,* **2005**, *26*(6), 880-881.

[18a] Jung, D.W.; Yang, D.A.; Kim, J.; Kim, J.; Ahn, W.S. Facile synthesis of MOF-177 by a sonochemical method using 1-methyl-2-pyrrolidinone as a solvent. *Dalton Trans.,* **2010**, *39*(11), 2883-2887.
[PMID: 20200716]

[19]   Horcajada, P.; Chalati, T.; Serre, C.; Gillet, B.; Sebrie, C.; Baati, T.; Eubank, J.F.; Heurtaux, D.; Clayette, P.; Kreuz, C.; Chang, J.S.; Hwang, Y.K.; Marsaud, V.; Bories, P.N.; Cynober, L.; Gil, S.; Férey, G.; Couvreur, P.; Gref, R. Porous metal-organic-framework nanoscale carriers as a potential platform for drug delivery and imaging. *Nat. Mater.,* **2010**, *9*(2), 172-178.
[http://dx.doi.org/10.1038/nmat2608] [PMID: 20010827]

[20]   Choi, J.Y.; Kim, J.; Jhung, S.H.; Kim, H.; Chang, J.; Chae, H.K. Microwave synthesis of a porous metal-organic framework, zinc terephthalate MOF-5. *Bull. Korean Chem. Soc.,* **2006**, *27*(10), 1523-1524.
[http://dx.doi.org/10.5012/bkcs.2006.27.10.1523]

[21]   Lu, C.M.; Liu, J.; Xiao, K.; Harris, A.T. Microwave enhanced synthesis of MOF-5 and its CO_2 capture ability at moderate temperatures across multiple capture and release cycles. *Chem. Eng. J.,* **2010**, *156*(2), 465-470.
[http://dx.doi.org/10.1016/j.cej.2009.10.067]

[22]   Schlesinger, M.; Schulze, S.; Hietschold, M.; Mehring, M. Evaluation of synthetic methods for microporous metal-organic frameworks exemplified by the competitive formation of [Cu_2 (btc) 3 (H_2O) 3]. *Microporous Mesoporous Mater.,* **2010**, *132*(1-2), 121-127.

[23]   Tonigold, M.; Lu, Y.; Bredenkötter, B.; Rieger, B.; Bahnmüller, S.; Hitzbleck, J.; Langstein, G.; Volkmer, D. Heterogeneous catalytic oxidation by MFU-1: a cobalt(II)-containing metal-organic framework. *Angew. Chem. Int. Ed.,* **2009**, *48*(41), 7546-7550.
[http://dx.doi.org/10.1002/anie.200901241] [PMID: 19746371]

[24]   Mueller, U.; Schubert, M.; Teich, F.; Puetter, H.; Schierle-Arndt, K.; Pastré, J. Metal-organic frameworks—prospective industrial applications. *J. Mater. Chem.,* **2006**, *16*(7), 626-636.
[http://dx.doi.org/10.1039/B511962F]

[25]   Mueller, U.; Puetter, H.; Hesse, M.; Wessel, H.; Patent, PCT *049,892,* **2005**.

[26]   Joaristi, A.M.J. Juan-Alcaniz, P. Serra-Crespo, F. Kapteijn. *J. Gascon. Cryst. Growth Des.,* **2012**, *12*, 3489-3498.

[27]   Tranchemontagne, D.J.; Hunt, J.R.; Yaghi, O.M. Room temperature synthesis of metal-organic frameworks: MOF-5, MOF-74, MOF-177, MOF-199, and IRMOF-0. *Tetrahedron,* **2008**, *64*(36), 8553-8557.
[http://dx.doi.org/10.1016/j.tet.2008.06.036]

[28]   Qiu, L.G.; Li, Z.Q.; Wu, Y.; Wang, W.; Xu, T.; Jiang, X. Facile synthesis of nanocrystals of a microporous metal-organic framework by an ultrasonic method and selective sensing of organoamines. *Chem. Commun. (Camb.),* **2008**, (31), 3642-3644.
[http://dx.doi.org/10.1039/b804126a] [PMID: 18665285]

[29]   Li, Z.Q.; Qiu, L.G.; Xu, T.; Wu, Y.; Wang, W.; Wu, Z.Y.; Jiang, X. Ultrasonic synthesis of the microporous metal-organic framework Cu3(BTC)2 at ambient temperature and pressure: An efficient and environmentally friendly method. *Mater. Lett.,* **2009**, *63*(1), 78-80.
[http://dx.doi.org/10.1016/j.matlet.2008.09.010]

[30]   Pichon, A.; Lazuen-Garay, A.; James, S.L. Solvent-free synthesis of a microporous metal-organic framework. *CrystEngComm,* **2006**, *8*(3), 211-214.
[http://dx.doi.org/10.1039/b513750k]

[31]   Pichon, A.; James, S.L. An array-based study of reactivity under solvent-free mechanochemical conditions—insights and trends. *CrystEngComm,* **2008**, *10*(12), 1839-1847.
[http://dx.doi.org/10.1039/b810857a]

[32]   Beldon, P.J.; Fábián, L.; Stein, R.S.; Thirumurugan, A.; Cheetham, A.K.; Friščić, T. Rapid room temperature synthesis of zeoliticimidazolate frameworks by using mechanochemistry. *Angew. Chem.,* **2010**, *122*(50), 9834-9837.
[http://dx.doi.org/10.1002/ange.201005547]

[33]   Ipadeola, A.K.; Ozoemena, K.I. Alkaline water-splitting reactions over Pd/Co-MOF-derived carbon obtained *via* microwave-assisted synthesis. *RSC Advances,* **2020**, *10*(29), 17359-17368.
[http://dx.doi.org/10.1039/D0RA02307H] [PMID: 35521459]

[34]   Wei, J.Z.; Gong, F.X.; Sun, X.J.; Li, Y.; Zhang, T.; Zhao, X.J.; Zhang, F.M. Rapid and low-cost electrochemical synthesis of UiO-66-NH$_2$ with enhanced fluorescence detection performance. *Inorg. Chem.,* **2019**, *58*(10), 6742-6747.
[http://dx.doi.org/10.1021/acs.inorgchem.9b00157] [PMID: 31026150]

[35]   Pirzadeh, K.; Ghoreyshi, A.A.; Rahimnejad, M.; Mohammadi, M. Electrochemical synthesis, characterization and application of a microstructure Cu$_3$(BTC)$_2$ metal organic framework for CO$_2$ and CH$_4$ separation. *Korean J. Chem. Eng.,* **2018**, *35*(4), 974-983.
[http://dx.doi.org/10.1007/s11814-017-0340-6]

[36]   Abdollahi, N.; Masoomi, M.Y.; Morsali, A.; Junk, P.C.; Wang, J. Sonochemical synthesis and structural characterization of a new Zn(II) nanoplate metal-organic framework with removal efficiency of Sudan red and Congo red. *Ultrason. Sonochem.,* **2018**, *45*, 50-56.
[http://dx.doi.org/10.1016/j.ultsonch.2018.03.001] [PMID: 29705324]

[37]   Masoomi, M.Y.; Morsali, A. Sonochemical synthesis of nanoplates of two Cd(II) based metal-organic frameworks and their applications as precursors for preparation of nano-materials. *Ultrason. Sonochem.,* **2016**, *28*, 240-249.
[http://dx.doi.org/10.1016/j.ultsonch.2015.07.017] [PMID: 26384904]

[38]   Singh, N.K.; Hardi, M.; Balema, V.P. Mechanochemical synthesis of an yttrium based metal-organic framework. *Chem. Commun. (Camb.),* **2013**, *49*(10), 972-974.
[http://dx.doi.org/10.1039/C2CC36325A] [PMID: 23128845]

[39]   Lv, D.; Chen, Y.; Li, Y.; Shi, R.; Wu, H.; Sun, X.; Xiao, J.; Xi, H.; Xia, Q.; Li, Z. Efficient

mechanochemical synthesis of MOF-5 for linear alkanes adsorption. *J. Chem. Eng. Data,* **2017**, *62*(7), 2030-2036.
[http://dx.doi.org/10.1021/acs.jced.7b00049]

[40] Keskin, S.; Kızılel, S. Biomedical applications of metal organic frameworks. *Ind. Eng. Chem. Res.,* **2011**, *50*(4), 1799-1812.
[http://dx.doi.org/10.1021/ie101312k]

[41] Seo, Y.K.; Hundal, G. , **2009**.

[42] Zhou, Y.Y.; Yan, X.P.; Kim, K.N.; Wang, S.W.; Liu, M.G. Exploration of coordination polymer as sorbent for flow injection solid-phase extraction on-line coupled with high-performance liquid chromatography for determination of polycyclic aromatic hydrocarbons in environmental materials. *J. Chromatogr. A,* **2006**, *1116*(1-2), 172-178.
[http://dx.doi.org/10.1016/j.chroma.2006.03.061] [PMID: 16616177]

[43] Huo, S.H.; Yan, X.P. Facile magnetization of metal-organic framework MIL-101 for magnetic solid-phase extraction of polycyclic aromatic hydrocarbons in environmental water samples. *Analyst (Lond.),* **2012**, *137*(15), 3445-3451.
[http://dx.doi.org/10.1039/c2an35429b] [PMID: 22695791]

[44] Ma, R.; Hao, L.; Wang, J.; Wang, C.; Wu, Q.; Wang, Z. Magnetic porous carbon derived from a metal-organic framework as a magnetic solid☐phase extraction adsorbent for the extraction of sex hormones from water and human urine. *J. Sep. Sci.,* **2016**, *39*(18), 3571-3577.
[http://dx.doi.org/10.1002/jssc.201600347] [PMID: 27470965]

[45] Tahmasebi, E.; Masoomi, M.Y.; Yamini, Y.; Morsali, A. Application of mechanosynthesized azine-decorated zinc(II) metal-organic frameworks for highly efficient removal and extraction of some heavy-metal ions from aqueous samples: a comparative study. *Inorg. Chem.,* **2015**, *54*(2), 425-433.
[http://dx.doi.org/10.1021/ic5015384] [PMID: 25548873]

[46] Ge, D.; Lee, H.K. Water stability of zeolite imidazolate framework 8 and application to porous membrane-protected micro-solid-phase extraction of polycyclic aromatic hydrocarbons from environmental water samples. *J. Chromatogr. A,* **2011**, *1218*(47), 8490-8495.
[http://dx.doi.org/10.1016/j.chroma.2011.09.077] [PMID: 22018717]

[47] Ma, J.; Yao, Z.; Hou, L.; Lu, W.; Yang, Q.; Li, J.; Chen, L. Metal organic frameworks (MOFs) for magnetic solid-phase extraction of pyrazole/pyrrole pesticides in environmental water samples followed by HPLC-DAD determination. *Talanta,* **2016**, *161*, 686-692.
[http://dx.doi.org/10.1016/j.talanta.2016.09.035] [PMID: 27769466]

[48] Hu, Y.; Huang, Z.; Liao, J.; Li, G. Chemical bonding approach for fabrication of hybrid magnetic metal-organic framework-5: high efficient adsorbents for magnetic enrichment of trace analytes. *Anal. Chem.,* **2013**, *85*(14), 6885-6893.
[http://dx.doi.org/10.1021/ac4011364] [PMID: 23758552]

[49] Wang, G.H.; Lei, Y.Q.; Song, H.C. Evaluation of $Fe_3O_4@SiO_2$-MOF-177 as an advantageous adsorbent for magnetic solid-phase extraction of phenols in environmental water samples. *Anal. Methods,* **2014**, *6*(19), 7842-7847.
[http://dx.doi.org/10.1039/C4AY00822G]

[50] Li, Q.L.; Wang, X.; Chen, X.F.; Wang, M.L.; Zhao, R.S. In situ hydrothermal growth of ytterbium-based metal-organic framework on stainless steel wire for solid-phase microextraction of polycyclic aromatic hydrocarbons from environmental samples. *J. Chromatogr. A,* **2015**, *1415*, 11-19.
[http://dx.doi.org/10.1016/j.chroma.2015.08.036] [PMID: 26346186]

[51] Zhang, Z.; Huang, Y.; Ding, W.; Li, G. Multilayer interparticle linking hybrid MOF-199 for noninvasive enrichment and analysis of plant hormone ethylene. *Anal. Chem.,* **2014**, *86*(7), 3533-3540.
[http://dx.doi.org/10.1021/ac404240n] [PMID: 24576104]

[52] Rocío-Bautista, P.; Martínez-Benito, C.; Pino, V.; Pasán, J.; Ayala, J.H.; Ruiz-Pérez, C.; Afonso, A.M. The metal-organic framework HKUST-1 as efficient sorbent in a vortex-assisted dispersive micro

solid-phase extraction of parabens from environmental waters, cosmetic creams, and human urine. *Talanta,* **2015**, *139*, 13-20.
[http://dx.doi.org/10.1016/j.talanta.2015.02.032] [PMID: 25882402]

[53]   Karmakar, A.; Samanta, P.; Dutta, S.; Ghosh, S.K. Fluorescent "Turn-on" Sensing Based on Metal-Organic Frameworks (MOFs). *Chem. Asian J.,* **2019**, *14*(24), 4506-4519.
[http://dx.doi.org/10.1002/asia.201901168] [PMID: 31573139]

[54]   Lian, X.; Yan, B. Phosphonate MOFs composite as off-on fluorescent sensor for detecting purine metabolite uric acid and diagnosing hyperuricuria. *Inorg. Chem.,* **2017**, *56*(12), 6802-6808.
[http://dx.doi.org/10.1021/acs.inorgchem.6b03009] [PMID: 28358494]

[55]   Bhogal, S.; Kaur, K.; Malik, A.K.; Sonne, C.; Lee, S.S.; Kim, K.H. Core-shell structured molecularly imprinted materials for sensing applications. *Trends Analyt. Chem.,* **2020**, *133*, 116043.
[http://dx.doi.org/10.1016/j.trac.2020.116043]

[56]   Wang, D.; Tan, Q.; Liu, J.; Liu, Z. A stable europium metal-organic framework as a dual-functional luminescent sensor for quantitatively detecting temperature and humidity. *Dalton Trans.,* **2016**, *45*(46), 18450-18454.
[http://dx.doi.org/10.1039/C6DT03812C] [PMID: 27830848]

[57]   Liu, Y.L.; Zhao, X.J.; Yang, X.X.; Li, Y.F. A nanosized metal-organic framework of Fe-MIL-88NH2 as a novel peroxidase mimic used for colorimetric detection of glucose. *Analyst (Lond.),* **2013**, *138*(16), 4526-4531.
[http://dx.doi.org/10.1039/c3an00560g] [PMID: 23775015]

[58]   Sun, Z.; Wu, S.; Ma, J.; Shi, H.; Wang, L.; Sheng, A.; Yin, T.; Sun, L.; Li, G. Colorimetric Sensor Array for Human Semen Identification Designed by Coupling Zirconium Metal-Organic Frameworks with DNA-Modified Gold Nanoparticles. *ACS Appl. Mater. Interfaces,* **2019**, *11*(40), 36316-36323.
[http://dx.doi.org/10.1021/acsami.9b10729] [PMID: 31522499]

[59]   Wang, C.; Gao, J.; Cao, Y.; Tan, H. Colorimetric logic gate for alkaline phosphatase based on copper (II)-based metal-organic frameworks with peroxidase-like activity. *Anal. Chim. Acta,* **2018**, *1004*, 74-81.
[http://dx.doi.org/10.1016/j.aca.2017.11.078] [PMID: 29329711]

[60]   Zhu, Q.; Chen, Y.; Wang, W.; Zhang, H.; Ren, C.; Chen, H.; Chen, X. A sensitive biosensor for dopamine determination based on the unique catalytic chemiluminescence of metal-organic framework HKUST-1. *Sens. Actuators B Chem.,* **2015**, *210*, 500-507.
[http://dx.doi.org/10.1016/j.snb.2015.01.012]

[61]   Luo, F.; Lin, Y.; Zheng, L.; Lin, X.; Chi, Y. Encapsulation of hemin in metal-organic frameworks for catalyzing the chemiluminescence reaction of the H_2O_2-luminol system and detecting glucose in the neutral condition. *ACS Appl. Mater. Interfaces,* **2015**, *7*(21), 11322-11329.
[http://dx.doi.org/10.1021/acsami.5b01706] [PMID: 25928385]

[62]   Mi, L.; Sun, Y.; Shi, L.; Li, T. Hemin-Bridged MOF Interface with Double Amplification of G-Quadruplex Payload and DNAzyme Catalysis: Ultrasensitive Lasting Chemiluminescence MicroRNA Imaging. *ACS Appl. Mater. Interfaces,* **2020**, *12*(7), 7879-7887.
[http://dx.doi.org/10.1021/acsami.9b18053] [PMID: 31983198]

[63]   Li, Y.; Huangfu, C.; Du, H.; Liu, W.; Li, Y.; Ye, J. Electrochemical behavior of metal-organic framework MIL-101 modified carbon paste electrode: An excellent candidate for electroanalysis. *J. Electroanal. Chem. (Lausanne),* **2013**, *709*, 65-69.
[http://dx.doi.org/10.1016/j.jelechem.2013.09.017]

[64]   Yuan, S.; Qin, J.S.; Lollar, C.T.; Zhou, H.C. Stable metal-organic frameworks with group 4 metals: current status and trends. *ACS Cent. Sci.,* **2018**, *4*(4), 440-450.
[http://dx.doi.org/10.1021/acscentsci.8b00073] [PMID: 29721526]

[65]   Wang, Y.; Wang, L.; Chen, H.; Hu, X.; Ma, S. Fabrication of highly sensitive and stable hydroxylamine electrochemical sensor based on gold nanoparticles and metal-metalloporphyrin

framework modified electrode. *ACS Appl. Mater. Interfaces,* **2016**, *8*(28), 18173-18181.
[http://dx.doi.org/10.1021/acsami.6b04819] [PMID: 27351460]

[66] Yang, J.; Yang, L.; Ye, H.; Zhao, F.; Zeng, B. Highly dispersed AuPd alloy nanoparticles immobilized on UiO-66-NH 2 metal-organic framework for the detection of nitrite. *Electrochim. Acta,* **2016**, *219*, 647-654.
[http://dx.doi.org/10.1016/j.electacta.2016.10.071]

[67] Chuang, C.H.; Kung, C.W. Metal−Organic Frameworks toward Electrochemical Sensors: Challenges and Opportunities. *Electroanalysis,* **2020**, *32*(9), 1885-1895.
[http://dx.doi.org/10.1002/elan.202060111]

[68] Zhang, H.; Liu, D.; Yao, Y.; Zhang, B.; Lin, Y.S. Stability of ZIF-8 membranes and crystalline powders in water at room temperature. *J. Membr. Sci.,* **2015**, *485*, 103-111.
[http://dx.doi.org/10.1016/j.memsci.2015.03.023]

[69] Ahmed, S.A.; Bagchi, D.; Katouah, H.A.; Hasan, M.N.; Altass, H.M.; Pal, S.K. enhanced Water Stability and photoresponsivity in Metal-organic framework (Mof): A potential tool to combat Drug-resistant Bacteria. *Sci. Rep.,* **2019**, *9*(1), 19372.
[http://dx.doi.org/10.1038/s41598-019-55542-8] [PMID: 31852949]

[70] Ren, K.; Guo, X.F.; Tang, Y.J.; Huang, B.H.; Wang, H. Size-controlled synthesis of metal-organic frameworks and their performance as fluorescence sensors. *Analyst (Lond.),* **2020**, *145*(22), 7349-7356.
[PMID: 32930197]

[71] Guo, Z.; Song, L.; Xu, T.; Gao, D.; Li, C.; Hu, X.; Chen, G. CeO2-CuO bimetal oxides derived from Ce-based MOF and their difference in catalytic activities for CO oxidation. *Mater. Chem. Phys.,* **2019**, *226*, 338-343.
[http://dx.doi.org/10.1016/j.matchemphys.2019.01.057]

[72] Zacho, S.L.; Mielby, J.; Kegnæs, S. Hydrolytic dehydrogenation of ammonia borane over ZIF-67 derived Co nanoparticle catalysts. *Catal. Sci. Technol.,* **2018**, *8*(18), 4741-4746.
[http://dx.doi.org/10.1039/C8CY01500G]

[73] Long, J.; Shen, K.; Chen, L.; Li, Y. Multimetal-MOF-derived transition metal alloy NPs embedded in an N-doped carbon matrix: highly active catalysts for hydrogenation reactions. *J. Mater. Chem. A Mater. Energy Sustain.,* **2016**, *4*(26), 10254-10262.
[http://dx.doi.org/10.1039/C6TA00157B]

[74] Yu, G.; Sun, J.; Muhammad, F.; Wang, P.; Zhu, G. Cobalt-based metal organic framework as precursor to achieve superior catalytic activity for aerobic epoxidation of styrene. *RSC Advances,* **2014**, *4*(73), 38804-38811.
[http://dx.doi.org/10.1039/C4RA03746D]

[75] Chen, Y.Z.; Cai, G.; Wang, Y.; Xu, Q.; Yu, S.H.; Jiang, H.L. Palladium nanoparticles stabilized with N-doped porous carbons derived from metal-organic frameworks for selective catalysis in biofuel upgrade: the role of catalyst wettability. *Green Chem.,* **2016**, *18*(5), 1212-1217.
[http://dx.doi.org/10.1039/C5GC02530C]

[76] Zhong, W.; Liu, H.; Bai, C.; Liao, S.; Li, Y. Base-free oxidation of alcohols to esters at room temperature and atmospheric conditions using nanoscale Co-based catalysts. *ACS Catal.,* **2015**, *5*(3), 1850-1856.
[http://dx.doi.org/10.1021/cs502101c]

[77]   Wang, X.; Zhao, S.; Zhang, Y.; Wang, Z.; Feng, J.; Song, S.; Zhang, H. CeO_2 nanowires self-inserted into porous Co_3O_4 frameworks as high-performance "noble metal free" hetero-catalysts. *Chem. Sci. (Camb.),* **2016**, *7*(2), 1109-1114.
[http://dx.doi.org/10.1039/C5SC03430B] [PMID: 29896375]

[78]   Zhang, J.; An, B.; Hong, Y.; Meng, Y.; Hu, X.; Wang, C.; Lin, J.; Lin, W.; Wang, Y. Pyrolysis of metal-organic frameworks to hierarchical porous Cu/Zn-nanoparticle@carbon materials for efficient CO_2 hydrogenation. *Mater. Chem. Front.,* **2017**, *1*(11), 2405-2409.
[http://dx.doi.org/10.1039/C7QM00328E]

CHAPTER 6

Nanostructure Impregnated MOFs for Photo-catalytic and Sensing Applications

Aman Grover[1a], **Irshad Mohiuddin**[1a], **Shikha Bhogal**[1], **Ashok Kumar Malik**[1,**] and **Jatinder Singh Aulakh**[1,*]

[1] *Department of Chemistry, Punjabi University, Patiala-147002, Punjab, India*

Abstract: Metal-organic frameworks (MOFs), due to their high porosity, enhanced surface area, rich topology, diverse structures and controllable chemical structures, have recently emerged as an exciting class of porous crystalline materials. The integration of nanostructures with MOFs generates MOF composites with synergistic properties and functions, attracting the broad application prospect. In this chapter, the primary strategies guiding the design of these materials, including MOFs, are described as host materials that contain and stabilize guest nanoparticles. A detailed discussion about the recent progress of nanostructure-impregnated MOFs based on diverse photocatalytic (*e.g.*, environmental remediation, oxidation of alcohols, CO_2 reduction, and H_2 generation) and sensing (organic pollutants, gaseous pollutants, and heavy metal ions) applications has been provided. With a deeper knowledge of nanostructure-impregnated MOFs, this book chapter will provide better guidance for the rational design of high-performance MOF-based materials and is likely to shed new light on future research in this promising field.

Keywords: Composites, Environmental remediation, Metal-organic frameworks (MOFs), Porous material, Photocatalysis, Sensing.

INTRODUCTION

Nanotechnology refers to a multi-disciplinary area in which the structure of materials is modified at the nanometer scale through chemical, physical, and biological paths. Nano-structured materials are materials that have at least one dimension in the nanoscale range (1-100 nm), or their basic units are in a three-dimensional space. Compared with conventional materials, nanomaterials are

* **Corresponding author Jatinder Singh Aulakh:** Department of Chemistry, Punjabi University, Patiala-147002, Punjab, India; E-mail: chemiaulakh@gmail.com
** **Co-corresponding author Ashok Kumar Malik:** Department of Chemistry, Punjabi University, Patiala-147002, Punjab, India; malik_chem2002@yahoo.co.uk
aThese authors contributed equally as first authors.

Karamjit Singh Dhaliwal (Ed.)

garnering a lot of attention in the scientific community because of their amazing properties (*e.g.*, small size, large surface area to volume ratio, high surface reactivity, adaptive surface functionalities, short diffusion, high sensitivity, quantum confinement, and compactness) [1]. With the advancements in nanotechnology, nanomaterials have been recognized as an excellent potential candidate in various fields owing to their size-dependent properties and high chemical, thermal, mechanical, optical, electrical, and biological properties. They have been extensively studied in diverse fields, such as photocatalytic, adsorption, sensing, medicine and health, aerospace, energy resources, environment, biotechnology, *etc*. The exploitation of nanostructured materials has led to fundamental changes in analytical sciences during the past few decades [2].

Recently, metal-organic frameworks (MOFs), a novel type of porous crystalline compounds, have sparked a lot of curiosity due to their plethora of advantages (*e.g.*, ultrahigh porosity and surface area, fully exposed active sites, uniform but tunable pores, plentiful compositions, and tunable morphologies), which makes them extremely useful in a variety of applications [3]. The structure of MOF can be tuned according to targeted properties by converting the geometry of the metal ions or clusters and multifunctional organic linkers, which are put together with the help of well-defined coordination linkages. For many advanced applications in nanotechnology, it is mandatory that the MOF must be impregnated with nanostructured materials. Through the integration of diverse building blocks, the impregnation of nanostructured materials within MOF has been pursued to develop new multifunctional composites/hybrids that demonstrate advantages of all of them that are superior to those of the individual components.

In this chapter, efforts were made to emphasize the promising applications of nanostructured impregnated MOF materials for photocatalytic (removal of environmental pollutants, CO_2 reduction, the hydrogen evolution reaction (HER), and organic photo-redox reactions) and target-specific sensing (organic pollutants sensing, gaseous pollutants sensing, and heavy metal ions sensing). A few reports concerning the investigation of such composites for photocatalytic and sensing applications have appeared, but there is still plenty of scope for improvement and optimization. Finally, the bottlenecks and perspectives for future work on nanostructured impregnated MOF materials in this highly significant area are critically proposed. Our purpose is to encourage the attention of researchers toward the application of nanostructure-impregnated MOF materials and thus further promote their development.

SYNTHESIS OF NANOSTRUCTURED IMPREGNATED MOFS

Due to their unique functional and structural features, MOFs are now recognized as a key category of porous compounds. These frameworks are fabricated by bridging organic linkers and metal ions or their clusters. The chosen primary building blocks are another important element playing a vital role in the final structure and properties of MOFs. Nonetheless, various additional synthetic processes and parameters, such as pressure, pH, temperature, reaction time, and solvent, must be taken into account. The "ship in bottle" strategy, "bottle around ship" approach, and one-step synthesis approach are three important proven approaches for immobilizing functional nanoscale objects inside the pores of MOFs. The "ship in bottle" method combines the encapsulation of MOF cavities with active small molecules or nanoparticle (NPs) precursors, followed by further processing to provide the desired functional structure (Fig. **1a**) [4]. Various methods have been used to incorporate NP precursors into MOFs, such as vapor deposition, solution infiltration, and solid grinding [5, 6]. However, it is quite difficult to precisely manage the structure, composition, location, and morphology of integrated guests using this synthetic technique.

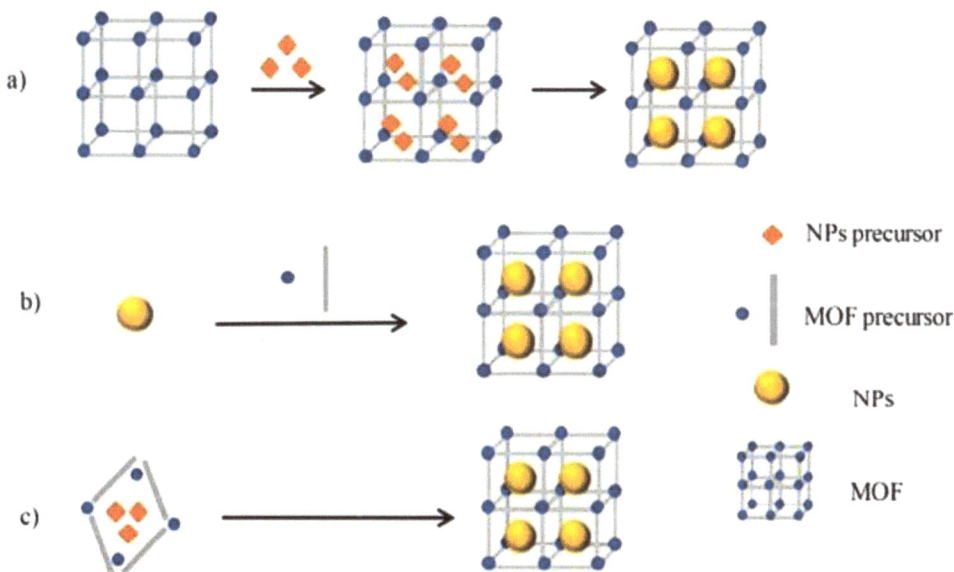

Fig. (1). Main approaches for the fabrication of nanostructure-impregnated MOF composites **(a)** Ship in the bottle; **(b)** bottle around the ship; and **(c)** in situ one-step synthesis [4].

In addition, the development of precursors and products on the exterior surface of MOFs must be taken into account. To prevent NPs accumulation on the exterior surface of MOFs, a double-solvent method (DSM) was successfully devised to rationally insert precursors within MOF cavities, followed by further treatment to make NPs@MOF composites [7]. The DSM is based on a hydrophilic and hydrophobic solvent and, in certain MOFs, on large cages with hydrophilic surroundings and high pore volumes. The quantitative volume (smaller than the MOF pore volume) of the aqueous precursor solution can be easily incorporated into the pores of MOF suspended in a substantial amount of low-boiling point hydrophobic solvent via capillary force and hydrophilic interactions. By combining DSM with reducing agents, such as H_2 and $NaBH_4$, metallic NPs, such as Pt, Pd, Rh, Ni, and AuNi, RuNi, AuCo, and CuCo bimetallic NPs were successfully immobilized within the pores of MOFs without aggregation on the external surface of the framework [8 - 11]. The "bottle around ship" strategy, also known as template synthesis, entails synthesizing the functional molecules or NPs separately and stabilizing them with capping agents or surfactants (Fig. **1b**). Then the pre-synthesized nanoscale artifacts subsequently add to a synthetic solution containing precursors of MOF to assemble the MOF. The nanoscale objects do not occupy the pore space but are surrounded by grown MOF materials. The problems of the aggregation of NPs on the external surface of MOFs can be reduced by using this synthesis approach as the morphology, size, and structure of trapped NPs can be easily controlled since they are performed before the MOF system is assembled [12]. The insertion of NPs can cause problems with the subsequent growth of the MOF due to the obvious high interfacial energy barrier between the two types of materials. Furthermore, the inclusion of capping agents [*e.g.*, polyvinyl pyrrolidone (PVP)] can reduce the entire exposure of active sites and potentially change or influence NPs efficiency.

Combining the precursor solutions of NPs and MOFs directly together, followed by the simultaneous development of NPs and MOFs, and assembling them all into a nanostructure is a one-step synthesis process (Fig. **1c**). This strategy is straightforward and simple compared to the above stepwise methods, but it usually needs to balance the rates of the self-nucleation and growth of the NPs and MOFs. The choice of functional groups in organic linkers or solvents is critical for trapping NP precursors and stabilizing in situ-formed NPs, as well as facilitating hetero-nucleation of MOFs on the surface of NPs [7, 13].

The difference in synthesis protocols of a specific catalyst might have a significant effect on its catalytic performance. For example, the uniform Pt NPs were successfully supported on or encapsulated within MOF particles to prepare $Pt/UiO-66-NH_2$ and $Pt@UiO-66-NH_2$ catalysts through the solution infiltration technique (the "ship in bottle" approach) and "bottle around ship" approach, respectively [14]. The different synthesis procedure causes a difference in the Pt location relative to MOF, which leads to a considerable difference in its photocatalytic activity. In the Pt@MOF, the internal Pt NPs greatly shorten the electron transfer path from MOF to Pt NPs in comparison to supported Pt. As a result, the utilization of Pt@UiO-66-NH$_2$ showcased a much better charge carrier and magnificently enhanced photocatalytic activity towards hydrogen production compared to Pt/UiO-66-NH$_2$. Moreover, Pt@UiO-66-NH$_2$ exhibited excellent stability and recycling performance due to the great confinement of Pt NPs in the MOF.

PHOTOCATALYTIC APPLICATIONS OF NANOSTRUCTURED IMPREGNATED MOFs

A variety of research attempts have been made to encapsulate MOFs with realistic guest NPs (including noble metal NPs, electrocatalysts, semiconducting metal/metal-oxide NPs, and metal-free semiconductors) in order to construct guest@MOF composites employing core/shell or surface films [15, 16]. The development of guest@MOF as a photocatalyst has three technological goals: faster light absorption by modulating the photo-responsive structure, improved separation of photogenerated charge (e$^-$/h$^+$) carriers and electron accumulation by redox-active metal node regulation, and selection of suitable MOF topologies for guest NP encapsulation with ample cavity to achieve efficient turnover. The functionalized guest NPs may serve as a co-catalyst to enhance visible photon absorption and/or increase active sites to improve e$^-$/h$^+$ separation rate during photocatalysis by producing hierarchical MOF composite structures.

Photocatalytic Removal of Environmental Pollutants

Photocatalytic Degradation of some Industrial and Emerging Contaminants

Zhang and co-workers fabricated Fe$_3$O$_4$NPs encapsulated MIL-100 via the in-situ solvothermal method and employed it for the photodegradation of methylene blue (MB) under visible light irradiation [17] (Fig. **2**). The encapsulated Fe$_3$O$_4$ NPs can be conveniently recovered with good recyclability (by applying a magnetic field) of Fe$_3$O$_4$@MIL-100.

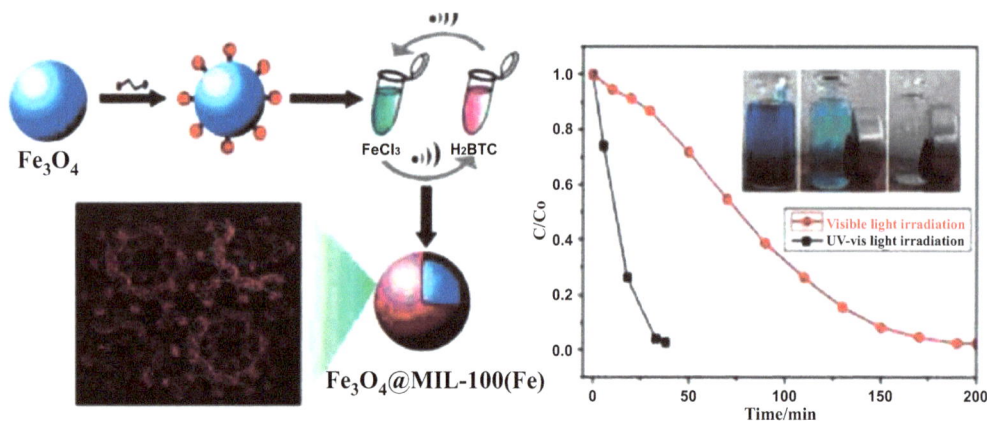

Fig. (2). Illustration of synthesis of Fe_3O_4@MIL-100 (Fe) core-shell microspheres (left) and their application for magnetically separable photocatalytic degradation of MB under UV-vis and visible light [17].

ZnO has also been integrated into various MOFs, *e.g.*, shell ZnO@MOF-46 (E_g = 2.44 eV) and ZnO@ZIF-8 (E_g = 3.24 eV), for the photocatalytic treatment of wastewater (organic dye and Cr (VI) reductions) [16, 18, 19]. Due to the creation of heterostructures, ZnO NPs trapped in MOF pores could narrow the bandgap compared to pristine MOFs. During photoexcitation, under visible irradiation, the reaction kinetics of the photocatalysis was significantly enhanced due to the migration of photo-induced from the LUMO level of the MOF to active sites (electron trap) of ZnO. Similarly, CdS, TiO_2, and BiOBr were impregnated in MOFs like UiO-66 (NH_2), MIL-68(Fe), MIL-125 (Ti), MIL-88B (Fe), CAU-17 (Bi-based MOF: Christian-Albrechts-University), and ZIF-67, to augment the photoresponse activities under visible light for organic pollutants degradation [16]. In the same way, Ag@MIL125 (Ag NPs 40 nm) synthesized via hydrothermal and photoreduction-mediated synthesis routes catalyzes the complete decomposition of RhB dye (*i.e.* 100% removal by 3 wt.% Ag-MIL-125 after 40 min) in the presence of visible light without apparent deactivation over 5 cycles [20]. The noticeable improvement in photodegradation efficiency is possible due to the improved catalytic ability of Ag NPs, which act as a catalyst for the efficient generation of $O_2\bullet^-$ and $\bullet OH$ oxidative radicals to facilitate electron transfer to oxygen. Additionally, Ag-MIL-125-AC (AC = acetylacetone) was synthesized by reacting MIL-125-NH_2 with AC vapors under N_2 for 24 h, which also acted as a reducing agent for $Ag^\pm$ ions in the solution mixture (acetonitrile/water;4:1 v/v) [21]. The developed Ag-MIL-125-AC (Ag NP of 5–10 nm) showed a considerable shift to a visible-light-induced bandgap of 2.09 eV (593 nm) with good photocatalytic activity toward the decolorization of MB dye (99.5% in 90 minutes) during 5 cycles of reuse. The photocatalytic efficiency of NH_2-MIL125 for textile dyes degradation (*e.g.*, RhB dye) was improved under visible light by encapsulating with BiOBr NPs (BiOBr/NH$_2$ $-$MIL-125: photocatalytic rate of 0.1 mmol g^{-1} h^{-1}) [22]. Similarly, the photocatalytic performance of virgin MIL-125 was enhanced by combining it with other semiconducting catalysts in a heterostructured nanocomposite. For example, the development of g-C$_3$N$_4$/MIL-125 [23] and In$_2$S$_3$/MIL-125 core-shell [24] exhibited an improved photodegradation efficiency in visible light irradiation towards RhB dye (0.25 mmol g^{-1} h^{-1}) and tetracycline (0.22 mmol g^{-1} h^{-1}), respectively. In these circumstances, the encapsulated catalysts (*e.g.*, g-C$_3$N$_4$, In$_2$S$_3$, and BiOBr) by increasing the number of active catalytic sites synergistically permit a prolonged lifetime of photogenerated e$^-$/h$^\pm$ pairs (*e.g.*, through inter-valence electron transfer: Ti$^{3\pm}$-Ti$^{4\pm}$). The photocatalytic efficiency of the heterostructure composite (in comparison to the pristine MIL-125) was substantially improved under visible light.

An incipient wetness impregnation procedure was also employed to introduce Ag mono-substituted oxometallate (POM: Ag/PW11O39)-doped AgCl NPs into MIL-101-NH$_2$, followed by photo-reduction to construct visible-driven Ag/POM-AgCl@MIL-101-NH$_2$ photocatalysts [25]. Ag/POM-AgCl@MIL-101-NH$_2$ under visible irradiation was able to complete the photocatalytic reduction of 0.12 mmol/L 4-nitrophenol (*i.e.* 100% in 20 min). Furthermore, Ag/POM-AgCl@MI--101-NH$_2$ showed a high photodegradation efficiency towards 20 ppm RhB dye (88%: 3 runs stability) in 150 min of visible light illumination in comparison to Ag/AgCl@MIL-101-NH$_2$ (28%). This demonstrates that the POM (no visible light absorption) promotes Ag/AgCl particle photoactivity by storing electrons on excited Ag NPs and inhibiting e$^-$/h$^\pm$ recombination.

For photocatalytic reduction of 4-nitrophenol in an aqueous solution, a novel Ce/Tb-doped MOF-76 nanocomposite impregnated with gold and silver nanoparticles (GNPs and SNPs) was produced [26]. Indeed, in contrast to commercial photocatalysts (Degussa P25 TiO$_2$ nanoparticles), these nanocomposites exhibited excellent photoactivity for the reduction of 4-nitrophenol. Among them, GNPs@MOF76 (Ce) (MOF-76(1a)) displayed the maximum enactment for photoreduction of 4-nitrophenol (96.3% removal with an apparent rate constant (K'$_{app}$) = 0.33 min^{-1}) and acceptable recyclability over 3 cycles.

Photocatalytic Removal of Gaseous Air Pollutants

Nanostructured impregnated MOF photocatalysts have been acknowledged as a viable technical option to eliminate toxic air pollutants in indoor environments (*e.g.*, gaseous NO$_x$ and warfare agents). This segment will address the photocatalytic removal of gaseous nitrogen oxides (NO$_x$: NO and NO$_2$) due to their high emission rate from fossil fuels and detrimental impacts on the environment and human health (*e.g.*, respiratory disease and immune system problems) [27]. Heterogeneous MOF photocatalysis performed well in the reduction of NO_x at high pollution levels in urban areas by photocatalytic oxidation of gaseous NO_x to nitrate (NO^{3-}). N/C-QDs/MIL-125 has been applied for photocatalytic removal of NO and possesses the conversion rate of 0.123 mmol g^{-1} h^{-1} for gaseous NO_x to NO^{3-} ion under visible irradiation. Moreover, N/C-QDs/MIL-125 exhibited stable photocatalytic activity of up to 5 sequential reusable cycles [28]. Similar to N/CQDs/MIL-125, to speed up visible-driven photo-conversion of NO_x to NO^{3-} ions, ZIF-8 was impregnated with CQDs co-catalyst (CQDs/ZIF-8(Zn)) using a surface soaking method [29]. CQDs/ZIF8(Zn) composite photocatalytic performance against 420 ppb NO_x was improved by increasing CQDs loading from 0.1 to 0.5 vol.% (*i.e.* 43% conversion compared to 7.1% by bare ZIF-8). Particularly, after three successive cycles, photoactivity of 0.5- CQD/ZIF-8 to NO_x oxidation was reduced by 9%. This lower efficiency could be due to the deactivation of active sites by adsorbed NO^{3-} and/or changes in the ZIF-8 framework property (texture and morphology) during photoreactions. CQDs were critical in boosting photocatalytic conversion of NO_x over ZIF-8 (Zn_4N tetrahedral site) during photocatalysis by improving photocurrent responsiveness (1.6 times greater than pristine ZIF-8) under visible-irradiation (up to 600 nm) and minimizing $e^-/h^{\pm}$ recombination (*i.e.* increasing the number of photogenerated charge carriers).

Photocatalytic Oxidation of Alcohols

The selective oxidation of alcohols, generally regarded as the key reaction in organic chemistry, is effectively catalyzed with nanostructured incorporated MOF composites. For instance, Fu and co-workers developed NH_2-MIL-125 doped with Ni NPs to improve visible-light-driven aerobic oxidation of benzyl alcohol via surface localized light sensitization of catalytically active Ni NPs (Ni^{2+} to Ni^0) [30]. Li and others investigated the catalytic efficiency in fluid-phase aerobic oxidation of alcohols by using a series of MOF composites encapsulated with different metal NPs (Au, Pd and Pt) [31 - 33]. The prepared Au/MIL-101, Pd/UIO-67, and Pt/DUT-5 composite catalysts showed excellent photo-oxidation for a variety of alcohols under base-free conditions, with conversions of over 99% and up to 100% selectivity to cinnamyl aldehyde. As a result, the improved photocatalytic activity was attributed to the designed MOF's synergistic effects of electron donation and confined nano-size characteristics.

Diverse composite structures, such as Au NPs, metal nanoclusters, and CdS, have also been found to have strong catalytic activity for the selective oxidation of various alcohol substrates to aldehydes [34 - 36]. Liu and co-workers, in 2016, reported the fabrication of precise nanoclusters @MIL-101 for the first time by using MOFs as size-confining templates [36]. Highly dispersed $Au_{13}Ag_{12}$@MOF structures showed promising catalytic activity in the oxidation of benzyl alcohol toward benzaldehyde (*i.e.* 75% conversion and 100% selectivity).

Photocatalytic Hydrogen Evolution

The hydrogen evolution reaction (HER) mainly depends on the number of active sites, light absorption capability, bandgap energy, and charge transfer/separation properties of MOF photocatalysts. Further H_2 generation from MOF photocatalysts also depends on the HOMO and LUMO levels relative to water-splitting potential energy. Due to this, co-catalysts and photosensitizers are frequently used to boost the light-harvesting capacity of MOF-based composites and the rate of H_2 generation.

Generally, Au, Pt, Ni, Ag, Cu, and Pd NPs possessing high Fermi energy levels have been usually employed as active co-catalysts to increase electron transfer [37 - 39]. Encapsulated metal NPs (MNPs) promote photocatalysis by improving the e^-/h^+ separation [16]. The selection of appropriate reducing and capping agents is critical during the fabrication of MNPs@MOFs in order to control the morphologies and distributions of MNPs while avoiding aggregation on the MOFs' outer surfaces. These capping agents impact MNPs and MOFs' binding sites while maintaining the porosity of MOFs [40]. For instance, the double-solvent approach is used to make a Pt@NH$_2$-MIL-125 composite photocatalyst [41]. This procedure involves the formation of an immiscible mixture by slowly adding an aqueous K_2PtCl solution in methanol-water (9:1 v/v, hydrophilic solvent) to a suspension of NH$_2$-MIL-125 in 10 ml hexane solvent (hydrophobic solvent). The hydrophilic solvent aids the uptake of hydrophilic precursor (Pt) by the hydrophilic interior pores of MOFs during sonication, allowing Pt to be impregnated. Due to the significant electron spillover induced by Pt NPs (4 nm), which favourably lowered the photo-charge recombination rate, Pt@NH$_2$-MI--125 displayed prolonged visible light absorption up to 500 nm with enhanced photocurrent stability (80% for 1 h) during H_2 production. Pt, Au, and Pd NPs were encapsulated in pristine MIL125(Ti) without the use of stabilizing/reducing agents during the synthesis process [42]. The photo-reduction of MIL-125(Ti) under N_2 and UV–Vis. light (320–780 nm) was used to prepare $Ti^{3\pm}$-MIL125 ($Ti^{3\pm}$ is a strong reducer with -1.37 V *vs.* SHE). Due to the strong redox power of $Ti^{3\pm}$, direct *in-situ* redox reactions between $Ti^{3\pm}$-MIL-125 (reductive) and noble metal salts (oxidative) were performed in the dark. Thus, M/MIL-125 (M = Pt, Au, and Pd NPs) has improved photocatalysis in terms of H_2 generation (up to 40 μmol for

5 h) and selective oxidation of benzyl alcohol (conversion of 26.4–36%) to benzene aldehyde relative to MIL-125 ($\leq$ 5 μmol H_2 and 18.9% conversion). Furthermore, under visible light, Ni/NiO$_x$ particles were photo-deposited on MIL-101 with erythrosine B sensitization for improved H_2 production (82 μmol h^{-1}) (Liu *et al*. 2015). The density functional theory (DFT) calculations revealed that NiMo@MIL101 nanocomposites retained good photocatalytic stability for H_2 production (740.2 μmol h^{-1}) with a high quantum efficiency (QE) of 75.7% at 520 nm illumination at pH 7 [43].

Carbondioxide (CO₂) Photoreduction

ZIF-8 impregnated with Ti/TiO$_2$ nanotubes (Ti/TiO$_2$@ZIF-8) was synthesized for both adsorption/photo-electrocatalytic conversion of CO_2 to methanol/ethanol under UV-visible irradiation at room temperature [44]. The Ti/TiO$_2$@ZIF-8 complex showcased high adsorption of CO_2 through carbamate formation due to the presence of the imidazole ligand, as confirmed by spectroscopic and voltammetric analysis. In the presence of 0.1 M Na$_2$SO$_4$, ZIF-8 served as a co-catalyst under UV–vis irradiation, mediating energy transfer from excited Ti/TiO$_2$ NTs to CO_2 during photo-electrocatalysis for the production of ethanol (10 mM) and methanol (0.7 mM). The rod-like Ag NP-impregnated Co-ZIF-9 also reported a photocatalytic conversion of CO_2 to CO under visible light illumination [45]. The Ag@Co-ZIF-9 MOF showed double enhancement in photocatalytic efficiency at a rate of 5.68×104 μmol g^{-1} h^{-1} and good photo-stability compared to pristine Co-ZIF-9 after 30 minutes of illumination. The increased photocatalytic reduction was attributed to the possible electron trap effects of doped Ag NPs, which speed up the electron transfer to CO_2 during photocatalysis.

NH$_2$-MIL-125(Ti) encapsulation with noble metal NPs (Pt- and Au-NPs, as electron trap) has also been examined to show increased activation of the redox sites for CO_2 reduction to formate in the presence of TEOA [46]. The measured formate was found to decrease in the order of Pt/NH$_2$-MIL125 (32.4 μmol g^{-1} h^{-1}) $\geq$ NH$_2$–MIL-125 (26.3 μmol g^{-1} h^{-1}) $\geq$ Au/NH$_2$–MIL-125 (22.5 μmol g^{-1} h^{-1}) after 8 h of visible light irradiation. Furthermore, Pt- and Au- doped NH$_2$-MIL-125(Ti) had a synergetic effect on HER (H_2 evolution rate $\approx$ 0.56 and 100 μmol g^{-1} h^{-1}, respectively), resulting in significant CO_2 reduction. The catalytic inferiority of Pt/NH$_2$-MIL-125 may be accountable for the hydrogen spillover mechanism over Pt NPs (serving as redox sites) along with a stronger Pt-O bond (391.6 kJ mol^{-1}) than that of Au-O bonds (221.8 kJ mol^{-1}) within the framework [47].

SENSING APPLICATIONS OF NANOSTRUCTURED IMPREGNATED MOFS

Sensing of Organic Pollutants

Nanostructure-impregnated MOFs are promising candidates for analytical sensing applications due to their unique intrinsic properties. Feng *et al.* prepared 1D carbon fiber/ZIF-67 composites and investigated the effect of the structure of carbon fibers on the catalytic activity of ZIF-67 [48] (Fig. **3**). The porous carbon fibers (PCFs) with hollow structures formed by simply carbonization of natural catkins significantly limited the growth and aggregation of ZIF-67 nanocrystals compared to solid carbon fibers (SCF), allowing ZIF-67/PCF to have more exposed active sites, better conductivity and greater channels of transport. Thus, ZIF-67/PCF composites exhibited remarkable electrocatalytic sensing of L-Cysteine and nitrobenzene, which included broad linear detection, low overpotential, high sensitivity, and excellent stability, selectivity and anti-interference ability.

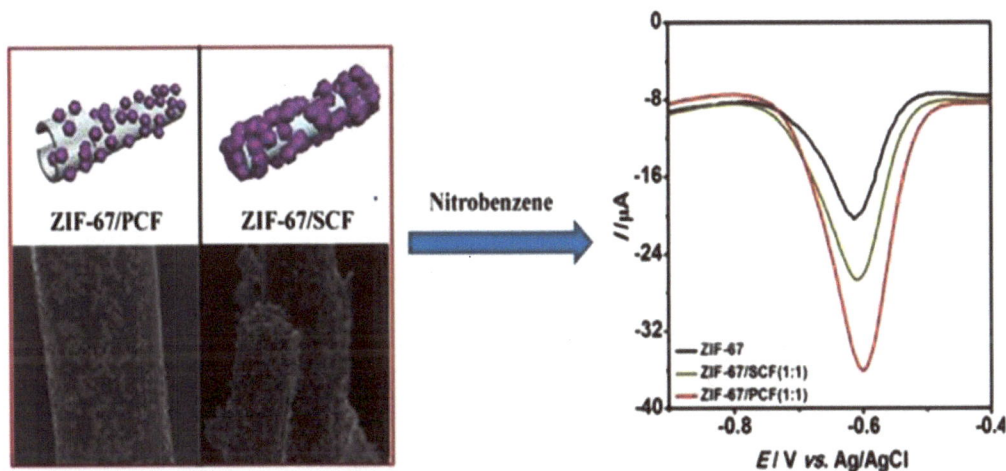

Fig. (3). Morphology of ZIF-67/carbon fiber composites and the detection of nitrobenzene [48].

For 2D graphene nanosheet/MOF composites, various modified forms of graphene, such as graphene nanoribbons (GNRs) [49], chemically reduced GO (rGO) [50], graphene oxide (GO) [51], and electrochemically reduced GO (ERGO) [52] have been applied in electrochemical sensing. To date, various techniques, such as solution diffusion [53], *in situ* growth [54], ultrasound dispersion [55], and electrodeposition [56], have been proposed to synthesize MOFs/GN composites for electrochemical sensor fabrication. For example, Wang *et al.* proposed a simple ultrasonication procedure for the synthesis of Cu-TPA

MOF/GO composites and employed it to detect dopamine and acetaminophen [51]. The binding mechanism of Cu-TPA and GO was assumed to be the synergistic interaction of hydrogen bonding, p-p stacking interactions, and Cu–O coordination.

Liu *et al.* also prepared a high sensitivity/selective HKUST-1/EGRO electrode for the determination of dihydroxybenzene isomers (DBI) [57]. Zhang and co-workers developed MIL-101(Cr)/rGO composites for 4-nonylphenol sensing. However, the MOF/GN composites synthesized via a simple ultrasonic method were not stable and effective enough for electron transfer [50]. To eradicate these disadvantages, Wan *et al.* reported the preparation of HKUST-1/GN composites by *in situ* growing HKUST-1 on liquid-phase exfoliated graphene nanosheets at room temperature [54]. This method provided the *in-situ* immobilization and uniform dispersion of ultra-small HKUST-1 nanoparticles on the surface of GN nanosheets, with excellent electron transfer capabilities and significantly improved adsorptiveness towards 8-hydroxy-2-deoxyguanosine, displaying high response signals and detecting sensitivities. Furthermore, Wu *et al.* developed ZIF-67/GN hybrids on acidified GN nanosheets through *in situ* liquid-phase growth of the ZIF-67 (Fig. **4**) [58]. Polyhedral ZIF-67 nanoparticles having a size of approximately 300 nm were loaded on each side of the GN nanosheets to form a ZIF-67/GN sandwiched heterostructure. Compared with ZIF-67 and the physical mixture of GN + ZIF-67, the as-synthesized ZIF-67/GN hybrids exhibited the largest electrode active area and the highest electrochemical activity toward glucose oxidation. These data further showed that the acidified graphene and ZIF-67 had been more evenly and precisely attached to each other using the in-situ synthetic procedure, leading to better electron conduction than the simple physical mixing approach. As a result, the as-synthesized sandwich-like ZIF-67/GN heterostructure was used to fabricate a highly sensitive electrochemical sensor for glucose oxidation with a linear detection range of 1–805.5l M, the detection limit was 0.36l M (S/N = 3), and the sensor exhibited remarkable selectivity and stability. Zheng *et al.* prepared ZIF-8/3D graphene composites via a step-by-step route by unceasingly immersing 3D graphene in $ZnCl_2$ solutions and ligand solutions at room temperature, and this process produced sensors with enhanced electroanalytical performances towards the detection of dopamine [53]. Xie *et al.* further introduced polypyrrole as linkers to enhance the binding between the graphene aerogels (GAs) and ZIF-8 and prepared 3D heterostructured PPy@ZIF-8/GAs composites, exhibiting ultrasensitive detection of dichlorophenol down to 0.1 nM [59].

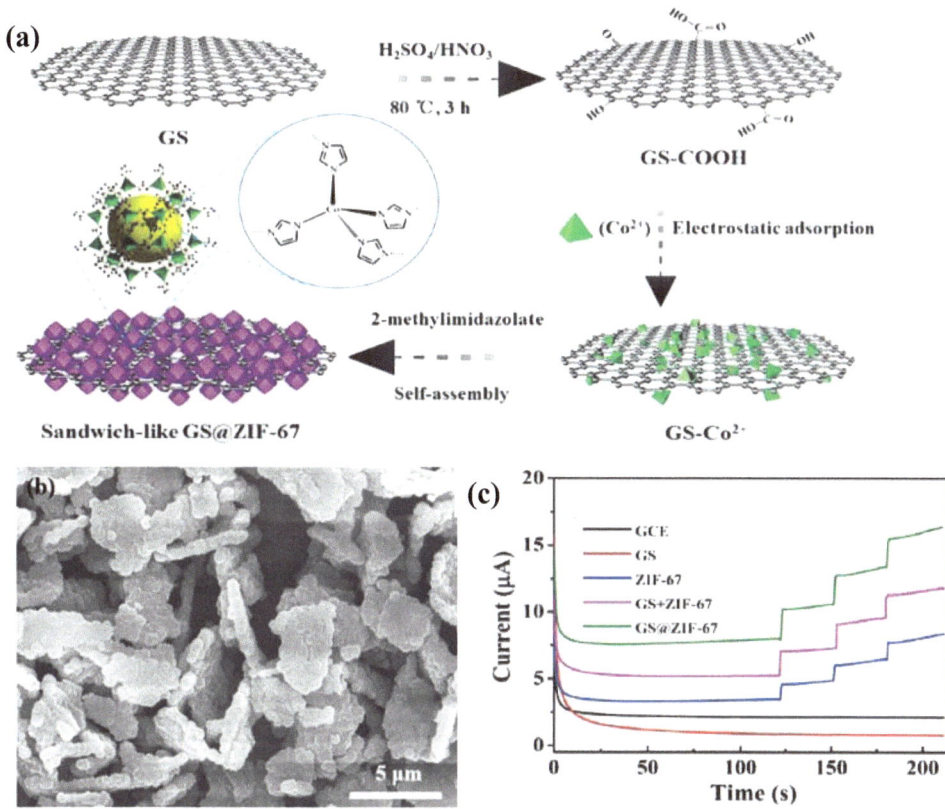

Fig. (4). (a) Preparation, **(b)** SEM image of the sandwich-like ZIF-67/GN hybrids, and **(c)** Current response of different electrodes towards successive injection of glucose [58].

Sensing of Gaseous Pollutants

By tuning the topological structure of the MOFs, the sensitiveness of pristine MOF-based electrochemical sensors can be enhanced. However, these pristine MOF sensors continue to suffer from certain constraints, such as extreme aggregation, poor conductivity, and lower electrocatalytic abilities. In order to overcome these shortcomings, an efficient way is to build MOFs based hybrid structures, such as hybrid MOFs, with carbon materials to achieve high-conductivity or hybrid MOFs with noble metal nanoparticles to achieve synergistic capabilities of catalytic action. Hybrid nanostructures will provide a short path between each component, which facilitate the transfer of electron and prevent the addition of individual components, creating more exposed active sites. MOF-based nanocomposite immobilized with noble metal NPs or optical fiber may demonstrate high performance in the detection of hydrogen or ethanol [60 -

62]. Recently, Surya *et al.* reported an Ag@UiO-66 composite for efficient H_2S gas sensing [63].

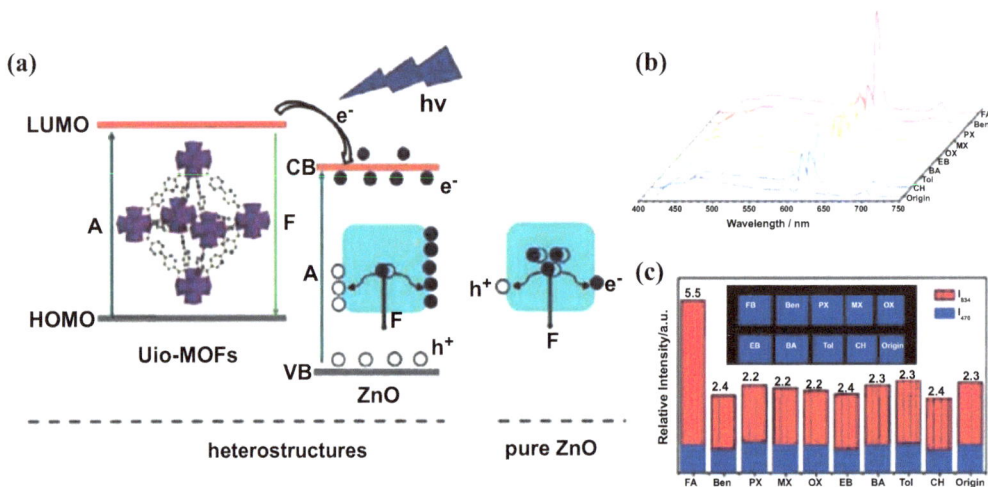

Fig. (5). (a) A simplified schematic diagram of the energy transfer between MOFs and ZnO in ZnO@UiO-MOF heterostructures and the effect of the presence or absence of UiO-MOF on the possible fluorescence emission mechanism of ZnO. **(b)** After exposure to various VOC gases in vehicles (lex ¼ 365 nm), the photoluminescence emission spectra and **(c)** intensities (I_{614} and I_{470}) of prepared Eu^{3+}@ZUM test paper. The inset shows the images under 365 nm UV light irradiation, and the numbers correspond to the ratio of I_{614} to I_{470} [64].

Xu and Yan developed a ppb-level sensing platform using Eu(III)-functionalized ZnO@MOF heterostructures [64]. ZnO nanoparticles were doped onto UiO-MOF via a simple solvothermal process, resulting in a ZnO@MOF heterostructure and post-synthesis of Eu^{3+}(Eu^{3+}@ZUM). In this way, a novel fluorescence sensor based on the heterostructure was fabricated, which was able to identify volatile aldehyde gas at room temperature. UiO-MOF has been developed using solvothermal approaches by mixing $ZrCl_4$, H_2bpydc and glacial acetic acid. MOFs acted as gas pre-concentrators, whereas ZnO NPs acted as reaction and charge centres through intense reactions between the reactive oxygen species on their surface and aldehyde molecules. Eu^{3+} acted as charge transfer centres, transferring charge from reaction centres to fluorescence sensing signals (Fig. **5a**). Eu^{3+}@ZUM was integrated into the filter paper for single VOCs detection by the sensor; it was exposed to 10 ppm of the main VOC gases (xylene, formaldehyde, toluene, benzene, butyl acetate, and cyclohexane) at 25°C for 30 min before photoluminescence measurements. For the detection of a mixed system, a mixture of different aldehydes (10 ppm) and other interfering gases (20 ppm) was exposed to Eu^{3+}@ZUM test paper at 25°C for 30 min. The results are shown in (Fig. **5b** and **c**). The Eu^{3+}@ZUM sensor exhibited better anti-interference in the mixed gas atmosphere, and the photoluminescence of formaldehyde displayed substantial reversible changes. The sensor exhibited high selectivity, sensitivity (the formaldehyde detection limit was 42 ppb), and reusability (the fluorescence intensity remained approximately unchanged after five consecutive runs). Moreover, detecting multiple pollutants simultaneously has a more realistic implementation benefit than detecting a single pollutant.

The Pd/ZIF composites integrating selective and responsive Pd nanowires with the molecule sieving MOF are the most representative example of chemiresistor. The selective detection of Pd against H_2 depends on the particular change in resistance resulting from the formation of a solid Pd/H or palladium hydride. The high surface area and well-formed interconnection of Pd nanotubes allow for a highly sensitive response toward H_2 at room temperature (R1000 ppm= 1000%), with a response time of ~3.17 min due to the slower speed of Pd/H compared to palladium hydride, which occurs in the range of 1–2% at room temperature. ZIF-8/MAF-4 is a subfamily of MOFs possessing good chemical and thermal stability, as well as versatile channels for gas adsorption and separation [65]. Accordingly, I.-D. Kim's group and R. M. Penner's group fabricated Pd/ZIF-8 NW bilayered thick film [66] (Fig. **6a**). Surprisingly, the acceleration effect of highly porous ZIF-8 gives rise to a remarkable response (~0.22 min) and recovery speed (~0.10 min) of Pd/ZIF-8 sensors toward 1000 ppm of H_2 (Fig. **6b, c** and **d**). Furthermore, the micropores (0.34 nm) of ZIF-8 may act as a molecular sieving layer in terms of kinetic parameters. Although the difference in kinetic diameter between H_2 (0.298 nm) and interfering O_2 (0.346 nm) makes rapid diffusion of H_2 than O_2 in ZIF-8, ZIF-8 does not separate a gas with a kinetic diameter smaller than 0.7 nm. Some related works on MOF-metal composites, such as Pt@Cu-HHTP and Au@ZnO@ZIF-8, were also designed for the detection of other gaseous NO_2 and HCHO [56, 67, 68], respectively. The Au@ZnO@ZIF-8 sensor with catalytic Au and porous MOF shell was illuminated by visible light irradiation and demonstrated a selective response to HCHO and strong anti-interference properties towards gasses, such as H_2O and toluene.

Sensing of Heavy Metal Ions

Semiconductor quantum dots (QDs) are a type of zero-dimensional nanomaterial with a standard size of 2-20 nm. Due to their unique features, such as wide excitation band, tunable emission, stable fluorescence, and large Stokes shift, QDs have been used in the field of fluorescence. By incorporating QDs into the pores of MOF, the composite materials displayed superior characteristics to individual QDs. For example, $CH_3NH_3PbBr_3$ QDs were encapsulated in MOF-5 to obtain $CH_3NH_3PbBr_3$@MOF-5 composite material with excellent water resistance and thermal stability, and the material showcased green fluorescence under 365 nm UV light illumination [55]. The pores of the MOF-5 framework provide excellent protection for QDs, hence $CH_3NH_3PbBr_3$@MOF-5 have better stability and a wider pH range than $CH_3NH_3PbBr_3$ QDs. $CH_3NH_3PbBr_3$@MOF-5 composites are stable fluorescence probes for the detection of Bi^{3+}, Al^{3+}, Cu^{2+}, Co^{2+}, Fe^{3+}, and Cd^{2+} ions in aqueous solution due to their different quenching luminescence performances. The material temperature has a strong linear relationship with the fluorescence intensity, which can be used to indicate the temperature in the range of 30–230 °C. The indicator of this material has good reversibility and provides more advantages than other normal thermometers.

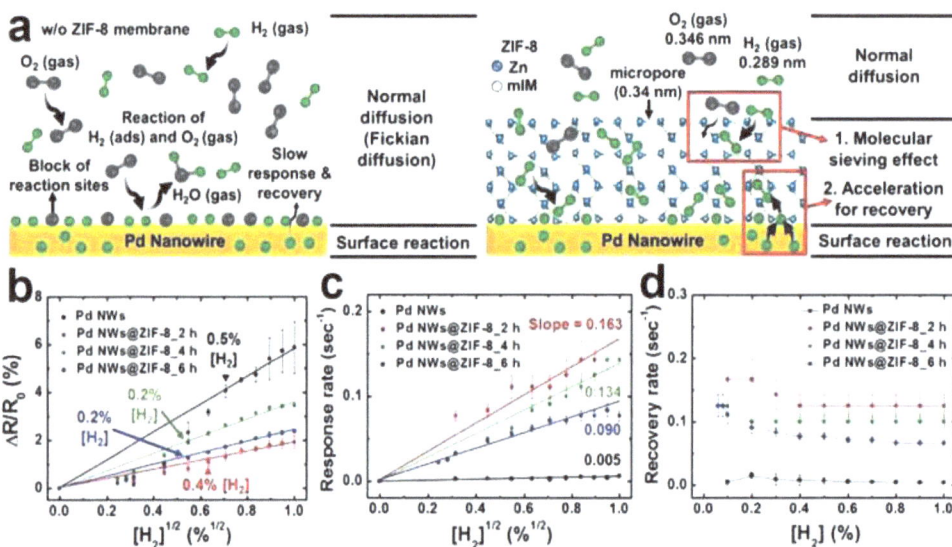

Fig. (6). (a) Sensing model for Pd NWs without ZIF-8 membrane and Pd NWs/ZIF-8. (b) Response versus $[H_2^{1/2}]$ in air, (c) response rate versus $[H_2^{1/2}]$ in air, and (d) recovery rate versus $[H_2]$ [66].

CONCLUSION AND OUTLOOK

In this analysis, we summarized recent developments of nano-structured impregnated MOFs with clearly defined and tunable crystal structure, chemical flexibility, excellent biocompatibility, appropriate and physiological stability, uniform pore structures, high surface areas, and specially confined nanopore microenvironments. However, MOFs are structurally flexible materials that can be readily tuned to synthetic conditions, such as pressure, temperature, salinity, pH, and light intensity. Owing to the intrinsic properties of smaller nanostructured impregnated MOFs, they exhibit great potential in photocatalyst and sensing applications. Many recent studies have confirmed the outstanding performance of nanostructured-impregnated MOFs in diverse fields. The progress of nanostructured impregnated MOFs has been achieved by a plethora of research efforts. However, there is still a long way to go to see more rational and general approaches to synthesizing nanostructured impregnated MOFs in diverse fields and thus to serve for enhanced performances. In this regard, future research should be directed towards the development of novel nanostructured impregnated MOFs that can help to scale up exciting results from the laboratory to real-world applications.

REFERENCES

[1]   Lu, F.; Astruc, D. Nanomaterials for removal of toxic elements from water. *Coord. Chem. Rev.,* **2018**, *356*, 147-164.
[http://dx.doi.org/10.1016/j.ccr.2017.11.003]

[2]   Rawtani, D.; Rao, P.K.; Hussain, C.M. Recent advances in analytical, bioanalytical and miscellaneous applications of green nanomaterial. *Trends Analyt. Chem.,* **2020**, *133*, 116109.
[http://dx.doi.org/10.1016/j.trac.2020.116109]

[3]   Li, H.Y.; Zhao, S.N.; Zang, S.Q.; Li, J. Functional metal–organic frameworks as effective sensors of gases and volatile compounds. *Chem. Soc. Rev.,* **2020**, *49*(17), 6364-6401.
[http://dx.doi.org/10.1039/C9CS00778D] [PMID: 32749390]

[4]   Xiang, W.; Zhang, Y.; Lin, H.; Liu, C. Nanoparticle/metal–organic framework composites for catalytic applications: current status and perspective. *Molecules,* **2017**, *22*(12), 2103.
[http://dx.doi.org/10.3390/molecules22122103] [PMID: 29189744]

[5]   Yu, J.; Mu, C.; Yan, B.; Qin, X.; Shen, C.; Xue, H.; Pang, H. Nanoparticle/MOF composites: Preparations and applications. *Mater. Horiz.,* **2017**, *4*, 557-569.
[http://dx.doi.org/10.1039/C6MH00586A]

[6]   Zhu, Q.L.; Xu, Q. Metal–organic framework composites. *Chem. Soc. Rev.,* **2014**, *43*(16), 5468-5512.
[http://dx.doi.org/10.1039/C3CS60472A] [PMID: 24638055]

[7]   Yang, Q.; Xu, Q.; Jiang, H.L. Metal-organic frameworks meet metal nanoparticles: synergistic effect for enhanced catalysis. *Chem. Soc. Rev.,* **2017**, *46*(15), 4774-4808.
[http://dx.doi.org/10.1039/C6CS00724D]

[8]   Roy, S.; Pachfule, P.; Xu, Q. High Catalytic Performance of MIL-101-Immobilized NiRu Alloy Nanoparticles towards the Hydrolytic Dehydrogenation of Ammonia Borane. *Eur. J. Inorg. Chem.,* **2016**, *2016*, 4353-4357.
[http://dx.doi.org/10.1002/ejic.201600180]

[9]   Li, J.; Zhu, Q.L.; Xu, Q. Highly active AuCo alloy nanoparticles encapsulated in the pores of metal–organic frameworks for hydrolytic dehydrogenation of ammonia borane. *Chem. Commun. (Camb.),* **2014**, *50*(44), 5899-5901.
[http://dx.doi.org/10.1039/c4cc00785a] [PMID: 24760206]

[10]   Li, J.; Zhu, Q.L.; Xu, Q. Non-noble bimetallic CuCo nanoparticles encapsulated in the pores of metal-organic frameworks: Synergetic catalysis in the hydrolysis of ammonia borane for hydrogen generation. *Catal. Sci. Technol.,* **2015**, *5*, 525-530.
[PMID: 10.1039/C4CY01049C]

[11]   Zhen, W.; Gao, F.; Tian, B.; Ding, P.; Deng, Y.; Li, Z.; Gao, H.; Lu, G. Enhancing activity for carbon dioxide methanation by encapsulating (1 1 1) facet Ni particle in metal–organic frameworks at low temperature. *J. Catal.,* **2017**, *348*, 200-211.
[http://dx.doi.org/10.1016/j.jcat.2017.02.031]

[12]   Aguilera-Sigalat, J.; Bradshaw, D. Synthesis and applications of metal-organic framework-quantum dot (QD@MOF) composites. *Coord. Chem. Rev.,* **2016**, *307*, 267-291.
[PMID: 10.1016/j.ccr.2015.08.004]

[13]   Chen, L.; Luque, R.; Li, Y. Controllable design of tunable nanostructures inside metal–organic frameworks. *Chem. Soc. Rev.,* **2017**, *46*(15), 4614-4630.
[http://dx.doi.org/10.1039/C6CS00537C] [PMID: 28516998]

[14]   Xiao, J.D.; Shang, Q.; Xiong, Y.; Zhang, Q.; Luo, Y.; Yu, S.H.; Jiang, H.L. Boosting photocatalytic hydrogen production of a metal-organic framework decorated with platinum nanoparticles: The platinum location matters. *Angew. Chem. Int. Ed.,* **2016**, *55*(32), 9389-9393.
[http://dx.doi.org/10.1002/anie.201603990] [PMID: 27321732]

[15]   Zhang, T.; Jin, Y.; Shi, Y.; Li, M.; Li, J.; Duan, C. Modulating photoelectronic performance of metal–organic frameworks for premium photocatalysis. *Coord. Chem. Rev.,* **2019**, *380*, 201-229.
[http://dx.doi.org/10.1016/j.ccr.2018.10.001]

[16]   Jiang, D.; Xu, P.; Wang, H.; Zeng, G.; Huang, D.; Chen, M. Strategies to improve metal-organic frameworks photocatalyst's performance for degradation of organic pollutants. *Coord. Chem. Rev.,* **2018**, *376*, 449-466.
[PMID: 10.1016/j.ccr.2018.08.005]

[17]   Zhang, C.F.; Qiu, L.G.; Ke, F.; Zhu, Y.J.; Yuan, Y.P.; Xu, G.S.; Jiang, X. A novel magnetic recyclable photocatalyst based on a core–shell metal–organic framework Fe3O4@MIL-100(Fe) for the decolorization of methylene blue dye. *J. Mater. Chem. A Mater. Energy Sustain.,* **2013**, *1*(45), 14329-14334.
[http://dx.doi.org/10.1039/c3ta13030d]

[18]   Rad, M.; Dehghanpour, S. ZnO as an efficient nucleating agent and morphology template for rapid, facile and scalable synthesis of MOF-46 and ZnO@MOF-46 with selective sensing properties and enhanced photocatalytic ability. *RSC Advances,* **2016**, *6*(66), 61784-61793.
[http://dx.doi.org/10.1039/C6RA12410K]

[19]   Wang, X.; Liu, J.; Leong, S.; Lin, X.; Wei, J.; Kong, B.; Xu, Y.; Low, Z.X.; Yao, J.; Wang, H. Rapid Construction of ZnO@ZIF-8 Heterostructures with Size-Selective Photocatalysis Properties. *ACS Appl. Mater. Interfaces,* **2016**, *8*(14), 9080-9087.
[http://dx.doi.org/10.1021/acsami.6b00028] [PMID: 26998617]

[20]   Guo, H.; Guo, D.; Zheng, Z.; Weng, W.; Chen, J. Visible-light photocatalytic activity of Ag@MIL-125(Ti) microspheres. *Appl. Organomet. Chem.,* **2015**, *29*(9), 618-623.
[http://dx.doi.org/10.1002/aoc.3341]

[21]   Abdelhameed, R.M.; Simões, M.M.Q.; Silva, A.M.S.; Rocha, J. Enhanced photocatalytic activity of MIL-125 by post-synthetic modification with Cr(III) and Ag nanoparticles. *Chemistry,* **2015**, *21*(31), 11072-11081.
[http://dx.doi.org/10.1002/chem.201500808] [PMID: 26095013]

[22]   Zhu, S.R.; Liu, P.F.; Wu, M.K.; Zhao, W.N.; Li, G.C.; Tao, K.; Yi, F.Y.; Han, L. Enhanced photocatalytic performance of BiOBr/NH$_2$-MIL-125(Ti) composite for dye degradation under visible light. *Dalton Trans.,* **2016**, *45*(43), 17521-17529.
[http://dx.doi.org/10.1039/C6DT02912D] [PMID: 27747336]

[23]   Wang, H.; Yuan, X.; Wu, Y.; Zeng, G.; Chen, X.; Leng, L.; Li, H. Synthesis and applications of novel graphitic carbon nitride/metal-organic frameworks mesoporous photocatalyst for dyes removal. *Appl. Catal. B,* **2015**, *174-175*, 445-454.
[http://dx.doi.org/10.1016/j.apcatb.2015.03.037]

[24]   Wang, H.; Yuan, X.; Wu, Y.; Zeng, G.; Dong, H.; Chen, X.; Leng, L.; Wu, Z.; Peng, L. In situ synthesis of In2S3@MIL-125(Ti) core–shell microparticle for the removal of tetracycline from wastewater by integrated adsorption and visible-light-driven photocatalysis. *Appl. Catal. B,* **2016**, *186*, 19-29.

[http://dx.doi.org/10.1016/j.apcatb.2015.12.041]

[25]   Yan, J.; Zhou, W.Z.; Tan, H.; Feng, X.J.; Wang, Y.H.; Li, Y.G. Ultrafine Ag/polyoxometalate-doped AgCl nanoparticles in metal–organic framework as efficient photocatalysts under visible light. *CrystEngComm,* **2016**, *18*(45), 8762-8768.
        [http://dx.doi.org/10.1039/C6CE02021F]

[26]   Singh, K.; Kukkar, D.; Singh, R.; Kukkar, P.; Bajaj, N.; Singh, J.; Rawat, M.; Kumar, A.; Kim, K.H. In situ green synthesis of Au/Ag nanostructures on a metal-organic framework surface for photocatalytic reduction of p-nitrophenol. *J. Ind. Eng. Chem.,* **2020**, *81*, 196-205.
        [http://dx.doi.org/10.1016/j.jiec.2019.09.008]

[27]   Sivaramakrishnan, K. Investigation on performance and emission characteristics of a variable compression multi fuel engine fuelled with Karanja biodiesel–diesel blend. *Egyptian Journal of Petroleum,* **2018**, *27*(2), 177-186.
        [http://dx.doi.org/10.1016/j.ejpe.2017.03.001]

[28]   Chen, M.; Wei, X.; Zhao, L.; Huang, Y.; Lee, S.; Ho, W.; Chen, K. Novel N/Carbon Quantum Dot Modified MIL-125(Ti) Composite for Enhanced Visible-Light Photocatalytic Removal of NO. *Ind. Eng. Chem. Res.,* **2020**, *59*(14), 6470-6478.
        [http://dx.doi.org/10.1021/acs.iecr.9b06816]

[29]   Wei, X.; Wang, Y.; Huang, Y.; Fan, C. Composite ZIF-8 with CQDs for boosting visible-light-driven photocatalytic removal of NO. *J. Alloys Compd.,* **2019**, *802*, 467-476.
        [http://dx.doi.org/10.1016/j.jallcom.2019.06.086]

[30]   Fu, Y.; Sun, L.; Yang, H.; Xu, L.; Zhang, F.; Zhu, W. Visible-light-induced aerobic photocatalytic oxidation of aromatic alcohols to aldehydes over Ni-doped NH2-MIL-125 (Ti). *Appl. Catal. B,* **2016**, *187*, 212-217.
        [PMID: 10.1016/j.apcatb.2016.01.038]

[31]   Liu, H.; Chang, L.; Chen, L.; Li, Y. In situ one-step synthesis of metal–organic framework encapsulated naked Pt nanoparticles without additional reductants. *J. Mater. Chem. A Mater. Energy Sustain.,* **2015**, *3*(15), 8028-8033.
        [http://dx.doi.org/10.1039/C5TA00030K]

[32]   Liu, H.; Liu, Y.; Li, Y.; Tang, Z.; Jiang, H. Metal−organic framework supported gold nanoparticles as a highly active heterogeneous catalyst for aerobic oxidation of alcohols. *J. Phys. Chem. C,* **2010**, *114*(31), 13362-13369.
        [http://dx.doi.org/10.1021/jp105666f]

[33]   Chen, L.; Chen, H.; Luque, R.; Li, Y. Metal−organic framework encapsulated Pd nanoparticles: towards advanced heterogeneous catalysts. *Chem. Sci. (Camb.),* **2014**, *5*(10), 3708-3714.
        [http://dx.doi.org/10.1039/C4SC01847H]

[34]   Shen, L.; Liang, S.; Wu, W.; Liang, R.; Wu, L. CdS-decorated UiO–66(NH2) nanocomposites fabricated by a facile photodeposition process: an efficient and stable visible-light-driven photocatalyst for selective oxidation of alcohols. *J. Mater. Chem. A Mater. Energy Sustain.,* **2013**, *1*(37), 11473.
        [http://dx.doi.org/10.1039/c3ta12645e]

[35]   Luan, Y.; Qi, Y.; Gao, H.; Zheng, N.; Wang, G. Synthesis of an amino-functionalized metal–organic framework at a nanoscale level for gold nanoparticle deposition and catalysis. *J. Mater. Chem. A Mater. Energy Sustain.,* **2014**, *2*(48), 20588-20596.
        [http://dx.doi.org/10.1039/C4TA04311A]

[36]   Liu, L.; Song, Y.; Chong, H.; Yang, S.; Xiang, J.; Jin, S.; Kang, X.; Zhang, J.; Yu, H.; Zhu, M. Size-confined growth of atom-precise nanoclusters in metal–organic frameworks and their catalytic applications. *Nanoscale,* **2016**, *8*(3), 1407-1412.
        [http://dx.doi.org/10.1039/C5NR06930K] [PMID: 26669234]

[37]   Bouhadoun, S.; Guillard, C.; Dapozze, F.; Singh, S.; Amans, D.; Bouclé, J.; Herlin-Boime, N. One step synthesis of N-doped and Au-loaded TiO2 nanoparticles by laser pyrolysis: Application in

photocatalysis. *Appl. Catal. B,* **2015**, *174-175*, 367-375.
[http://dx.doi.org/10.1016/j.apcatb.2015.03.022]

[38]   Nguyen, N.T.; Altomare, M.; Yoo, J.; Schmuki, P. Efficient photocatalytic H2 evolution: controlled dewetting–dealloying to fabricate site-selective high-activity nanoporous Au particles on highly ordered TiO2 nanotube arrays. *Adv. Mater.,* **2015**, *27*(20), 3208-3215.
[http://dx.doi.org/10.1002/adma.201500742] [PMID: 25872758]

[39]   Zhou, W.; Li, T.; Wang, J.; Qu, Y.; Pan, K.; Xie, Y.; Tian, G.; Wang, L.; Ren, Z.; Jiang, B.; Fu, H. Composites of small Ag clusters confined in the channels of well-ordered mesoporous anatase TiO$_2$ and their excellent solar-light-driven photocatalytic performance. *Nano Res.,* **2014**, *7*(5), 731-742.
[http://dx.doi.org/10.1007/s12274-014-0434-y]

[40]   Dhakshinamoorthy, A.; Asiri, A.M.; Garcia, H. Metalorganic frameworks as versatile hosts of Au nanoparticles in heterogeneous catalysis. *ACS Catal.,* **2017**, *7*(4), 2896-2919.
[http://dx.doi.org/10.1021/acscatal.6b03386]

[41]   Hou, C.; Xu, Q.; Wang, Y.; Hu, X. Synthesis of Pt@NH2-MIL-125(Ti) as a photocathode material for photoelectrochemical hydrogen production. *RSC Advances,* **2013**, *3*(43), 19820-19823.
[http://dx.doi.org/10.1039/c3ra43188f]

[42]   Shen, L.; Luo, M.; Huang, L.; Feng, P.; Wu, L. A clean and general strategy to decorate a titanium metal-organic framework with noble-metal nanoparticles for versatile photocatalytic applications. *Inorg. Chem.,* **2015**, *54*(4), 1191-1193.
[http://dx.doi.org/10.1021/ic502609a] [PMID: 25594784]

[43]   Zhen, W.; Gao, H.; Tian, B.; Ma, J.; Lu, G. Fabrication of low adsorption energy Ni–Mo cluster cocatalyst in metal–organic frameworks for visible photocatalytic hydrogen evolution. *ACS Appl. Mater. Interfaces,* **2016**, *8*(17), 10808-10819.
[http://dx.doi.org/10.1021/acsami.5b12524] [PMID: 27070204]

[44]   Cardoso, J.C.; Stulp, S.; de Brito, J.F.; Flor, J.B.S.; Frem, R.C.G.; Zanoni, M.V.B. MOFs based on ZIF-8 deposited on TiO$_2$ nanotubes increase the surface adsorption of CO$_2$ and its photoelectrocatalytic reduction to alcohols in aqueous media. *Appl. Catal. B,* **2018**, *225*, 563-573.
[http://dx.doi.org/10.1016/j.apcatb.2017.12.013]

[45]   Chen, M.; Han, L.; Zhou, J.; Sun, C.; Hu, C.; Wang, X.; Su, Z. Photoreduction of carbon dioxide under visible light by ultra-small Ag nanoparticles doped into Co-ZIF-9. *Nanotechnology,* **2018**, *29*(28), 284003.
[http://dx.doi.org/10.1088/1361-6528/aabdb1] [PMID: 29648546]

[46]   Sun, D.; Liu, W.; Fu, Y.; Fang, Z.; Sun, F.; Fu, X.; Zhang, Y.; Li, Z. Noble metals can have different effects on photocatalysis over metal-organic frameworks (MOFs): a case study on M/NH$_2$-MI--125(Ti) (M=Pt and Au). *Chemistry,* **2014**, *20*(16), 4780-4788.
[http://dx.doi.org/10.1002/chem.201304067] [PMID: 24644131]

[47]   Zhu, J.; Li, P.Z.; Guo, W.; Zhao, Y.; Zou, R. Titanium-based metal–organic frameworks for photocatalytic applications. *Coord. Chem. Rev.,* **2018**, *359*, 80-101.
[http://dx.doi.org/10.1016/j.ccr.2017.12.013]

[48]   Feng, X.; Lin, S.; Li, M.; Bo, X.; Guo, L. Comparative study of carbon fiber structure on the electrocatalytic performance of ZIF-67. *Anal. Chim. Acta,* **2017**, *984*, 96-106.
[http://dx.doi.org/10.1016/j.aca.2017.07.020] [PMID: 28843573]

[49]   Kung, C.W.; Li, Y.S.; Lee, M.H.; Wang, S.Y.; Chiang, W.H.; Ho, K.C. In situ growth of porphyrinic metal–organic framework nanocrystals on graphene nanoribbons for the electrocatalytic oxidation of nitrite. *J. Mater. Chem. A Mater. Energy Sustain.,* **2016**, *4*(27), 10673-10682.
[http://dx.doi.org/10.1039/C6TA02563C]

[50]   Zhang, Y.; Yan, P.; Wan, Q.; Yang, N. Integration of chromium terephthalate metal-organic frameworks with reduced graphene oxide for voltammetry of 4-nonylphenol. *Carbon,* **2018**, *134*, 540-547.

[http://dx.doi.org/10.1016/j.carbon.2018.02.072]

[51]   Wang, X.; Wang, Q.; Wang, Q.; Gao, F.; Gao, F.; Yang, Y.; Guo, H. Highly dispersible and stable copper terephthalate metal-organic framework-graphene oxide nanocomposite for an electrochemical sensing application. *ACS Appl. Mater. Interfaces,* **2014,** *6*(14), 11573-11580.
        [http://dx.doi.org/10.1021/am5019918] [PMID: 25000168]

[52]   Yang, Y.; Wang, Q.; Qiu, W.; Guo, H.; Gao, F. Covalent immobilization of Cu3 (btc) 2 at chitosan–electroreduced graphene oxide hybrid film and its application for simultaneous detection of dihydroxybenzene isomers. *J. Phys. Chem. C,* **2016,** *120*(18), 9794-9803.
        [http://dx.doi.org/10.1021/acs.jpcc.6b01574]

[53]   Zheng, Y.Y.; Li, C.X.; Ding, X.T.; Yang, Q.; Qi, Y.M.; Zhang, H.M.; Qu, L.T. Detection of dopamine at graphene-ZIF-8 nanocomposite modified electrode. *Chin. Chem. Lett.,* **2017,** *28*(7), 1473-1478.
        [http://dx.doi.org/10.1016/j.cclet.2017.03.014]

[54]   Cao, G.; Wu, C.; Tang, Y.; Wan, C. Ultrasmall HKUST-1 nanoparticles decorated graphite nanosheets for highly sensitive electrochemical sensing of DNA damage biomarker 8-hydroxy-2'-deoxyguanosine. *Anal. Chim. Acta,* **2019,** *1058,* 80-88.
        [http://dx.doi.org/10.1016/j.aca.2019.01.031] [PMID: 30851856]

[55]   Zhang, D.; Xu, Y.; Liu, Q.; Xia, Z. Encapsulation of $CH_3NH_3PbBr_3$ perovskite quantum dots in MOF-5 microcrystals as a stable platform for temperature and aqueous heavy metal ion detection. *Inorg. Chem.,* **2018,** *57*(8), 4613-4619.
        [http://dx.doi.org/10.1021/acs.inorgchem.8b00355] [PMID: 29600846]

[56]   Wang, D.; Li, Z.; Zhou, J.; Fang, H.; He, X.; Jena, P.; Zeng, J.B.; Wang, W.N. Simultaneous detection and removal of formaldehyde at room temperature: Janus Au@ ZnO@ ZIF-8 nanoparticles. *Nano-Micro Lett.,* **2018,** *10*(1), 4.
        [http://dx.doi.org/10.1007/s40820-017-0158-0] [PMID: 30393653]

[57]   Li, J.; Xia, J.; Zhang, F.; Wang, Z.; Liu, Q. An electrochemical sensor based on copper-based metal-organic frameworks-graphene composites for determination of dihydroxybenzene isomers in water. *Talanta,* **2018,** *181,* 80-86.
        [http://dx.doi.org/10.1016/j.talanta.2018.01.002] [PMID: 29426545]

[58]   Chen, X.; Liu, D.; Cao, G.; Tang, Y.; Wu, C. In Situ Synthesis of a sandwich-like graphene@ ZIF-67 heterostructure for highly sensitive nonenzymatic glucose sensing in human serums. *ACS Appl. Mater. Interfaces,* **2019,** *11*(9), 9374-9384.
        [http://dx.doi.org/10.1021/acsami.8b22478] [PMID: 30727733]

[59]   Xie, Y.; Tu, X.; Ma, X.; Xiao, M.; Liu, G.; Qu, F.; Dai, R.; Lu, L.; Wang, W. In-situ synthesis of hierarchically porous polypyrrole@ZIF-8/graphene aerogels for enhanced electrochemical sensing of 2, 2-methylenebis (4-chlorophenol). *Electrochim. Acta,* **2019,** *311,* 114-122.
        [http://dx.doi.org/10.1016/j.electacta.2019.04.132]

[60]   Weber, M.; Kim, J.H.; Lee, J.H.; Kim, J.Y.; Iatsunskyi, I.; Coy, E.; Drobek, M.; Julbe, A.; Bechelany, M.; Kim, S.S. High-Performance Nanowire Hydrogen Sensors by Exploiting the Synergistic Effect of Pd Nanoparticles and Metal–Organic Framework Membranes. *ACS Appl. Mater. Interfaces,* **2018,** *10*(40), 34765-34773.
        [http://dx.doi.org/10.1021/acsami.8b12569] [PMID: 30226042]

[61]   Nguyen, D.K.; Lee, J.H.; Nguyen, T.B.; Le Hoang Doan, T.; Phan, B.T.; Mirzaei, A.; Kim, H.W.; Kim, S.S. Realization of selective CO detection by Ni-incorporated metal-organic frameworks. *Sens. Actuators B Chem.,* **2020,** *315,* 128110.
        [http://dx.doi.org/10.1016/j.snb.2020.128110]

[62]   Hromadka, J.; Tokay, B.; Correia, R.; Morgan, S.P.; Korposh, S. Highly sensitive volatile organic compounds vapour measurements using a long period grating optical fibre sensor coated with metal organic framework ZIF-8. *Sens. Actuators B Chem.,* **2018,** *260,* 685-692.
        [http://dx.doi.org/10.1016/j.snb.2018.01.015]

[63]   Surya, S.G.; Bhanoth, S.; Majhi, S.M.; More, Y.D.; Teja, V.M.; Chappanda, K.N. A silver nanoparticle-anchored UiO-66(Zr) metal–organic framework (MOF)-based capacitive H$_2$S gas sensor. *CrystEngComm,* **2019**, *21*(47), 7303-7312.
[http://dx.doi.org/10.1039/C9CE01323G]

[64]   Xu, X.Y.; Yan, B. Eu(III)-functionalized ZnO@MOF heterostructures: integration of pre-concentration and efficient charge transfer for the fabrication of a ppb-level sensing platform for volatile aldehyde gases in vehicles. *J. Mater. Chem. A Mater. Energy Sustain.,* **2017**, *5*(5), 2215-2223.
[http://dx.doi.org/10.1039/C6TA10019H]

[65]   Huang, X.C.; Lin, Y.Y.; Zhang, J.P.; Chen, X.M. Ligand-directed strategy for zeolite-type metal-organic frameworks: zinc(II) imidazolates with unusual zeolitic topologies. *Angew. Chem. Int. Ed.,* **2006**, *45*(10), 1557-1559.
[http://dx.doi.org/10.1002/anie.200503778] [PMID: 16440383]

[66]   Koo, W.T.; Qiao, S.; Ogata, A.F.; Jha, G.; Jang, J.S.; Chen, V.T.; Kim, I.D.; Penner, R.M. Accelerating palladium nanowire H$_2$ sensors using engineered nanofiltration. *ACS Nano,* **2017**, *11*(9), 9276-9285.
[http://dx.doi.org/10.1021/acsnano.7b04529] [PMID: 28820935]

[67]   Koo, W.T.; Kim, S.J.; Jang, J.S.; Kim, D.H.; Kim, I.D. Catalytic Metal Nanoparticles Embedded in Conductive Metal–Organic Frameworks for Chemiresistors: Highly Active and Conductive Porous Materials. *Adv. Sci. (Weinh.),* **2019**, *6*(21), 1900250.
[http://dx.doi.org/10.1002/advs.201900250] [PMID: 31728270]

[68]   Li, P.; Zhan, H.; Tian, S.; Wang, J.; Wang, X.; Zhu, Z.; Dai, J.; Dai, Y.; Wang, Z.; Zhang, C.; Huang, X.; Huang, W. Sequential ligand exchange of coordination polymers hybridized with *in situ* grown and aligned au nanowires for rapid and selective gas sensing. *ACS Appl. Mater. Interfaces,* **2019**, *11*(14), 13624-13631.
[PMID: 30888141]

CHAPTER 7

Metal-Organic Frame Works (MOFs) for Smart Applications

Manju[1,2], Megha Jain[1,2], Sanjay Kumar[3], Ankush Vij[4, *] and Anup Thakur[1, **]

[1] *Department of Basic and Applied Sciences, Advanced Materials Research Lab, Punjabi University, Patiala-147 002, Punjab, India*

[2] *Department of Physics, Punjabi University, Patiala-147002, Punjab, India*

[3] *Department of Chemistry, Multani Mal Modi College, Patiala-147 001, Punjab, India*

[4] *Department of Physics, University of Petroleum and Energy Studies, Dehradun-248 007, Uttarakhand, India*

Abstract: Metal-organic framework (MOF) is a class of materials, which is formed by combining metal/inorganic and organic linkers, resulting in the formation of a framework with high surface area and permanent porosity. The freedom to vary inorganic and organic linkers stimulated the synthesis of thousands of MOF structures, for their utility in various applications. The presence of high porosity, high surface area and high free volume made these materials a perfect choice among the class of solid adsorbents. The metal nodes, tunable pore, versatile structure and functionalized surface allow various types of chemical interactions, *viz.* electrostatic interactions, π complexation, H-bonding, coordination bonding, van der Waals interactions, hydrophobic/hydrophilic interactions. All these features made MOF a customizable material to be utilized for targeted applications. This chapter involves a discussion about the usage of versatile MOFs in smart applications, such as gas storage, gas separation and drug delivery, along with a brief discussion about the synthesis of MOFs.

Keywords: Alkane separation, Carbon dioxide storage, Drug delivery, Gas storage, Gas separation, Hydrogen storage, Methane storage, Molecular building approach, Membrane, Nano synthesis, Porous coordination framework, Reticular chemistry, Toxic gas storage.

---

* **Corresponding author Ankush Vij:** Department of Physics, University of Petroleum and Energy Studies, Dehradun-248 007, Uttarakhand, India; E-mail: vij_anx@yahoo.com

** **Co-corresponding author Anup Thakur:** Department of Basic and Applied Sciences, Advanced Materials Research Lab, Punjabi University, Patiala-147 002, Punjab, India; E-mail: dranupthakur@gmail.com

Karamjit Singh Dhaliwal (Ed.)
All rights reserved-© 2023 Bentham Science Publishers

INTRODUCTION

Coordination compounds engulf the class of materials formed by the uni-dimensional or cross-linked array of repeating units, to give macrostructures of various fashions [1]. As a subclass of coordination polymers, metalorganic framework (MOF) represents a class of compounds having inorganic clusters integrated with organic units, leading in the formation of a crystalline network with high porosity and surface area [2]. MOF possesses very high porosity with ~90% free volume, a high surface area of ~7000 m^2/g, and low densities of ~0.13 g cm^{-3} [3]. MOFs might resemble to sponge but they differ from sponge form in the aspect that MOFs have uniform, programmable, controllable pores, which can be used and altered as per desired functionality.

The reticular synthesis is responsible for the creation of strong bonds between organic-inorganic counterparts. By choosing a desirable metal-organic linker combination, the properties of MOF can be tailored as per desire, which is the driving force backing the hot market of MOF in present times. This field experienced unprecedented growth compared to any other field, as a result, vast literature is available on this class of material [4].

The presence of permanent and tunable porosity, functionally alterable surface and high surface area worked as key features for the exploitation of MOFs in gas storage [5], as good quality adsorbents [6], catalysis [7], biomedical imaging [8], drug delivery [9], optoelectronics [10], *etc.*, as shown in Fig. (**1**). Apart from the experimental work, a plentiful interest is also shown by computational researchers in predicting the new MOFs for modifying the properties/performance of currently existing systems. Having a customized MOF is not the ultimate goal, the designed MOF should be chemically, mechanically and thermally stable to withstand the conditions encountered during practical exploitation. Different applications demand a different type of stability of MOF, such as the need of tolerable mechanical stability for electrocatalysis. Metal nodes in MOF are chemically weak points,which poses serious concerns about the chemical stability of MOF. Moreover, weak node-linker bonds are prone to thermal degradation and hydrothermal instability of framework. The boon of porosity in MOF is actually the bane in the context of its mechanical stability, which is detrimental to the palletization process at the application stage. The stability of MOF is judged by variation in properties (structure, surface area or adsorption isotherm) upon exposure to such an environment. So, it can be stated that key features of MOFs may somehow invite some sort of instability in the structure. However, these factors did not create any serious dents on the popularity and employability of MOFs and great attention is also devoted to addressing the stability issues in MOFs. Several measures, like the presence of anionic or nitrogen-containing linkers, absence of polar groups or small apertures, catenation, strong metal-oxyanion bonds, less or no defects, metal centres with high valency *etc.*, are found to be helpful in gaining reasonable chemical, thermal and hydrothermal stability in MOF structures [3]. The pore size, functionality, metal centre and its environment are all customizable, which makes it an interesting playground for researchers. The interesting features of pores are observed in the form of the breathing effect and gate phenomenon, which are the result of the ability of pores to change their dimensions without rupturing the chemical bonds [11].

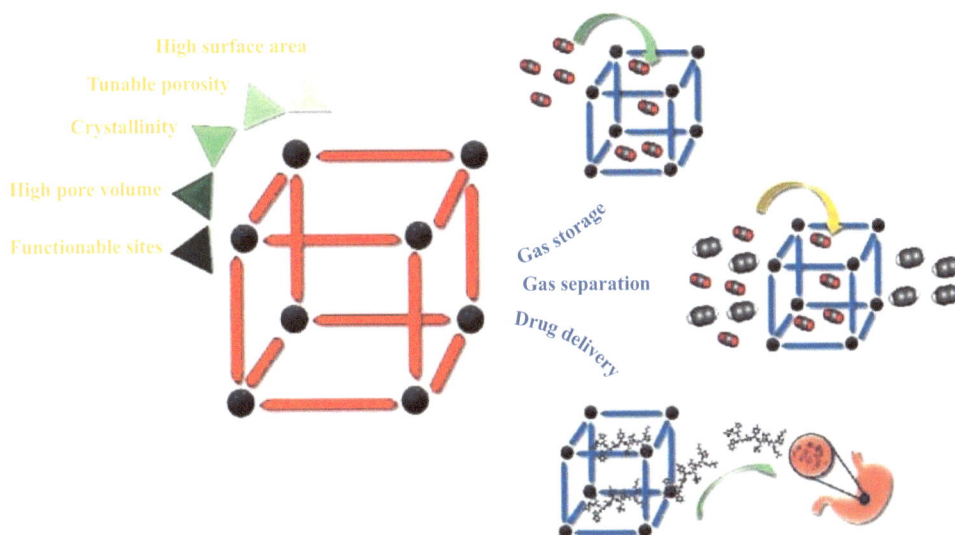

Fig. (1). A diagram representing general properties of MOFs for their application in gas storage, gas separation and drug delivery, discussed in the chapter.

Among porous solid adsorbents, zeolites have also been extensively explored owing to their high surface area and enough active sites available for chemical interactions. These were found to be reducing the power consumption compared to the conventional distillation process. The ability to synthesize zeolites as powder as well as membrane posed them as a good choice for gas storage and separation applications. Zeolites also provide uniform pore distribution, but less room for chemical tailoring to modify or customize it, as in the case of MOFs. On the other hand, MOFs were found to be exhibiting higher surface area than zeolites, holding the position of most porous materials known to date. The presence of metal nodes directly in the framework enhances the chemical interactions to a bigger extent. MOFs outperformed zeolites in selectivity, regeneration, reusability and cost. Moreover, structural versatility and pore tuning are more feasible in MOFs as compared to zeolites, which resulted in a sudden shift of research interest from later to the former [12 - 14].

This chapter is intended to discuss the utility of MOFs in smart applications. As it is impractical to summarize all its properties in a single chapter, the presented work is a brief discussion about the applicability of MOFs in gas storage, gas separation and drug delivery. The text also provides some glimpses about the synthesis of MOFs through reticular chemistry, in nano and membrane form. The reader would gain basic ideas from properties and basic interaction principles to their correlation to the targeted application.

SYNTHESIS

It is usually desirable for MOF to have a high surface area accompanied by significant pore volume, pore shape and its functionality for their utility in practical applications. To attain an optimum yield of targeted MOF, some conditions are required to be achieved: (a) light and abundant elements constituted inorganic unit, (b) organic unit with desirable shape, size and functionality through expansion and features, (c) platform to permit isoreticular access to MOF having high surface area, tunable pores with a large volume and (d) topology which prohibits interpenetration upon MOF growth. Various techniques are reported for the synthesis of MOFs, which are critical for designing the final set of properties of synthesized MOFs. Some of the techniques are discussed below:

Molecular Building Approach

In this approach, desired geometrical and structural attributes are introduced into molecular building blocks (MBB), which ultimately assemble to give complete MOF with incorporated properties. Thus, using isoreticular chemistry, MBB approach successfully fabricates targeted MOF with desired properties. This technique is altering dimensionality and functionality without impacting the topology, and hence provides a pathway for the generation of functional materials with typical properties [15]. Such selective and targeted generation of MOF of the desired topology by implementing isoreticular chemistry is a key to potentially exploit material to address the need to application.

Membrane Form

For gas separation, membrane form is highly adaptive and various techniques are worked out for this purpose. The synthesis of MOF membranes for gas storage is expanding rapidly by adapting various methods and techniques to functionalize the membrane for adsorption of targeted gas. For the construction of 2D MOF nanosheets, both top-down and bottom-up approaches can be employed.

Sonication treatment, as a top-down approach, can result in the formation of ~5 Å thick sheets, with acetone as solvent being more suitable for this purpose.

The first MOF membrane was developed for the separation of H_2 from CO_2. Yang *et al.* [16] commenced the synthesis with poly[Zn_2(benzimidazole)$_4$] (Zn_2(bim)$_4$, (bim$^-$= benzimidazolate)), because its bulk structure is comprised of Zn(bim) stacked sheets *via* week van der Waals forces. After that, exfoliation was done at low-speed wet ball milling by optimizing the solvents (methanal and propanol in this case) to maintain structural integrity. The ultrathin membrane was deposited using the hot drop coating technique, preventing the stacking of nanosheets, which could otherwise block the sieve pores. Hence, the resulted Zn_2(bim)$_4$ membrane was found to be having cages of 0.21 nm, which exhibited excellent H_2 selectivity and permittivity, with the thermal stability of over 120 h.

The bottom-up technique involves the synthesis of 2D MOF nanosheets directly by a judicial balance of solvent, contact mode and capping agents. It can be done through diffusion-mediated modulation, which is represented with an example of copper 1,4-benzenedicarboxylate (CuBDC) MOF lamellae [17]. In this technique, three-column of liquids *viz.* $Cu(NO_3)_2$ and BDCA (benzene 1,4-dicarboxylic acid) linkers were placed in the top and bottom columns, separated by a solvent layer. The diffusion of Cu^{2+} and BDCA ligands was allowed to take place under static conditions, resulting in the formation of MOF nanosheets. This method is easy and efficient in synthesizing MOF nanosheets.

Apart from synthesizing the membranes, the pore size of MOF membranes is re-tuned to adjust the permeability and selectivity for a particular gas mixture. This can be done through three ways:

- The first approach contains the formation of a mixed linker framework, starting with two MOF of similar structure/topology but different ligands (polar and non-polar), resulting in the formation of two membranes. Thus, the ratio of mixed ligands and ratio of polar to non-polar ligands can tune the pore size of ligands for targeted gas. This technique is worked out for CO_2/CH_4, *n*-butane/*i*-butane, H_2/CH_4, H_2/N_2 and H_2/CO_2 separation [18].
- The second approach involves exchange of the metal centre of two MOFs, which resulted in the elimination of void interfaces, reduction in gas channel and enhanced gas selectivity. An example is propylene/propane separation by growing ZIF-67 on ZIF-8 seed layers [19].
- The third approach is the introduction of guest molecules into channels of MOF as an additional step to synthesis. Adding guest molecules adjust the pore size and enhance the interactions of MOF with targeted gas molecules, making it a

suitable way to address the selectivity and permeability issues in MOF membranes [20].

Besides pore tuning, MOF membranes can be combined with other species to functionalize the membrane for particular gas mixture separation. Having the same metal source in substrate and MOF can improve the interfacing bonding between both, which regulates the growth of the layer and consequently result in making a thinner and continuous membrane. The introduction of other species may also act as a repair agent to cure the crystal defects present in the MOF, which subsequently help in improving the sieving properties of the membrane, as shown in Fig. (**2**), for the covalent organic framework (COF)-MOF composite membranes. Transmission electron microscopic (TEM) images in the Figure are indicating the amorphous part filling the gaps between the crystalline part (shown by fringes). This technique improved the membrane selectivity and permeability for H_2/CO_2 mixture [21].

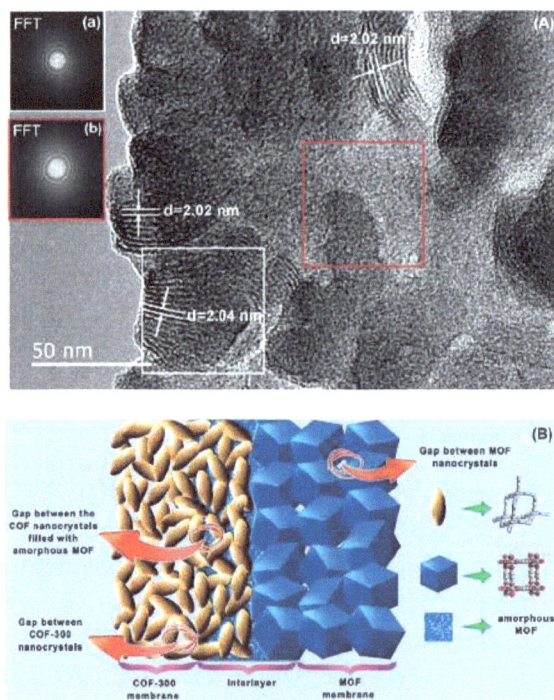

Fig. (2). (a) TEM image and FFT analysis (shown in insets a and b) of the [COF300] [$Zn_2(bdc)_2(dabco)$] composite membrane. Inset (a) is the FFT of the white marked area, and inset (b) is the FFT of the red marked area. **(b)** Schematic of the interlayer formed by amorphous MOF, which has a similar pore size as a crystalline MOF, occupying the gaps between the COF nanocrystals and the interface between the COF and MOF crystalline layers. Reprinted with permission from [21]. Copyright 2016 American Chemical Society.

Nano Form

Controlling particle size is another desirable factor when designing a research problem involving MOFs. In some applications, where nanoscale MOFs are required, a variety of synthetic conditions are already worked out, (Fig. **3**), summarises strategies for manipulating the morphology of MOF crystals, some of which are briefly discussed below:

- **Solvothermal Route**: This technique demands control of reaction time, concentration, temperature, pressure and pH to adjust the reaction kinetics and hence growth of nano MOFs. The use of blocking agents and inhibitors can help in controlling the size in desired limits [23, 24]. The polymers and/or surfactants used in the synthesis reaction adjoins with organic linkers. A noticeable inverse dependence of molecular weight of polymer and size of MOF is observed in various reports [25, 26].
- **Reverse Microemulsions**: It is a very flexible method to control the morphology and size of materials by adjusting the factor, w (water to surfactant molar ratio). Thus, $w = 5$ can yield nanorods of 100-125 nm long and 40 nm wide, whereas $w = 10$ can result in 1-2 μm long nanorods with width of 100 nm [27].
- **Ultrasonic Method**: It is a widely employed and low-cost method to synthesize nano MOF of tunable sizes and dimensionality. The size of the final product can be controlled by adjusting the reaction time of synthesis [28].
- **Microwave Irradiation**: It is considered the best technique to obtain nano MOFs having narrow size distribution, controlled morphology, rapid and high yield. Thus, nano MOFs of various sizes can be synthesized by judicial adjustment of temperature and heating rates. This method is efficient in cutting down the synthesis time, and is a considerable route for mass synthesis of nano MOFs [29].

APPLICATIONS

The extremely high porosities of MOFs, their tendency to accommodate gas molecules, high and tunable pore sizes, and pore surfaces having desirable functional sites have made them a very suitable choice to be used for gas storage and separation. Brunauer Emmett Teller (BET) surface area of up to ~7000 m^2/g has been reported for MOFs. Owing to their functional and versatile pores, sieving of gas molecules and their tendency to differentiate the interactions with gas molecules is unmatched.

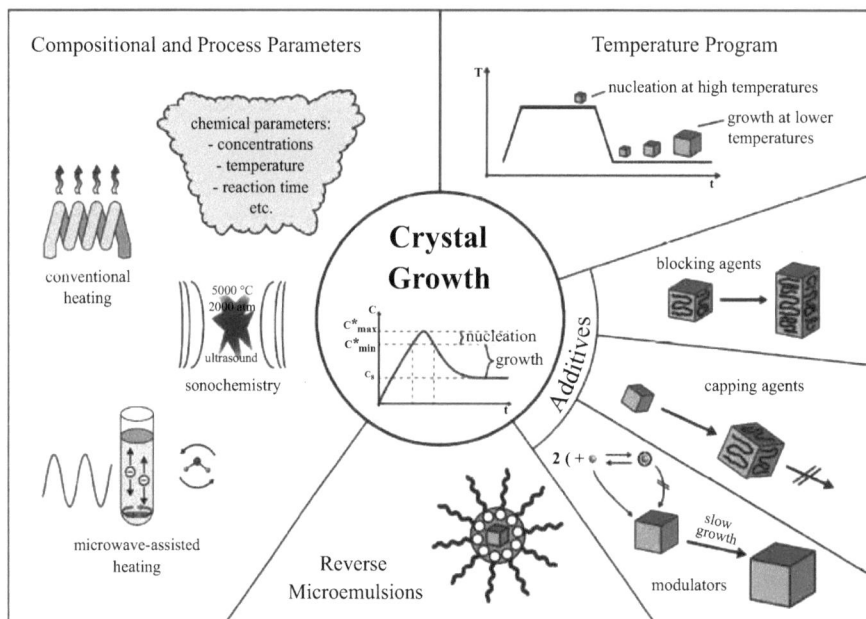

Fig. (3). Summary of strategies for manipulating the crystal growth, *i.e.*, size and morphology of MOF crystals. Reprinted with permission from [22]. Copyright 2012 American Chemical Society.

Gas Storage

The permanent porosity of MOFs is a key factor for their utility as gas storage materials. Gas storage and gas separation are two different issues to be addressed differently. For storage, the material not only needs to store much, but it should also have good deliverable capacity at low pressures, as pressures are the main issue of cost. The working capacity of the gas storage unit or adsorbent material is a key factor to determine its performance. The storage reservoirs usually have some permanent inventory for maintaining the pressure and deliverability during withdrawal sessions, which is termed as base gas. Thus, the working capacity is defined as the difference of total storage capacity and base gas. For systems having large total uptakes, the inability to desorb a significant proportion of gas at operational desorption pressure would be detrimental to its overall performance. The following text is discussing methane, carbon dioxide, hydrogen and toxic gas storage in MOFs due to their obvious importance.

Hydrogen Storage

Enough attention has been grabbed by H_2 production through shale gas, which posed the need for utilization of H_2 as fuel due to its ultrahigh combustion heat and mild by-products [30]. It can serve as a major solution for climate problems caused by fossil fuels. However, the low boiling point (20 K) and critical temperature (38 K) poses serious concerns over liquefication or compression of hydrogen. One method is to store hydrogen in liquid form, which demands a big share of energy to liquefy and store the hydrogen in tanks. Another method is to use high-pressure tanks, which reduce actual gravimetric capacity due to the usage of heavy apparatus. As a consequence of these constraints, the current hydrogen storage capacity is 14-28 kg m^{-3}. Thus, emphasis is drawn on MOF, as a large surface area and plentiful interaction sites can efficiently be employed to store hydrogen at low pressure and temperature [31].

United States Department of Energy (DOE) had set a target for the hydrogen storage system to store 9.0 wt% and 81 g/l by 2015, which is not achieved to date. The limited or less remarkable participation of open metal sites in H_2 binding is due to the high coordination number of the metal linkers, which could not take up enough H_2 molecules due to saturation of coordination tendency. Thus, $Mn_2(dsbdc)$ ($H_4dsbdc=$ 2,5disulfhydrylbenzene-1,4-dicarboxylic acid) was proposed, which is having low coordination numbered metal site. There are two crystallographic sites for Mn^{2+} with six coordinated sites with octahedral geometry and four coordinated sites with see-saw geometry. As anticipated, four coordinated open metal sites exhibited enhanced H_2 uptake.

When it comes to hydrogen storage, the weak van der Waals forces between hydrogen molecules and open metal sites of MOFs were found as a crucial factor that enhances the interaction between the two, consequently increasing H_2 adsorption isosteric heat to ~12 kJ /mol. From theoretical calculations, H_2 adsorption isosteric heat of 15 kJ/mol is required to meet the high hydrogen storage standards, non-achievement of which has created dormancy in further progress in MOFs for hydrogen storage [13]. MOF-filled tanks can retain hydrogen at 77 K and at pressures <100 bar [31]. It is known for MOFs that open metal sites and the surface of cage and/or channels can provide strong and weak interaction positions, respectively. As an example of this fact, MOF-74-Mg, which has the highest open metal site density, showed 1.7 wt% H_2 uptake, which is far less than other MOFs with low open metal site density [32]. So, if H_2 uptake takes place at weak interaction sites, a large pore size would be detrimental as it would reduce the attractive interaction between the pore surface and H_2. A theoretical estimate for pore width of 9 Å is given for carbon-based materials [33], which also holds good for MOFs as their cages and channels mainly consist of carbon. Apart from pore engineering, synthesis with interpenetration can also be helpful to divide large voids. It can be observed by comparing the H_2 uptake stats for two-fold interpenetrated PCN-6 and non-interpenetrated PCN-6' [31]. The record gravimetric H_2 uptake of 17.6 wt% at 77 K and 80 bar is observed in

MOF-210 [34]. These statistics are higher than contemporary carbon-based absorbing materials, zeolites and metal/chemical hydrides. However, researchers are thriving to attain optimum uptake values at more convenient temperatures.

Apart from MOF only, metal nanocrystals are also known for their good affinity in hydrogen-related storage and catalytic reaction, among which Pd holds an unraveling place. Pd is not a new name for hydrogen-related storage and catalytic reactions,as it provides easy dissociation and permeation of H_2 into the lattice of the metal. Thus, a blended approach to exploit Pd nanocrystal and porous MOF was envisaged as a good bet for H_2 storage. Li *et al.* [30] adopted a similar approach of covering Pd nanocrystals with HKUST-1 MOF, which resulted in twice H_2 storage capacity than the bare Pd nanocrystals.

Methane Storage

Because of environmental concerns and to find fuels with low carbon emission, methane presents itself as a good choice for such applications due to being the main component of natural gas and a constituent of two-thirds of fossil fuels. The challenge lies in technology to store and deliver large amounts of methane fuel at low pressures and room temperature. The main challenge in expanding methane fuel for automobile fuelling is its low energy density, which hinders deliverability and transportation. Various strategies are adopted to overcome this problem. In using liquefied natural gas, a very high cost is levied on cryogenic systems (working up to -162 °C) to compress the gas to 0.16 vol%. Another method is compressed natural gas, which needs high pressures (200-250 bar) in fuel tanks, compressing the methane to 1 vol%. Using compressed gas in vehicles has a major concern of safety issues. The third way is to use suitable adsorbents to get adsorbed natural gas. Such adsorbents are found to be working at low pressures and near room temperatures [31, 35]. Because of crystallinity, high permanent porosity, functionalization ability, abundant binding sites, structural diversity, MOFs have been a good choice for their utility for gas storage and separation [31, 36].

US DOE flagged a research program, named "Methane Opportunities for Vehicular Energy (MOVE)", to develop this fuelling technology at room temperature and optimized pressures (35-80 bar, and high up to 250 bar) to assist the commercial availability [37]. This initiative has attracted numerous corporate-academic collaborations to work on MOF materials for methane storage. The moderate interactions between methane and MOFs make them good candidates for methane storage at moderate temperatures and pressures. The pore size, its curvature and distribution are deciding factors in methane storage capabilities of various MOFs [13]. He *et al.* [38] established an empirical relation between gravimetric methane uptake and pore volume (V_p) for MOF with $V_p < 1.50$ cm^3 g^{-1}, given as follows

$$C = -126.69 \times V_p^2 + 381.62 \times V_p - 12.57 \tag{1}$$

Here, C represents excess gravimetric methane storage capacity at 300 K and 35 bar (cm^3 (STP) g^{-1}). Excess adsorption is defined as the amount of adsorbed gas in interaction with MOF, and absolute adsorption is described as the amount of gas interacting as well as staying in pores in absence of gas-solid interaction. The deliverable amount is calculated from the amount of adsorbed methane between 5 bar to upper working pressure (generally 35 bar). The working capacity in methane storage is decided by the release of most of the gas when the pressure drops to 5 bar. Co(bdp) and Fe(bdp) (H$_2$bdp = 1,4-benzenedipyrazolate) are two such MOFs that exhibited the release of >96% of methane at a lowering of pressure from 65 bar to 5 bar [39], with Co(bdp) holding record of highest working capacity even with low methane uptake among others. Such performance of these MOFs is credited to their flexible framework, which enables expansion and contraction of its channels on varying methane pressures, *i.e.*, expansion at high pressure to make way for incoming methane and contraction at low pressures to push out the residing methane.

Another popular MOF, HKUST-1, showed the highest volumetric uptake capacity (267 cc/cc at 65 bar and room temperature), representing the early examples of MOF for gas storage. Till now, few MOFs have shown the volumetric deliverable capacity to meet US DOE storage targets, such as MOF-519 (209 cc/cc), NJU-Bai-43 (198 cc/cc), Co(bdp) (197 cc/cc) and UTSA-76a (197 cc/cc) [36].

Apart from altering the pores, the effect of synthesis is also observed in the performance of MOFs. As it is already known that US DOE set the target of 263 cm^3 (STP) cm^{-3} of volumetric storage for adsorbed natural gas at 65 bar and room temperature, which is equivalent to the storage capacity of an empty tank at 250 bar, the quest to achieve which has initiated widespread exploration of MOFs for this work. Theoretically, HKUST-1 is the top-performing MOF to meet the maximum uptake of 270 cm^3 (STP) cm^{-3} at 65 bar. But, experimentally, the volumetric adsorption capacity was found to decrease up to 35% of its theoretical value, due to mechanical breakdown in pore structure upon densification and palletization operations. Tian *et al.* [40] synthesized monolithic HKUST-1 ($_{mono}$HKUST-1) by the sol-gel process without binders, which showed a methane capacity of 259 cm^3 (STP) cm^{-3} at 65 bar to become the first adsorbent to show such performance after subsequent densification and shaping. The hardness of $_{mono}$HKUST-1 was found to be exceedingly twice that of conventional HKUST-1 MOF.

On the other hand, varying the synthesis conditions can also help in varying the framework links, which ultimately reflects the overall performance of MOF. The SBU in MOF-519 is having 8 octahedrally coordinated Al atoms, cornered joined by doubly bridged OH⁻, 12 carboxylate BTB links for extended structure, and 4 terminal BTB ligands. On the other hand, MOF-520 is having similar structural parameters, but has 4 formate ligands instead of 4 terminal BTB ligands as in MOF-519, resulting in a larger void space in MOF-520 than in MOF-519, the effect of which was seen in varied methane storage performance in both (Table 1) [37].

Table 1. The methane uptake characteristics of various compounds at 298 K, along with topology, pore volume (V_p) and surface area.

Material	Topology	V_p	Surface Area		Gas Uptake		References
			BET	Langmuir	Uptake	Pressure	
		(cm^3/g)	(m^2/g)	(m^2/g)	$(cm^3\,cm^{-3})$	(bar)	
HKUST-1	tbo	0.69		1977	225	35	[37]
					272	80	
monoHKUST-1	tbo	0.51	1193		259	65	[40]
MOF-519	sum	0.94	2400	2660	200	35	[37]
					279	80	
MOF-520	sum	1.28	3290	3930	162	35	[37]
					231	80	
MOF-205	ith-d	2.16	6240	10400	120	35	[37]
					205	80	
PCN-14	fof/nbo	0.83		2360	200	35	[37]
					250	80	
MOF-177	qom	1.89	4500	5340	122	35	[37]
					205	80	
MOF-5	pcu	1.38	3320	4400	126	35	[37]
					198	80	
MAF-38	pcu	0.62	2022	2229	226	35	[31, 41]
					265	80	
MOF-905	ith-d	1.34	3490	3770	145	35	[35]
					228	80	
MOF-906-Me₂	ith-d	1.39	3640	3920	138	35	[35]
					211	80	
MOF-905-Naph	ith-d	1.25	3310	3540	146	35	[35]

(Table 1) cont.....

					217	80	
MOF-905-NO$_2$	**ith-d**	1.29	3380	3600	132	35	[35]
					203	80	
MOF-950	**ith-d**	1.30	3440	3650	145	35	[35]
					209	80	
Co(bdp)	**oab**	1.02	2911		163	35	[35, 36, 39]
					205	65	
Fe(bdp)	**tbo**	0.99	2780		156	35	[36, 39]
					196	65	
Fe-pbpta	**soc**	2.15	4937		139	35	[36]
					219	65	
Al-soc-MOF-1	**edq/soc**		5585		127	35	[35, 36]
					221	80	
UTSA-76a	**fof/nbo**	1.09	2820		211	35	[31, 35, 42]
					257	65	
PCN-250(Fe$_2$Co)	**soc**	0.573	21653		200	35	[31, 43, 44]
Ni-MOF-74	**etb**	0.51		1438	230	35	[37, 45]
					267	80	
Co-MOF-74	**etb**	0.51		1433	221	35	[31, 45, 46]
					249	65	
Mg-MOF-74	**etb**	0.91		2176	200	35	[31, 45, 46]
Zn-MOF-74	**etb**	0.91			188	35	[31, 45]

The isosteric enthalpies of adsorption (Q_{st}) at zero coverage indicate the methane-methane interactions. Open metal sites offer high values of Q_{st}, which is the reason for the high volumetric uptake achieved in MOF74, HKUST-1 and PCN-14. Usually, each metal site works as a binding entity for each methane molecule, which limits the storage capacity in MOF. Inorganic secondary building units (SBUs), with or without open metal sites, are primary sites of adsorption. The pore size should be large enough to accommodate the gas molecule and small enough to uphold/confine it. The storage capacity can be altered by playing with the pore space of the MOF. Jiang *et al.* stated that the usage of BTB (benzene1,3,5-tribenzoate) linkers in some MOFs (MOF-177, MOF-205, MOF-519, MOF-520, *etc.*) might result in large pore size, giving dead space for the gas molecules. The replacement of the peripheral phenylene ring of the BTB linker with double/triple bond spacer may reduce the dead volume. As in the zinc-based MOF series (MOF-905, MOF-905-Me$_2$, MOF-905-Naph, MOF905-NO$_2$ and MOF-950), the usage of a new tricarboxylate organic linker, BTAC

(H$_3$BTAC=benzene-1,3,5-tri-β-acrylic acid) reduced the dead space to achieve high volumetric storage capacity at 80 bar, shown in Fig. (**4**) [35]. Two main factors govern the methane storage in MOFs, namely electrostatic interaction and van der Waals interaction. Talking at rest about electrostatic interactions, these take place at open metal sites in MOFs. Among the M-MOF-74 (M=Mg, Co, Ni, Zn, Mn) series, Ni-MOF-74 has the highest total volumetric capacity (Table **1**) due to the strongest polarizing ability of Ni^{2+} ions, resulting in the strongest electrostatic interaction with methane. Each open Ni site is considered to be occupying one methane molecule. However, if electrostatic interaction exceeds a certain limit, it would cause considerable trapping of methane in the framework, hampering the working capacity [46 - 48]. The second considerable factor is van derWaals interactions, which can be controlled by pore size and shape. An example being Al-soc-MOF-1, which showed low methane adsorption at the surface of cages, indicating van der Waals interactions play a major role in methane storage [15]. A similar theory is also there for HKUST-1, UTSA76a and PCN-250 (Fe$_2$Co) [42]. MAF-38 is another such example, which has no open metal site but shows significantly high methane uptake due to van der Waals interaction at three sites of the framework. The exceptional performance of MAF-38 lies in its monotonically increasing -Q$_{st}$curves [41]. The introduction of functional groups in pores has been another strategy to increase the storage capacity, as was observed in the case of UTSA-76. This MOF along with the pyrimidine group exhibited volumetric storage of 260 cm^3 (STP) cm^{-3}, accompanied by a high working capacity of 200 cm^3 (STP) cm^{-3}. The reason for such record-high storage statistics with pyrimidine groups is their Lewis basic N sites and the dynamic freedom of these groups [49]. Apart from that, in the case of methane storage, open metal sites were having less impact on working capacity [13]. Similar Lewis basic N sites were found to be impacting the performance of Fe-based **soc** MOF, Fepbpta, which exhibited the highest gravimetric methane uptake values, comparable to Al-soc-MOF-1. **soc** topology provided well-defined cages and channels, with narrow pores of higher localized charge density. It showed the highest surface area among soc-MOF families so far. The presence of Lewis basic pyridine nitrogen atoms resulted in appreciable methane uptake and appropriate pore size helped in enhancing the methane-methane interaction, which collectively resulted in outstanding methane storage performance from this MOF [36].

As carbon-based materials are also found suitable for gas storage, a good portion of the research is devoted to blending MOF and graphene oxide (GO)/its derivatives to achieve a hybrid material having properties different than the individual components. Epoxy groups from basal planes of GO presented themselves as good linkers to improve the gas uptake properties than bare MOF. Al-Naddaf *et al.* [50] investigated the hybrid nanocomposite adsorbents by blending HKUST-1 with GO, reduced GO (rGO) and carboxyl-functionalized GO (fGO) yielding modified pores and uptake performance when compared to bare HKUST-1, though still less than DOE targets. Among all three hybrid MOFs, the one with rGO showed enhanced methane deliverability and uptake due to the availability of more surface defects in rGO and good bridging linkers in this derivative.

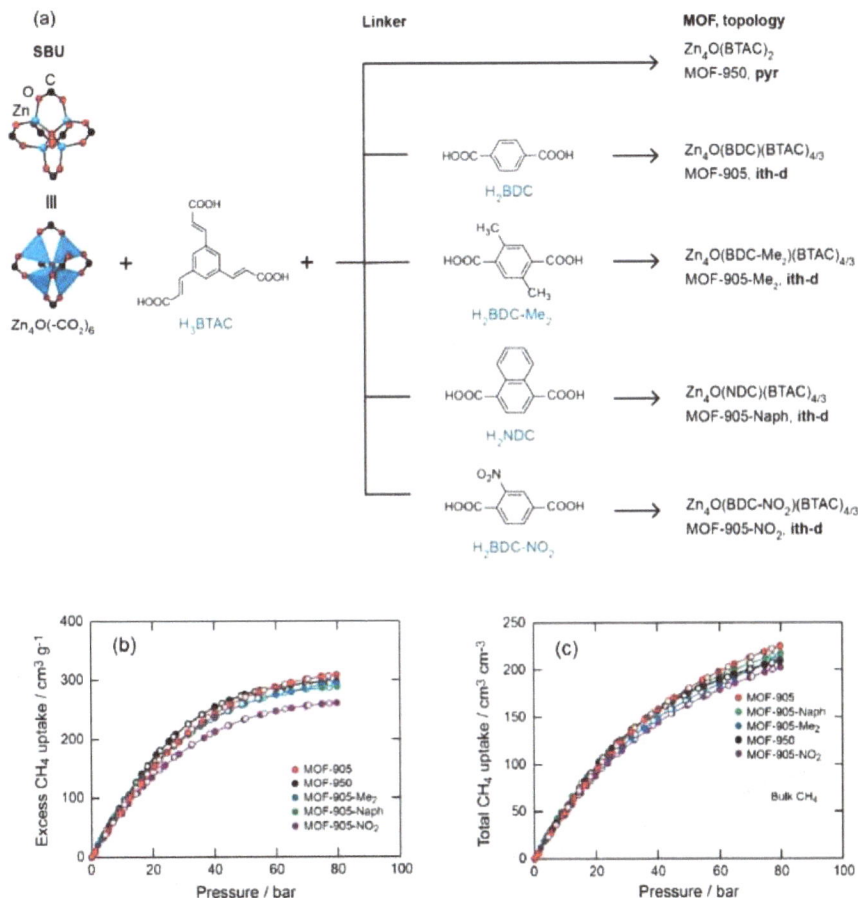

Fig. (4). (a) $Zn_4O(CO_2)_6$ secondary building units (SBUs) are connected with organic linkers to form MOF-950, MOF-905 and functionalized MOF-905. (b) Excess and (c) total CH_4 isotherms for MOFs measured at 298 K. Reprinted with permission from [35]. Copyright 2016 American Chemical Society.

Other than MOFs, zeolites, carbon nanotubes, activated carbon, COF, and porous polymer network are also working in competition with MOFs in methane storage. The low surface area and hydrophilic nature of zeolites are detrimental to this application, which limits the storage capacity to 106 cm³ (STP) cm⁻³ at room temperature and 35 bar. On the other hand, carbon material exhibits volumetric uptakes in the range 50-160 cm^3 (STP) cm^{-3}, showing comparable performance as MOFs. The other two, COF and porous polymer networks, exhibit high surface areas of up to ~6500 m^2/g, but the low densities of these materials are less favorable for desirable volume storage goals. Thus, MOFs are the leading methane storage industry due to the above-mentioned reasons [13].

Carbon Dioxide Storage

Due to the increase in concern about global warming, many efforts are being worked out in various domains for CO_2 emission control. The combustion of fossil fuels in the power sector, industry, and automobiles have been a major and unavoidable vector of CO_2 emission, leading to global warming. Transportation, being a distributed source, contributes one-half of total CO_2 emissions, with the rest from fossil fuel-based power stations, serving as concentrated sources. Aqueous alkanolamine solution-based CO_2 scrubbers are found to be effective in removing CO_2 from gas mixtures, but high regeneration energy supply is a concern owing to the major water content in their composition [5, 51, 52]. Amine functionalized adsorbents emerged as a good substitute to amine solution technologies, due to the former having a low heat capacity comparable to the later [53]. MOF, offering selective and functionalized sites, is the primary candidate for adsorbing targeted gas. The alkylamine functionalization ushers the chemisorption of CO_2 due to the formation of covalent C-N bonds and yields ammonium carbamate as a product of CO_2 uptake [54]. The mechanism of CO_2 capture by the amine group is a two-step process with the first step involving the formation of zwitter ion. Another amine group present in the vicinity of zwitter ion (formed in step one) would act as a base and will lead to the formation of ammonium carbamate (Fig. **5**). Thus, it is always desirable to have a high density of amine groups near to each other to have better CO_2 capture efficiency. However, in humid conditions, the second step may lead to the formation of ammonium bicarbonate [55, 59].

Generally, post-synthetic modification of MOF is more effective than amine functionalized pristine MOF. As in the case of Mg-dobpdc (H_4dobpdc=4,4'dihydroxy-(1,1'-biphenyl-3,3'-dicarboxylicacid)), the solvent attached to the metal centre can be removed by vacuum or high temperature giving unsaturated metal site, which can combine with one end of diamine group during post-synthesis modification. The other end of the diamine can bind CO_2 as per the mechanism explained above and shown in Fig. (**5**) [60, 61]. The absence of an amine group could leave a strongly acidic open metal site to be vulnerable to moisture as well, instead of its effectivity to bind CO_2. Amine functionalization provides moisture stability to MOF and hence improves overall CO_2 capturing performance. But not all metal sites can work well with amine functionalization, as in the case of mmen-Ni-dobpdc (mmen=N,N'dimethylethylenediamine), which exhibits strong bonding between Ni and amine ground and hence hampering the insertion of CO_2 in the metal-amine system [62].

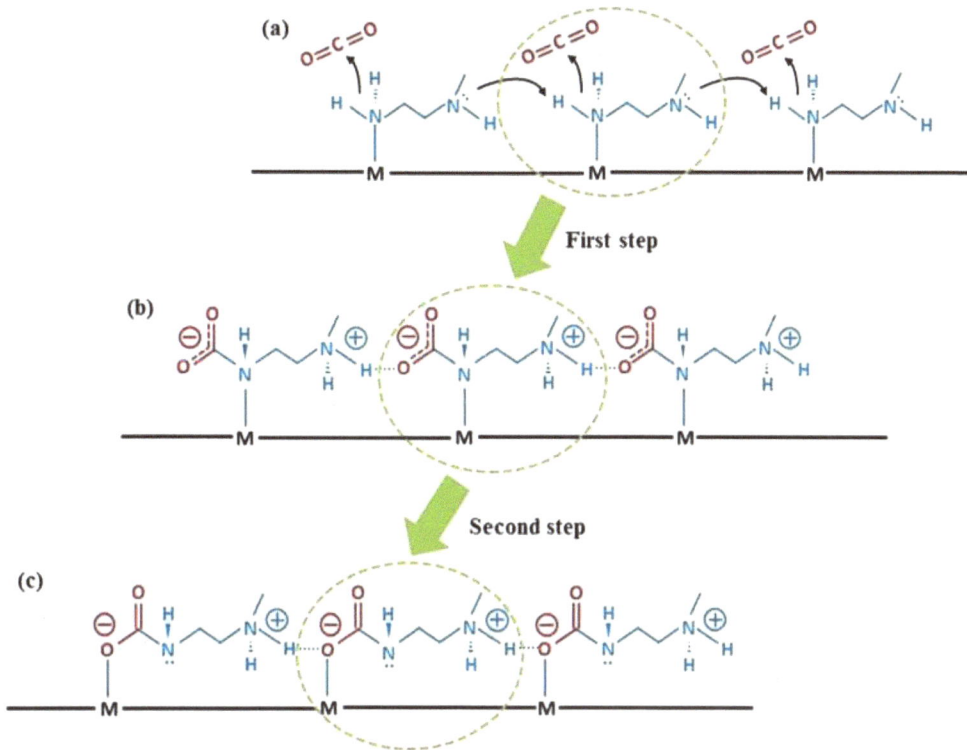

Fig. (5). (a-c) CO_2 adsorption mechanism of 1'-m-2-M_2(dpbpdc) along *c* axis. Reprinted with permission from [60]. Copyright 2020 American Chemical Society.

In terms of surface area and CO_2 capture, a linear proportionality between BET surface area and total gravimetric CO_2 capture capacities is observed. Examples are, MOF-5, MOF-177, NU-100 and MOF-210, having BET surface areas of 3800, 4500, 6143 and 6240 m^2 g^{-1} showing gravimetric CO_2 uptakes of 1072, 1656, 2315 and 2480 mg g^{-1}, at 40 bar and room temperature. Among these, gravimetric CO_2 uptake of 2870 mg g^{-1} is also observed for MOF-210 at 50 bar and room temperature [34, 42]. The performance of MOFs in CO_2 uptakes is appreciable in comparison to contemporary polymer networks (e.g., PPN-4 and PAF-1) [13].

CO_2 capture from combustion can be done in pre-combustion or post-combustion mode [31], as discussed below:

- **Pre-combustion Capture**: This technique is based on decarbonizing the fuels to separate CO_2 from gas mixtures to have zero CO_2 emissions after combustion. The basic idea of this separation lies in the different kinetic diameters of H_2 and CO_2, which are 2.89 and 3.30 Å, respectively [63]. So, such separation can be obtained with the help of molecular sieve-based MOF membranes. An example being $Zn_2(bim)_4$ with an aperture size of 2.9 Å, allowing fast passage for H_2 and impeding permeation of CO_2. Though, special care of synthesis temperature is required while coating the layers onto the substrate to avoid lamellar restacking of layers, which would restrict the pathway for H_2. This MOF showed remarkable H_2 passage and high H_2/CO_2 selectivity. However, the expensive step of mass gasification of fuels restricts any solid exploration for post-combustion CO_2 capture [64].
- **Post-combustion Capture**: The flue gas (73-77% N_2 and 15-16% CO_2 with a total pressure of ~1 bar) liberated from power plants interacts with the CO_2 scrubber at 40-60 °C [63]. Thus, the post-combustion process deals with separating CO_2 from this flue gas. The open metal sites of MOF interact strongly with CO_2, but moisture in flue gas reduces the CO_2 uptake by MOFs. Thus, the incorporation of alkylamine in MOF is desired to substitute alkanolamine solutions. An example being functionalization of $Mg_2(dpbpdc)$ with mmen which imparted a high density of amine groups in the MOF, resulting in applaudable CO_2 selective uptake, even with existing moisture. One end of mmen group is linked to the metal site and other dangling end captures CO_2. Apart from mmen, dmen (N, N'-dimethylethylenediamine), en (ethylenediamine), men (1- methylethylenediamine) and den (1,1dimethylethylenediamine) are also found to be effective in optimizing CO_2 uptake, working capacity, adjusting regeneration energy, moisture stability, *etc.* [58, 62, 65, 66].

Toxic Gases

The burning of fossil fuels and large-scale industrialization came with a major drawback of air pollution, which results in the increasing content of toxic gases in the environment. However, some of the otherwise poisonous/toxic gases are important chemical reagents, which need to be purified prior to their use. In the case of toxic gas adsorption, targeted surface chemistry is required to functionalize the pores or surface of the framework as per the gas to be adsorbed. Thus, open metal sites and functionalized surfaces may result in selective and efficient interactions *viz.* acid-base or electrostatic interactions, π complexations,

H-bonding, coordination bonding, *etc.* [67]. This section is covering the adsorption/purification of some toxic gases using MOFs.

- Carbon monoxide: Carbon monoxide (CO) is having a dual role in chemistry, from being a starting material for numerous basic chemical reactions to being a poisonous gas for catalysts used in ammonia production, and fuel cells as well as toxic for humans and other animals too. The former issue demands CO purification, which is done with cryogenic systems while the latter demands CO separation/uptakes at low pressure. Thus, for CO purification at ambient temperature, M-MOF-74 (M= Mg^{2+}, Mn^{2+}, Fe^{2+}, Co^{2+}, Ni^{2+}, Zn^{2+}) shows high CO uptakes, better Co/H_2, N_2 selectivity for this application. On the other hand, for removal/uptake of toxic CO at low pressure, M $\rightarrow$ CO π back-donation is a crucial factor for better CO uptake at low pressure. Weak field H_4dobdc ligands (H_4dobdc= 2,5-dioxido1,4-benzenedicarboxylic acid) in M-MOF-74 results in high spin and electron-poor open metal sites in these MOFs, which ultimately causes poor M $\rightarrow$ CO π back donation [68]. On the other hand, strong field azolate ligands in triazolate based Fe-BTTri exhibits strong M $\rightarrow$ CO π back donation, and hence confirming its high CO uptakes at very low pressure. But their performance is half than Ni-MOF-74 at high pressure [69].
- Ammonia: Ammonia (NH_3) is a toxic and corrosive gas, but it is also useful in the pharmaceutical, chemical, and fertilizer industries. Thus, structurally and functionally versatile MOFs are suitable adsorbents for both bulk NH_3 storage as well as removing trace NH_3. To remove trace NH_3, it is desirable to have a high affinity between NH_3 and adsorbent. On the other hand, for NH_3 storage, MOF must also have good resistance to NH_3, in order to sustain corrosivity over various adsorption cycles. Here again, triazolate-based MOFs are a better choice for NH_3 storage due to strong field σ-donors, such as M_2Cl_2(BTDD)(H_2O)$_2$ (M= Mn, Co Ni; H_2BTDD= 1H-1,2,3-triazolo [4,5b],[4',5'-i]dibenzo- [1, 4]dioxin) and UiO-66-NH_2 [70]. Open metal sites, as well as functional groups in the framework, are potential NH_3 capture sites. Acid-bearing MOFs having carboxylic and sulfonic groups show enhanced NH_3 absorptivity, example being Fe-MIL-101-SO_3H [71].
- Hydrogen sulphide: The open metal sites and donor-acceptor groups for H-bonding are suitable features for MOF for hydrogen sulphide (H_2S) adsorption. Series of MIL MOF (MIL-53 (Cr) and MIL 53 (V)) are found to be a promising candidate for H_2S adsorption due to their flexible framework structure and tendency to form strong hydrogen bonds with H_2S molecules at the pore surface. But, due to strong interactions, the desorption process is irreversible and limits the usage of this MOF series for practical use [72, 73]. Another series, M-CO-27 (M= Zn, Ni), also exhibited strong interaction with H_2S molecules with subsequent desorption in presence of moisture [74].

- Nitric oxide: For capture and desorption of nitric oxide (NO) for biological applications, $[Cu_3(btc)_2]$ (btc=1,3,5-benzenetricarboxylate) exhibited satisfactory performance than other MOFs in terms of beneficial availability of open metal sites. The post-synthesis functionalization of open metal sites with bifunctional 4-(methylamio)-pyridine results in the formation of secondary amines, which reacted with NO to form diazen-1-ium-1,2,diolate (NONOate) [75]. This strategy of NO interaction to form NONOate derivates is found to be helpful in NO capture in other systems also.

Gas Separation

The separation of gas mixtures holds utmost industrial importance, mainly involving hydrogen (H_2/N_2, H_2/CO, H_2/CO_2), air (N_2/O_2), natural gas (CO_2/CH_4) and hydrocarbon separation for pharmaceutical, fuel, plastic and polymer industries. Gas separation based on distillation holds big energy consumption in order to run one cycle of heating and cooling involving adsorption and desorption of gas. However, adsorbent-based technologies relying on porous and surface structures consume quite less energy as compared to the distillation technique [76].

Regarding gas separation, the material/linking site should be sensitive or responsive to a particular gas and not to other gas. Thus, MOFs are functionalized in such a way that they act as a filter to stick the gas to be separated and let pass other constituents. This ability to functionalize or program the MOF pores is the main stakeholder for gas storage property. The separation ability of gases lies in their unique and fundamental property of polarizability and their size. Polarizability is the factor that decides their stickiness with pores or functional sites of MOFs. The pore size, their hydrophobicity or hydrophilicity can be functionalized to make it reactive for a particular gas.

Membrane separation is a potential technique for gas separation, for which polymers and zeolites are exploited to great extent. However, optimum permeability and selectivity are found to be inversely related to polymers. On the other hand, zeolites, though having uniform pore distribution, do not provide vast room for chemical tailoring for expanding the selectivity. Thus, porous MOFs with flexible chemical and physical structures came into the picture for gas separation. The high crystallinity, versatile pore functionality and structural diversity make MOF an interesting candidate for this application. The pore or channels can be tuned by altering the type and length of organic linkers and introducing different functionalities in pore structure, so as to customize the framework for adsorption/separation of gas of interest [77 - 79]. Gas separation can be done by three techniques: cryogenic distillation, membrane-based separation and adsorptive techniques, a brief about which is discussed in this section.

A lot of interest is vested by researchers in fabricating 2D nanosheets, tuning pore size and integrating with other species to optimize the performance of MOF membranes. Membranes with ultrathin thickness and uniform pore sizes are desirable for the effective separation of gas molecules through this sieve [14, 80, 81]. A brief introduction to 2D membrane synthesis and functionalization is given in the synthesis section. A detailed study of MOF membrane synthesis can be found in reference [77]. Rigid MOF may show adsorption selectivity based on molecular sieving effect or by the virtue of adsorbent-adsorbate interaction strength. On the other hand, flexible MOFs exhibit type I isotherms. Molecular sieving-based separation works upon the fact that the size of pores and the size of targeted gas should be the same, but pore size should be small for other gases so as to block them. Materials with a cage containing 1D channels and separated by windows work well for selective adsorption if the size of the aperture is comparable to the size of the targeted gas. Like in case of $Mg_3(ndc)_3$ (ndc=2,6-naphthalenedicarboxylate), PCN-13, $Cu(F-pymo)_2$ (F-pymo=5uoropyrimidin-2-olate) *etc.*, which selectively adsorb H_2 from the H_2/N_2 mixture [82 - 85].

Apart from the pore structure, adsorbent-adsorbate interactions are also important for the efficient adsorption of targeted gas. In this case, adsorption is governed by properties such as polarity, polarizability, quadrupole moment, the tendency of H-bonding, $\pi-\pi$ interactions, hydrophobic/hydrophilic interactions, *etc.* Example being the selective adsorption of C_2H_2 from C_2H_2/CO_2 gas mixture by $Cu_2(pzdc)_2(pyz)$ (pzdc=2,3-pyrazinedicarboxylate and pyz=pyrazine).This MOF has 1D open channels with O on the surface of the channel. The O present on the surface acts as an active site for H-bonding between O and C_2H_2 which effectively separates C_2H_2 from the mixture. Detailed information in this context can be found in the table of reference [82].

Adsorptive separation is a process in which a gas mixture is passed through a column of adsorbents, which is selective in the adsorption of particular gas from the mixture until equilibrium, followed by desorption of an adsorbed component of the gas mixture so that the adsorbent can be reused. Desorption, known as adsorbent regeneration, can be done by either heating the adsorbent (thermal swing adsorption) or by the application of pressure (pressure swing adsorption). Adsorptive separation can be either bulk separation (>10% adsorption) or purification (<10% adsorption). Thus, a good adsorbent must possess good adsorption, promising selectivity, favourable adsorption kinetics and regeneration capabilities. The adsorption capacity and adsorption selectivity depend upon pressure, temperature, the nature of the material and its pores [82].

Various mechanisms regarding gas separation are proposed, which are briefly mentioned below:

- Molecular sieving effect: In this process, the pore size of the framework is the deciding factor, which would allow the passage of gas molecules of comparable size and would block the larger ones. But such sieving is challenging with gas mixture components having comparable properties or sizes.
- Thermodynamic equilibrium effect: It happens in case of MOF with large pore structures and when adsorption is dependent upon the magnitude of host-guest interactions. The isosteric adoption heat at zero coverage is deciding factor in this case, which can be optimized by incorporating strong binding sites in the framework. Such an effect is observed due to the difference in the adsorption affinity of a particular gas type with the adsorbent surface.
- Kinetic effect: As the name suggests, this phenomenon is based on the difference in diffusivity while passing through pores/channels. Some components diffuse faster into pores than others, which depends on the different diffusion rates of various gas molecules. Gas with high diffusivity would occupy the pore space before others, and hence can be separated before adsorption equilibrium.
- Conformational separation: When the pore structure is large enough and absorbent-absorbate interactions are similar, then packing efficiency is observed as a driving force in such cases.

The surface of the adsorbent also plays a crucial role in the selective adsorption of targeted gas. If target gas molecules have no polarity, but high polarizability/high dipole moment/high quadrupole moment, then adsorbent with a high surface area/with high polarized surface/with a surface having high electric field gradients will be suitable, respectively. Apart from that, hydrogen bonding or π bonds may also be beneficial in sticking the molecules of a gas of interest [86, 87].

Thus, based on the above factors, some exemplary works in the gas separation field with MOFs are discussed as follows:

Alkane Isomers

Catalytic isomerization reactions in the oil refining process generate various alkane isomers of C_5-C_6, having different degrees of branching. The degree of branching provides a premise for the separation of alkanes based on their research octane number (RON) (Table **2**). Higher RON of branched alkanes than linear ones incited the need for their separation to enhance the octane rating of gasoline. Dubbeldam *et al.* [88] showed through grand canonical Monte Carlo simulations for MOF-1 that among a multicomponent mixture of C_5-C_7, the isomers with lower RON showed strong adsorption, except for 2, 3-dimethylpentane, as seen in Fig. (**6**). As another aspect, hexane isomers exhibit negligible dipole moments. Being saturated hydrocarbon, these interact with other reagents/species *via* van der Waals interactions only. In addition to that, size-based differentiation is an effective approach for the separation of these isomers, which require pore size tuning. On the other hand, C_8 alkylaromatic compounds exhibit significant polarizability, which indicates the possibility of separation based on host-guest interactions [76]. Chen *et al.* [89] tuned the pore size of MOF-508 [Zn(BDC)(4,4'-Bipy)$_{0.5}$] (BDC= 1, 4-benzenedicarboxylic acid, Bipy= bipyridine) *via* double interpenetration to get the 1D channels of ~4.0 × 4.0 Å, which allowed the passing of linear chains of pentane and hexane, and blocked the branched isomers. The difference in van der Waals interactions of chain and branched isomers formed the basis of this separation, which was further assisted by rigid and flexible 4, 4'-Bipy linkers and paddle wheel clusters of variable distortion.

Table 2. Table listing the research octane number (RON) and kinetic Diameters of C_5-C_6 alkane isomers [12].

Alkane	RON	Kinetic Diameter (Å)
n-pentane	62.0	4.5
2-methylbutane	92.3	5.0
2,2-dimethylpropane	80.2	6.2
n-hexane	24.8	4.3
2-methylpentane	73.4	5.0-5.5
3-methlypentane	74.5	5.0-5.5
2,3-dimethylbutane	101.7	5.6
2,2-dimethylbutane	91.8	6.2

C_2H_2/C_2H_4

Ethylene (C_2H_4) is an important and essential raw component in polymer and chemical industries. During production, acetylene (C_2H_2) is usually found as a by-product, which is labeled as poison in the polymerization reaction of ethylene to form polyethylene. Similar molecular size, dipole moments, polarizability and other properties of C_2H_2 and C_2H_4 make it challenging to separate these two. The kinetic diameter of C_2H_4 and C_2H_2 is 4.2 and 3.3 Å, respectively, which indicates the scope of pore engineering to separate these gases [76]. Mixed MOF (M'MOF) is reported to be having chiral pore cavities with open metal centres, exhibiting small pores to selectively adsorb the C_2H_2. The activated M'MOFs could effectively decrease the C_2H_2 concentration to 40 ppm, leaving behind highly pure

C_2H_4 [90]. MMOF-74 series, due to having a high density of open metal sites, were also worked out for this separation. Very strong interactions of C_2H_4 with open metal sites of M-MOF-74 indicated good binding affinity, which resulted in low C_2H_2/C_2H_4 selectivity [91]. On the other hand, UTSA-100,[Cu(atbdc)] (H_2atbdc= 5-(5-amino-1H-tetrazol-1-yl)1-3,benzenedicarboxylic acid) [92], is a better choice for this separation. Activated UTSA-100a consists of 1D channels (4.3 Å) and cavities of 4.0 Å with 3.3 Å aperture, which resulted in better C_2H_2 selectivity as compared to other discussed MOFs.

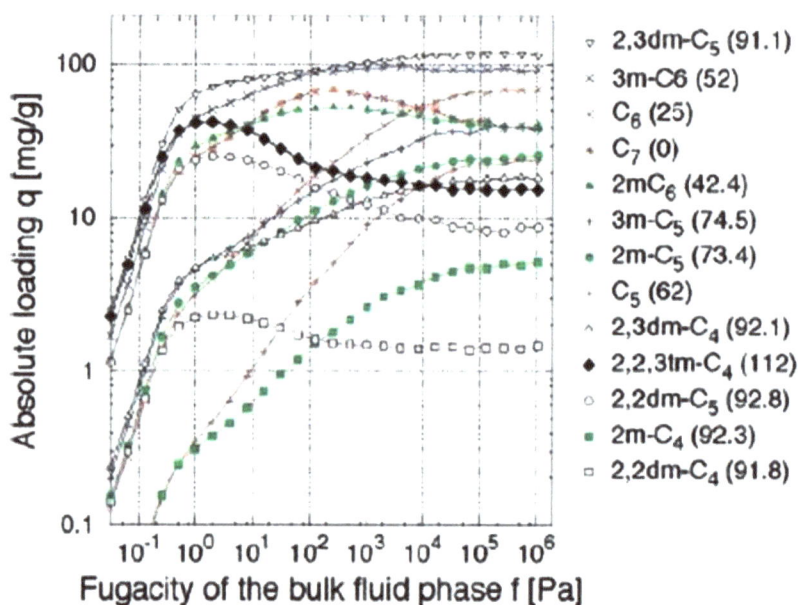

Fig. (6). Multicomponent isotherms at 298 K: an equimolar 13-component mixture of *n*-pentane, *n*-hexane, *n*-heptane, 2-methylbutane, 2-methylpentane, 2-methylhexane, 3-methylpentane, 3-methylhexane, 2,2-dimethylbutane, 2,3-dimethylbutane, 2,2-dimethylpentane, 2,3-dimethylpentane, and 2,2,3-trimethylbutane in MOF-1. Numbers in brackets denote the Research Octane Number. The order in the legend corresponds to the adsorption order at high loading. The statistical uncertainty of the computed loadings is smaller than the symbol size. Reprinted with permission from [88]. Copyright 2008 American Chemical Society.

C_2H_2/CO_2

As a starting reagent for many chemical reactions, pure C_2H_2 is always needed on industrial scale. However, CO_2 is usually found as an impurity during production. C_2H_2 and CO_2 exhibit almost similar physical and structural properties. However, differentiation can be done on the basis of their interaction with open metal sites and Lewis base sites. C_2H_2 forms strong covalent π interaction with open metal sites, while CO_2 approaches covalently through electronegative oxygen atoms. Similarly, C_2H_2 can bind through H-bonding interactions to Lewis base sites, but CO_2 exhibits weak electrostatic interactions due to the presence of electropositive carbon. Based upon these considerations, excellent C_2H_2/CO_2 separation was observed from Zn-UTSA-74a [Zn_2(dobdc)(H_2O)].$0.5H_2O$, which has plentiful Zn open sites to bind two C_2H_2 and one CO_2 molecule per metal site [93]. Though for complete separation, UTSA-300 [Zn(dps)$_2$(SiF$_6$)] (dps= 4,4'dipyridylsulfide) [94] is found to be the only MOF so far to achieve this milestone. It exhibits multiple binding sites and a pore aperture of 3.3 Å. The pores undergo open-close transformation while activation and desolvation, respectively. The aperture size is so fine so as to adsorb C_2H_2 only, showing the highest uptake ratio reported so far for C_2H_2/CO_2 and selectivity of 743 at 1 bar and 298 K. The adsorption takes place *via* C-H··· F and π-π stacking interactions. Thus, C_2H_2 interacts with SiF to open the pores, but CO_2 fails to do so at the same site due to the opposite quadrupole moment. Thus, such interaction-based selectivity makes UTSA-300 an outstanding MOF for C_2H_2/CO_2 separation.

C_3H_4/C_3H_6

The polymerization reaction of propylene (C_3H_6) is interfered with by unwanted propyne (C_3H_4) impurity and only 5 ppm of this impurity concentration is allowed. As in the case of acetylene/ethylene separation, C_3H_4/C_3H_6 separation is also challenging due to similarity in shape, size, and physical and chemical properties. The kinetic diameters of C_3H_4 and C_3H_6 are 4.76 and 4.68 Å, respectively, due to which a lot of MOFs fail in this separation process. However, ELM-12, exhibiting two kinds of cavities that matched well with the shape and size of C_3H_4 to give strong interaction with it, broke the stereotype about flexible MOFs for gas separation. C_3H_4 was found to be strongly interacting with two oxygen atoms of OTf⁻ group through C-H··· O hydrogen bonds, which results in remarkable selectivity in this MOF [95].

Drug Delivery

The quest for suitable carriers in the field of drug delivery has always been a challenge, which has welcomed new materials resulting in an expansion in this technology day by day. Uncontrolled release and lower drug loading are general problems encountered in this technology. MOFs are found to be a better substitute for other porous counterparts due to obvious advantages *viz.* tunable and porous structure, versatile surface offering chemical functionalities, high surface area and large pore sizes for efficient loading and encapsulation of medicine molecules [96]. The drug delivery technology demands some desired and defined set of properties a material must possess before its employment for the said application, such as bearable toxicology, controlled degradation of the material, high loading encapsulation, controlled release and easy surface engineering. These are briefly discussed below:

- For toxicological compatibility, the metal and organic parts must be chosen so as to befriend body cycles. In context to the metal part, Ca, Cu, Mn, Zn, Fe, Mg, Ti and Zr are to be chosen, which are hitherto present in the body in some or other forms. Apart from that, their oral lethal dose is in a safe range. Similarly, for an organic part, both exogenous and endogenous linkers are preferred. The exogenous linkers are made from natural components, which offer tolerable toxicity (Table **3**), for example being polycarboxylates, imidazolates, amines, pyridyls, *etc.* On the other hand, endogenous ones are constituents of body compositions and are likely to be reused in the body after the delivery of the drug. These could be amino acids, nucleobases or saccharide-based organic parts [97 - 99].
- The drug and MOF interaction should be stable enough to remain intact before reaching the target and should be unstable enough to detach the drug from the framework to give high drug release efficiency. This can be gained by tuning the composition, structure and pre-noting the effect on MOF-drug composite in a targeted environment.
- The judicial control over size is also desirable. The drug-loaded MOF must have a large enough size for stable suspension, but a small enough size to pass through capillaries without aggregation. Various routes to synthesize nanosized MOF for addressing this problem are discussed briefly in the synthesis section.
- For obstruction-free delivery of a drug to its target, the surface of MOF is so engineered so as to arrange an easy passage without interacting with biological media encountered in its transit. Silica coating or hydrophilic polymer (poly(ethylene glycol)) based coating is usually practiced for this [100, 101].

Table 3. Oral lethal dose (LD$_{50}$) of some metals and organic compounds [97].

Metal	LD$_{50}$	Organic compounds	LD$_{50}$
Ca	1 g/kg	Terephthalic acid	5 g/kg
Cu	25 μg/kg	Trimesic acid	8.4 g/kg
Mn	1.5 g/kg	2,6napthalenedicarboxylic acid	5 g/kg
Mg	8.1 g/kg	1-methylimidazole	1.13 g/kg
Zn	350 μg/kg	2-methylimidazole	1.4 g/kg
Fe	30 μg/kg	Isonicotinic acid	5 g/kg
Ti	25 g/kg	5-aminoisophthalic acid	1.6 g/kg
Zr	4.1 g/kg	-	-

While delivering the drug to tumor cells, the drug distribution and its selectivity for tumor cells are of paramount interest. The observance of enhanced permeability and retention in MOFs at the nano scale makes these a suitable candidate for efficient nano-drug delivery [25]. However, the whole process of drug loading and delivery contains various parts and strategies, which are discussed below:

- Encapsulation strategy: The loading of the drug in pores/voids of MOF *via* covalent/non-covalent interactions is covered under this strategy (Fig. 7). Drug loading can be done by impregnating porous adsorber into a solution containing drug molecules, which results in the loading of drug depending on pore structure/dimensions of the adsorber. Thus, there is a key need to have a pore size comparable to the size of drug molecules. The small pores would not be able to store the drug, while large pores would result in a burst release of the drug. While designing or opting for any MOF for a particular drug, the loading capacity and release rate are the factors of consideration. MOF having an interactive surface and open metal sites along with tunable pores are worked out for different drugs. To decipher this fact, let us take the example of two MOFs of MIL series *i.e.*, MIL-100 (Cr) and MIL-53(Fe). The toxic Cr based MIL-100 exhibits giant pores (~30 Å), very high surface area (3100-5900 m^2g^{-1}) and large pore volumes. Thus, for the test drug Ibuprofen (IBU), MIL-100 exhibited a loading capacity of 1.376 g of IBU/g, which is higher than zeolites. But due to the weak bonding of the drug with MOF, the first release of the drug was observed in the first 2 hours and rest of all were released after 3 days. On the other hand, less toxic Fe based MIL-53 has small pores and comparatively less loading capacity of 0.210 g of IBU/g. However, a fascinating delivery of the entire IBU after 3 weeks indicated its suitability in long therapies. The hydrogen bonding between hydroxyl groups of MIL-53 and carboxylic groups of IBU is the reason for the slow release of drugs from this compound. Thus, the example of MIL-100 and MIL-53 indicated that tuning the pore size and drug-framework interactions can help in tuning the loading and delivery of drugs [102, 103]. Similarly, ZrUiO-66 amorphous MOF has shown drug release up to 30 days, by the virtue of trapping drugs through irreversibly collapsing porous structures [104].
- Direct assembly strategy: In some cases, the pre-mature release of the drug is also observed. To avoid this, a covalent attachment process between drug and solids is also practiced (Fig. 7). For the successful covalent attachment, functional groups must be kinetically available at the surface of MOF and extra strong interactions should be avoided so as to stop the creation of pro-drug [96]. This approach enables uniform drug delivery and high loading into the matrix. Various platinum-based chemotherapeutics (cisplatin, carboplatin and oxliplatin) have been loaded *via* this route. As an example, the formation of a coordination compound of Tb^{3+} with cisplatin pro-drug (c,c,t(diamminedichlorodisuccinato) Pt(IV)) (DSCP) ligands to form TbDSCP nanocomposite showed controlled

delivery and appreciable stability [105].

- Post-synthetic strategy: It is a two-step process-preparing desired nano MOF and then introducing cargo in a subsequent step to attach covalently at the outer surface of pre-synthesized nano MOF.

Fig. (7). Three kinds of cargo loading strategies for MOF classifying by the location and effect of cargos and the host-guest interactions within the MOF frameworks. The formation of coordination bonds between cargo molecules and unsaturated metal sites or ligands. The formation of coordination bonds between cargo molecules and unsaturated metal sites or ligand defect sites in the post-synthesis strategy are shown in purple dotted lines for highlight. For encapsulation strategy: cargos are located in the pores or channels of MOF *via* noncovalent interactions and do not change the framework structures. For direct assembly strategy: cargos are used as ligands to partly participate in the formation of MOFs *via* coordination bonds. For post-synthesis strategy: cargos are located in the surface, unsaturated or defect metal sites, and functions on linkers of pre-synthesized MOFs through the formation of coordination bonds and covalent bonds between metal nodes or organic linker and cargos. After the cargo loading, the structure, size and morphology of nanocarriers are well maintained. For mixed usage: it may contain features of the above strategies in one cargo delivery system. Reproduced from [25] with permission from The Royal Society of Chemistry.

Another approach of directly incorporating therapeutic molecules as metal nodes of MOF is also practiced. This approach reduces the concerns about the degradation of MOF and related toxicity. The association of nicotinic acid and non-toxic iron to construct therapeutic MOF exhibited more release of nicotinic acid as compared to drug-loaded MOF. Apart from that, antibacterial silver and zinc, antiarthritic gold, antiulcer bismuth, antidiabetic vanadium, *etc.*, are reported to be used as active metals for drug delivery [106, 107]. Due to the above-mentioned advantages, MOFs are worked out for a wide range of drugs. An example of MOF in the delivery of anticancer drugs is discussed here. Among anticancer/antiviral drugs, azidothymidine triphosphate (AZT-Tp), cidofovir (CDV) busulfan (Bu) and doxorubicin (Doxo) are widely studied. The main problem encountered while their usage is their poor solubility in a biological solvent, short half-lives and limited bypassing, which makes it challenging to incorporate and encapsulate these drugs into carriers. An applaudable job is done by MIL-100(Fe) in the efficient entrapping of these drugs, which is credited to the interaction between phosphate groups of drugs and open metal sites of MOF. Apart from that, another famous anticancer drug, 5- ourouacil (5-FU), was encapsulated in NH_2 modified terephthalic and copper-based MOF. The hydrogen bonding and π-π stacking between 5-FU and organic parts of MOF, along with tuning the pore sizes of MOF, resulted in the applaudable stepwise release of the drug [108, 109].

The pH stability of drug-carrier composite is highly desirable as the pH of the tumor cell's surroundings is less than blood and other tissues. pH-sensitive ZIF-8 is reported for its pH-dependent drug release performance for the 5-FU drug. ZIF-8 possesses large pores and a high surface area, but a small aperture as compared to the size of 5-FU. However, the flexible framework made it possible to load 5-FU into it. The drug release efficiency was more in acetate buffer (pH=5.0) than in phosphate-buffered saline (PBS) (pH=7.4), due to the degradation of ZIF-8 in the former. Hence, such pH dependent properties make such MOFs a suitable candidate for the delivery of anticancer drugs [110]. Thus, the above discussion provided an insight into versatility of MOF for various applications.

CONCLUSION

The presence of permanent porosity, high surface area, high density of open metal sites, tunable pores/surface and functional alterability made metal-organic framework (MOF) an outstanding candidate among solid adsorbents. Among a vast list of applications, the present chapter discussed the utility of MOFs for gas storage, gas separation and drug delivery. It was observed that the ability to tune and functionalize pore/surface sites allowed the customization of chemical interactions with gas molecules, resulting in high gas uptake and separation efficiency. Various MOF structures showed high values of isosteric enthalpies of adsorption at zero coverage, which was attributed to numerous open metal sites resulting in enhanced gas uptake through electrostatic interactions. In other cases, when electrostatic interactions were not possible or in a system with a low density of metal sites, surface or pores can be functionalized to interact with gas molecules. Apart from that, the ability to tune the size of MOF to nanoscale made it possible to use these frameworks as drug carriers. Excellent applications of MOFs in both drug-loading and drug-releasing efficiency are documented in the literature. Hence, it can be stated that MOFs are versatile and functionally modifiable materials, which can be customized for targeted applications. Owing to such a splendid set of properties, the field of exploration of MOFs is continuously attracting researchers.

REFERENCES

[1]   Bétard, A.; Fischer, R.A. Metal-organic framework thin films: from fundamentals to applications. *Chem. Rev.,* **2012**, *112*(2), 1055-1083.
[http://dx.doi.org/10.1021/cr200167v] [PMID: 21928861]

[2]   Makiura, R.; Motoyama, S.; Umemura, Y.; Yamanaka, H.; Sakata, O.; Kitagawa, H. Surface nano-architecture of a metal–organic framework. *Nat. Mater.,* **2010**, *9*(7), 565-571.
[http://dx.doi.org/10.1038/nmat2769] [PMID: 20512155]

[3]   Howarth, A.J.; Liu, Y.; Li, P.; Li, Z.; Wang, T.C.; Hupp, J.T.; Farha, O.K. Chemical, thermal and mechanical stabilities of metal–organic frameworks. *Nat. Rev. Mater.,* **2016**, *1*(3), 15018.
[http://dx.doi.org/10.1038/natrevmats.2015.18]

[4]   Yaghi, O.M.; O'Keeffe, M.; Ockwig, N.W.; Chae, H.K.; Eddaoudi, M.; Kim, J. Reticular synthesis and the design of new materials. *Nature,* **2003**, *423*(6941), 705-714.
[http://dx.doi.org/10.1038/nature01650] [PMID: 12802325]

[5]   Trickett, C.A.; Helal, A.; Al-Maythalony, B.A.; Yamani, Z.H.; Cordova, K.E.; Yaghi, O.M. The chemistry of metal–organic frameworks for CO_2 capture, regeneration and conversion. *Nat. Rev. Mater.,* **2017**, *2*(8), 17045.
[http://dx.doi.org/10.1038/natrevmats.2017.45]

[6]   Zhou, Y.; Lu, J.; Zhou, Y.; Liu, Y. Recent advances for dyes removal using novel adsorbents: A review. *Environ. Pollut.,* **2019**, *252*(Pt A), 352-365.
[http://dx.doi.org/10.1016/j.envpol.2019.05.072] [PMID: 31158664]

[7]   Goetjen, T.A.; Liu, J.; Wu, Y.; Sui, J.; Zhang, X.; Hupp, J.T.; Farha, O.K. Metal–organic framework (MOF) materials as polymerization catalysts: a review and recent advances. *Chem. Commun. (Camb.),* **2020**, *56*(72), 10409-10418.
[http://dx.doi.org/10.1039/D0CC03790G] [PMID: 32745156]

[8]   Luo, Z.; Fan, S.; Gu, C.; Liu, W.; Chen, J.; Li, B.; Liu, J. Metal-organic framework (MOF)-based nanomaterials for biomedicalapplications. *Curr. Med. Chem.,* **2019**, *26*(18), 3341-3369.
[http://dx.doi.org/10.2174/0929867325666180214123500] [PMID: 29446726]

[9]   Cao, J.; Li, X.; Tian, H. Metal-organic framework (MOF)-baseddrug delivery. *Curr. Med. Chem.,* **2020**, *27*(35), 5949-5969.
[http://dx.doi.org/10.2174/0929867326666190618152518] [PMID: 31215374]

[10]   Stavila, V.; Talin, A.A.; Allendorf, M.D. MOF-based electronic and opto-electronic devices. *Chem. Soc. Rev.,* **2014**, *43*(16), 5994-6010.
[http://dx.doi.org/10.1039/C4CS00096J] [PMID: 24802763]

[11]   Gascon, J.; Corma, A.; Kapteijn, F.; Llabrés i Xamena, F.X. Metal organic framework catalysis: quo vadis? ACS Catal., **2014**, *4*, 361.

[12]   Wang, H.; Li, J. Microporous metal-organic frameworks for adsorptive separation of C5-C6 alkane isomers. *Acc. Chem. Res.,* **2019**, *52*(7), 1968-1978.
[http://dx.doi.org/10.1021/acs.accounts.8b00658] [PMID: 30883088]

[13]   Li, B.; Wen, H.M.; Cui, Y.; Zhou, W.; Qian, G.; Chen, B. Emerging multifunctional metal-organic framework materials. *Adv. Mater.,* **2016**, *28*(40), 8819-8860.
[http://dx.doi.org/10.1002/adma.201601133] [PMID: 27454668]

[14]   Caro, J. Are MOF membranes better in gas separation than those made of zeolites? *Curr. Opin. Chem. Eng.,* **2011**, *1*(1), 77-83.
[http://dx.doi.org/10.1016/j.coche.2011.08.007]

[15]   Alezi, D.; Belmabkhout, Y.; Suyetin, M.; Bhatt, P.M.; Weseliński, Ł.J.; Solovyeva, V.; Adil, K.; Spanopoulos, I.; Trikalitis, P.N.; Emwas, A.H.; Eddaoudi, M. MOF crystal chemistry paving the way to gas storage needs: aluminum-based soc-MOF for CH_4, O_2, and CO_2 storage. *J. Am. Chem. Soc.,* **2015**, *137*(41), 13308-13318.

[http://dx.doi.org/10.1021/jacs.5b07053] [PMID: 26364990]

[16]   Li, Y.; Yang, W. Molecular sieve membranes: From 3D zeolites to 2D MOFs. *Chin. J. Catal.,* **2015**, *36*(5), 692-697.
[http://dx.doi.org/10.1016/S1872-2067(15)60838-5]

[17]   Rodenas, T.; Luz, I.; Prieto, G.; Seoane, B.; Miro, H.; Corma, A.; Kapteijn, F. iXamena, F. X. L., and Gascon, J. Metal-organic framework nanosheets in polymer composite materials forgas separation. *Nat. Mater.,* **2015**, *14*, 48.
[http://dx.doi.org/10.1038/nmat4113] [PMID: 25362353]

[18]   Kang, Z.; Xue, M.; Fan, L.; Huang, L.; Guo, L.; Wei, G.; Chen, B.; Qiu, S. Highly selective sieving of small gas molecules by using an ultra-microporous metal–organic framework membrane. *Energy Environ. Sci.,* **2014**, *7*(12), 4053-4060.
[http://dx.doi.org/10.1039/C4EE02275K]

[19]   Kwon, H.T.; Jeong, H.K.; Lee, A.S.; An, H.S.; Lee, J.S. Heteroepitaxially grown zeolitic imidazolate framework membranes with unprecedented propylene/propane separation performances. *J. Am. Chem. Soc.,* **2015**, *137*(38), 12304-12311.
[http://dx.doi.org/10.1021/jacs.5b06730] [PMID: 26364888]

[20]   Wang, N.; Mundstock, A.; Liu, Y.; Huang, A.; Caro, J. Amine-modified Mg-MOF-74/CPO-27-Mg membrane with enhanced H_2/CO_2 separation. *Chem. Eng. Sci.,* **2015**, *124*, 27-36.
[http://dx.doi.org/10.1016/j.ces.2014.10.037]

[21]   Fu, J.; Das, S.; Xing, G.; Ben, T.; Valtchev, V.; Qiu, S. Fabricationof COF-MOF composite membranes and their highly selectiveseparation of H_2/CO_2. *J. Am. Chem. Soc.,* **2016**, *138*(24), 7673-7680.
[http://dx.doi.org/10.1021/jacs.6b03348] [PMID: 27225027]

[22]   Stock, N.; Biswas, S. Synthesis of metal-organic frameworks (MOFs): routes to various MOF topologies, morphologies, and composites. *Chem. Rev.,* **2012**, *112*(2), 933-969.
[http://dx.doi.org/10.1021/cr200304e] [PMID: 22098087]

[23]   Chalati, T.; Horcajada, P.; Gref, R.; Couvreur, P.; Serre, C. Optimisation of the synthesis of MOF nanoparticles made of flexible porous iron fumarate MIL-88A. *J. Mater. Chem.,* **2011**, *21*(7), 2220-2227.
[http://dx.doi.org/10.1039/C0JM03563G]

[24]   Cho, W.; Lee, H.J.; Oh, M. Growth-controlled formation of porous coordination polymer particles. *J. Am. Chem. Soc.,* **2008**, *130*(50), 16943-16946.
[http://dx.doi.org/10.1021/ja8039794] [PMID: 19007221]

[25]   Wang, L.; Zheng, M.; Xie, Z. Nanoscale metal–organic frameworks for drug delivery: a conventional platform with new promise. *J. Mater. Chem. B Mater. Biol. Med.,* **2018**, *6*(5), 707-717.
[http://dx.doi.org/10.1039/C7TB02970E] [PMID: 32254257]

[26]   Cai, X.; Xie, Z.; Li, D.; Kassymova, M.; Zang, S.Q.; Jiang, H.L. Nano-sized metal-organic frameworks: Synthesis and applications. *Coord. Chem. Rev.,* **2020**, *417*213366
[http://dx.doi.org/10.1016/j.ccr.2020.213366]

[27]   Rieter, W.J.; Taylor, K.M.L.; An, H.; Lin, W.; Lin, W. Nanoscale metal-organic frameworks as potential multimodal contrast enhancing agents. *J. Am. Chem. Soc.,* **2006**, *128*(28), 9024-9025.
[http://dx.doi.org/10.1021/ja0627444] [PMID: 16834362]

[28]   Qiu, L-G.; Li, Z-Q.; Wu, Y.; Wang, W.; Xu, T.; Jiang, X. Facile synthesis of nanocrystals of a microporous metal-organic framework by an ultrasonic method and selective sensing oforganoamines. *ChemComm,* **2008**, *31*, 3642.

[29]   Jhung, S.H.; Lee, J.H.; Yoon, J.W.; Serre, C.; Férey, G.; Chang, J-S. andChang, J.-S. Microwave synthesis of chromium terephthalate MIL-101and its benzene sorption ability. *Adv. Mater.,* **2007**, *19*(1), 121-124.

[http://dx.doi.org/10.1002/adma.200601604]

[30]   Li, G.; Kobayashi, H.; Taylor, J.M.; Ikeda, R.; Kubota, Y.; Kato, K.; Takata, M.; Yamamoto, T.; Toh,
        S.; Matsumura, S.; Kitagawa, H. Hydrogen storage in Pd nanocrystals covered with a metal–organic
        framework. *Nat. Mater.,* **2014**, *13*(8), 802-806.
        [http://dx.doi.org/10.1038/nmat4030] [PMID: 25017188]

[31]   Li, H.; Wang, K.; Sun, Y.; Lollar, C.T.; Li, J.; Zhou, H.C. Recent advances in gas storage and
        separation using metal–organic frameworks. *Mater. Today,* **2018**, *21*(2), 108-121.
        [http://dx.doi.org/10.1016/j.mattod.2017.07.006]

[32]   Britt, D.; Furukawa, H.; Wang, B.; Glover, T.G.; Yaghi, O.M. Highly efficient separation of carbon
        dioxide by a metal-organic framework replete with open metal sites. *Proc. Natl. Acad. Sci. USA,* **2009**,
        106(49), 20637-20640.
        [http://dx.doi.org/10.1073/pnas.0909718106] [PMID: 19948967]

[33]   Simonyan, V.V.; Diep, P.; Johnson, J.K. Molecular simulation of hydrogen adsorption in charged
        single-walled carbon nanotubes. *J. Chem. Phys.,* **1999**, *111*(21), 9778-9783.
        [http://dx.doi.org/10.1063/1.480313]

[34]   Furukawa, H.; Ko, N.; Go, Y.B.; Aratani, N.; Choi, S.B.; Choi, E.; Yazaydin, A.Ö.; Snurr, R.Q.;
        O'Keeffe, M.; Kim, J.; Yaghi, O.M. Ultrahigh porosity in metal-organic frameworks. *Science,* **2010**,
        329(5990), 424-428.
        [http://dx.doi.org/10.1126/science.1192160] [PMID: 20595583]

[35]   Jiang, J.; Furukawa, H.; Zhang, Y.B.; Yaghi, O.M. Highmethane storage working capacity in metal-
        organic frameworks withacrylate links. *J. Am. Chem. Soc.,* **2016**, *138*(32), 10244-10251.
        [http://dx.doi.org/10.1021/jacs.6b05261] [PMID: 27442620]

[36]   Verma, G.; Kumar, S.; Vardhan, H.; Ren, J.; Niu, Z.; Pham, T.; Wojtas, L.; Butikofer, S.; Echeverria
        Garcia, J.C.; Chen, Y.S.; Space, B.; Ma, S. A robust soc-MOF platform exhibiting high gravimetric
        uptake and volumetric deliverable capacity for on-board methane storage. *Nano Res.,* **2021**, *14*(2),
        512-517.
        [http://dx.doi.org/10.1007/s12274-020-2794-9]

[37]   Gándara, F.; Furukawa, H.; Lee, S.; Yaghi, O.M. High methane storage capacity in aluminum metal-
        organic frameworks. *J. Am. Chem. Soc.,* **2014**, *136*(14), 5271-5274.
        [http://dx.doi.org/10.1021/ja501606h] [PMID: 24661065]

[38]   He, Y.; Zhou, W.; Yildirim, T.; Chen, B. A series of metal–organic frameworks with high methane
        uptake and an empirical equation for predicting methane storage capacity. *Energy Environ. Sci.,* **2013**,
        6(9), 2735.
        [http://dx.doi.org/10.1039/c3ee41166d]

[39]   Mason, J.A.; Oktawiec, J.; Taylor, M.K.; Hudson, M.R.; Rodriguez, J.; Bachman, J.E.; Gonzalez, M.I.;
        Cervellino, A.; Guagliardi, A.; Brown, C.M.; Llewellyn, P.L.; Masciocchi, N.; Long, J.R. Methane
        storage in flexible metal–organic frameworks with intrinsic thermal management. *Nature,* **2015**,
        527(7578), 357-361.
        [http://dx.doi.org/10.1038/nature15732] [PMID: 26503057]

[40]   Tian, T.; Zeng, Z.; Vulpe, D.; Casco, M.E.; Divitini, G.; Midgley, P.A.; Silvestre-Albero, J.; Tan, J-C.;
        Moghadam, P.Z. andFairen-Jimenez, D. A sol-gel monolithic metal-organic frameworkwith enhanced
        methane uptake. *Nat. Mater.,* **2018**, *17*, 174.
        [http://dx.doi.org/10.1038/nmat5050] [PMID: 29251723]

[41]   Lin, J.M.; He, C.T.; Liu, Y.; Liao, P.Q.; Zhou, D.D.; Zhang, J.P.; Chen, X.M. A metal-organic
        framework with a poresize/shape suitable for strong binding and close packing of methane. *Angew.
        Chem.,* **2016**, *128*(15), 4752-4756.
        [http://dx.doi.org/10.1002/ange.201511006]

[42]   Li, B.; Wen, H.M.; Wang, H.; Wu, H.; Tyagi, M.; Yildirim, T.; Zhou, W.; Chen, B. A porous metal-
        organic framework with dynamic pyrimidine groups exhibiting record high methane storage working

capacity. *J. Am. Chem. Soc.,* **2014**, *136*(17), 6207-6210.
[http://dx.doi.org/10.1021/ja501810r] [PMID: 24730649]

[43]   Chen, Y.; Qiao, Z.; Huang, J.; Wu, H.; Xiao, J.; Xia, Q.; Xi, H.; Hu, J.; Zhou, J.; Li, Z. Unusual moisture-enhanced CO_2 capture within microporous PCN-250 frameworks. *ACS Appl. Mater. Interfaces,* **2018**, *10*(44), 38638-38647.
[http://dx.doi.org/10.1021/acsami.8b14400] [PMID: 30360051]

[44]   Barona, M.; Ahn, S.; Morris, W.; Hoover, W.; Notestein, J.M.; Farha, O.K.; Snurr, R.Q. Computational predictions andexperimental validation of alkane oxidative dehydrogenation by Fe_2MMOF nodes. *ACS Catal.,* **2020**, *10*(2), 1460-1469.
[http://dx.doi.org/10.1021/acscatal.9b03932]

[45]   Witman, M.; Ling, S.; Anderson, S.; Tong, L.; Stylianou, K.C.; Slater, B.; Smit, B.; Haranczyk, M. *In silico* design and screening of hypothetical MOF-74 analogs and their experimental synthesis. *Chem. Sci. (Camb.),* **2016**, *7*(9), 6263-6272.
[http://dx.doi.org/10.1039/C6SC01477A] [PMID: 30034767]

[46]   Mason, J.A.; Veenstra, M.; Long, J.R. Evaluating metal–organic frameworks for natural gas storage. *Chem. Sci. (Camb.),* **2014**, *5*(1), 32-51.
[http://dx.doi.org/10.1039/C3SC52633J]

[47]   Wu, H.; Zhou, W.; Yildirim, T. High-capacity methane storage in metal-organic frameworks M_2(dhtp): the important role of open metal sites. *J. Am. Chem. Soc.,* **2009**, *131*(13), 4995-5000.
[http://dx.doi.org/10.1021/ja900258t] [PMID: 19275154]

[48]   Dietzel, P.D.C.; Besikiotis, V.; Blom, R. Application of metal–organic frameworks with coordinatively unsaturated metal sites in storage and separation of methane and carbon dioxide. *J. Mater. Chem.,* **2009**, *19*(39), 7362.
[http://dx.doi.org/10.1039/b911242a]

[49]   He, Y.; Zhou, W.; Qian, G.; Chen, B. Methane storage in metal–organic frameworks. *Chem. Soc. Rev.,* **2014**, *43*(16), 5657-5678.
[http://dx.doi.org/10.1039/C4CS00032C] [PMID: 24658531]

[50]   Al-Naddaf, Q.; Al-Mansour, M.; Thakkar, H.; Rezaei, F. MOF-GOhybrid nanocomposite adsorbents for methane storage. *Ind. Eng. Chem. Res.,* **2018**, *57*(51), 17470-17479.
[http://dx.doi.org/10.1021/acs.iecr.8b03638]

[51]   Li, S.; Chung, Y.G.; Snurr, R.Q. High-throughput screeningof metal-organic frameworks for CO_2 capture in the presence of water. *Langmuir,* **2016**, *32*(40), 10368-10376.
[http://dx.doi.org/10.1021/acs.langmuir.6b02803] [PMID: 27627635]

[52]   Olajire, A.A. Synthesis chemistry of metal-organic frameworks for CO_2 capture and conversion for sustainable energy future. *Renew. Sustain. Energy Rev.,* **2018**, *92*, 570-607.
[http://dx.doi.org/10.1016/j.rser.2018.04.073]

[53]   Vitillo, J.G.; Savonnet, M.; Ricchiardi, G.; Bordiga, S. Tailoring metal-organic frameworks for CO_2 capture: the amino effect. *ChemSusChem,* **2011**, *4*(9), 1281-1290.
[http://dx.doi.org/10.1002/cssc.201000458] [PMID: 21922680]

[54]   Flaig, R.W.; Osborn Popp, T.M.; Fracaroli, A.M.; Kapustin, E.A.; Kalmutzki, M.J.; Altamimi, R.M.; Fathieh, F.; Reimer, J.A.; Yaghi, O.M. The chemistry of CO_2 capture in an aminefunctionalizedmetal-organic framework under dry and humid conditions. *J. Am. Chem. Soc.,* **2017**, *139*(35), 12125-12128.
[http://dx.doi.org/10.1021/jacs.7b06382] [PMID: 28817269]

[55]   Bollini, P.; Didas, S.A.; Jones, C.W. Amine-oxide hybrid materials for acid gas separations. *J. Mater. Chem.,* **2011**, *21*(39), 15100.
[http://dx.doi.org/10.1039/c1jm12522b]

[56]   Didas, S.A.; Sakwa-Novak, M.A.; Foo, G.S.; Sievers, C.; Jones, C.W. andJones, C. W. Effect of amine surface coverage on the co-adsorptionof CO_2 and water: spectral deconvolution of adsorbed species. *J.*

Phys. Chem. Lett., **2014**, *5*(23), 4194-4200.
[http://dx.doi.org/10.1021/jz502032c] [PMID: 26278953]

[57]   Mebane, D.S.; Kress, J.D.; Storlie, C.B.; Fauth, D.J.; Gray, M.L.; Li, K. Transport, zwitterions, and the role of waterfor CO_2 adsorption in mesoporous silica-supported amine sorbents. *J. Phys. Chem. C,* **2013**, *117*(50), 26617-26627.
[http://dx.doi.org/10.1021/jp4076417]

[58]   McDonald, T.M.; Mason, J.A.; Kong, X.; Bloch, E.D.; Gygi, D.; Dani, A.; Crocellà, V.; Giordanino, F.; Odoh, S.O.; Drisdell, W.S.; Vlaisavljevich, B.; Dzubak, A.L.; Poloni, R.; Schnell, S.K.; Planas, N.; Lee, K.; Pascal, T.; Wan, L.F.; Prendergast, D.; Neaton, J.B.; Smit, B.; Kortright, J.B.; Gagliardi, L.; Bordiga, S.; Reimer, J.A.; Long, J.R. Cooperative insertion of CO_2 in diamine-appended metal-organic frameworks. *Nature,* **2015**, *519*(7543), 303-308.
[http://dx.doi.org/10.1038/nature14327] [PMID: 25762144]

[59]   Darunte, L.A.; Walton, K.S.; Sholl, D.S.; Jones, C.W. CO_2 capture *via* adsorption in amine-functionalized sorbents. *Curr. Opin. Chem. Eng.,* **2016**, *12*, 82-90.
[http://dx.doi.org/10.1016/j.coche.2016.03.002]

[60]   Zhang, H.; Yang, L.M.; Ganz, E. Unveiling the molecularmechanism of CO_2 capture in *n*-methylethylenediamine-grafted M_2(dobpdc). *ACS Sustain. Chem.& Eng.,* **2020**, *8*(38), 14616-14626.
[http://dx.doi.org/10.1021/acssuschemeng.0c05951]

[61]   McDonald, T.M.; Lee, W.R.; Mason, J.A.; Wiers, B.M.; Hong, C.S.; Long, J.R. Capture of carbon dioxide from air and flue gas in the alkylamine-appended metal-organic framework mmen-Mg_2(dobpdc). *J. Am. Chem. Soc.,* **2012**, *134*(16), 7056-7065.
[http://dx.doi.org/10.1021/ja300034j] [PMID: 22475173]

[62]   Lee, W.R.; Jo, H.; Yang, L.M.; Lee, H.; Ryu, D.W.; Lim, K.S.; Song, J.H.; Min, D.Y.; Han, S.S.; Seo, J.G.; Park, Y.K.; Moon, D.; Hong, C.S. Exceptional CO_2 working capacity in a heterodiamine-grafted metal–organic framework. *Chem. Sci. (Camb.),* **2015**, *6*(7), 3697-3705.
[http://dx.doi.org/10.1039/C5SC01191D] [PMID: 28706716]

[63]   Sumida, K.; Rogow, D.L.; Mason, J.A.; McDonald, T.M.; Bloch, E.D.; Herm, Z.R.; Bae, T.H.; Long, J.R. Carbon dioxide capture in metal-organic frameworks. *Chem. Rev.,* **2012**, *112*(2), 724-781.
[http://dx.doi.org/10.1021/cr2003272] [PMID: 22204561]

[64]   Peng, Y.; Li, Y.; Ban, Y.; Jin, H.; Jiao, W.; Liu, X.; Yang, W. Metal-organic framework nanosheets as building blocks for molecular sieving membranes. *Science,* **2014**, *346*(6215), 1356-1359.
[http://dx.doi.org/10.1126/science.1254227] [PMID: 25504718]

[65]   Lee, W.R.; Hwang, S.Y.; Ryu, D.W.; Lim, K.S.; Han, S.S.; Moon, D.; Choi, J.; Hong, C.S. Diamine-functionalized metal–organic framework: exceptionally high CO_2 capacities from ambient air and flue gas, ultrafast CO_2 uptake rate, and adsorption mechanism. *Energy Environ. Sci.,* **2014**, *7*(2), 744-751.
[http://dx.doi.org/10.1039/C3EE42328J]

[66]   Jo, H.; Lee, W.R.; Kim, N.W.; Jung, H.; Lim, K.S.; Kim, J.E.; Kang, D.W.; Lee, H.; Hiremath, V.; Seo, J.G.; Jin, H.; Moon, D.; Han, S.S.; Hong, C.S. Fine-tuning of the carbondioxide capture capability of diamine-grafted metal-organic frameworkadsorbents through amine functionalization. *ChemSusChem,* **2017**, *10*(3), 541-550.
[http://dx.doi.org/10.1002/cssc.201601203] [PMID: 28004886]

[67]   Barea, E.; Montoro, C.; Navarro, J.A.R. Toxic gas removal – metal–organic frameworks for the capture and degradation of toxic gases and vapours. *Chem. Soc. Rev.,* **2014**, *43*(16), 5419-5430.
[http://dx.doi.org/10.1039/C3CS60475F] [PMID: 24705539]

[68]   Bloch, E.D.; Hudson, M.R.; Mason, J.A.; Chavan, S.; Crocellà, V.; Howe, J.D.; Lee, K.; Dzubak, A.L.; Queen, W.L.; Zadrozny, J.M.; Geier, S.J.; Lin, L.C.; Gagliardi, L.; Smit, B.; Neaton, J.B.; Bordiga, S.; Brown, C.M.; Long, J.R. Reversible CO binding enables tunable CO/H_2 and CO/N_2 separations in metal-organic frameworks with exposed divalent metal cations. *J. Am. Chem. Soc.,* **2014**, *136*(30), 10752-10761.

[http://dx.doi.org/10.1021/ja505318p] [PMID: 24999916]

[69]   Reed, D.A.; Xiao, D.J.; Gonzalez, M.I.; Darago, L.E.; Herm, Z.R.; Grandjean, F.; Long, J.R. Reversible CO scavenging viaadsorbate-dependent spin state transitions in an iron (ii)-triazolatemetal-organic framework. *J. Am. Chem. Soc.,* **2016**, *138*(17), 5594-5602. [http://dx.doi.org/10.1021/jacs.6b00248] [PMID: 27097297]

[70]   Rieth, A.J.; Tulchinsky, Y.; Dincă, M. High and reversibleammonia uptake in mesoporous azolate metal-organic frameworkswith open Mn, Co, and Ni sites. *J. Am. Chem. Soc.,* **2016**, *138*(30), 9401-9404. [http://dx.doi.org/10.1021/jacs.6b05723] [PMID: 27420652]

[71]   Van Humbeck, J.F.; McDonald, T.M.; Jing, X.; Wiers, B.M.; Zhu, G.; Long, J.R. Ammonia capture in porous organic polymers densely functionalized with Brønsted acid groups. *J. Am. Chem. Soc.,* **2014**, *136*(6), 2432-2440. [http://dx.doi.org/10.1021/ja4105478] [PMID: 24456083]

[72]   Hamon, L.; Serre, C.; Devic, T.; Loiseau, T.; Millange, F.; Férey, G.; Weireld, G.D. Comparative study of hydrogen sulfide adsorption in the MIL-53(Al, Cr, Fe), MIL-47(V), MIL-100(Cr), and MIL-101(Cr) metal-organic frameworks at room temperature. *J. Am. Chem. Soc.,* **2009**, *131*(25), 8775-8777. [http://dx.doi.org/10.1021/ja901587t] [PMID: 19505146]

[73]   Hamon, L.; Leclerc, H.; Ghoufi, A.; Oliviero, L.; Travert, A.; Lavalley, J.C.; Devic, T.; Serre, C.; Férey, G.; De Weireld, G.; Vimont, A.; Maurin, G. Molecular insight into the adsorptionof H_2S in the flexible MIL-53 (Cr) and rigid MIL-47 (V) MOFs: infraredspectroscopy combined to molecular simulations. *J. Phys. Chem. C,* **2011**, *115*(5), 2047-2056. [http://dx.doi.org/10.1021/jp1092724]

[74]   Allan, P.K.; Wheatley, P.S.; Aldous, D.; Mohideen, M.I.; Tang, C.; Hriljac, J.A.; Megson, I.L.; Chapman, K.W.; De Weireld, G.; Vaesen, S.; Morris, R.E. Metal–organic frameworks for the storage and delivery of biologically active hydrogen sulfide. *Dalton Trans.,* **2012**, *41*(14), 4060-4066. [http://dx.doi.org/10.1039/c2dt12069k] [PMID: 22378060]

[75]   Ingleson, M.J.; Heck, R.; Gould, J.A.; Rosseinsky, M.J. Nitric oxide chemisorption in a postsynthetically modified metal-organic framework. *Inorg. Chem.,* **2009**, *48*(21), 9986-9988. [http://dx.doi.org/10.1021/ic9015977] [PMID: 19795833]

[76]   Lin, R.B.; Xiang, S.; Xing, H.; Zhou, W.; Chen, B. Exploration of porous metal–organic frameworks for gas separation and purification. *Coord. Chem. Rev.,* **2019**, *378*, 87-103. [http://dx.doi.org/10.1016/j.ccr.2017.09.027]

[77]   Kang, Z.; Fan, L.; Sun, D. Recent advances and challenges of metal–organic framework membranes for gas separation. *J. Mater. Chem. A Mater. Energy Sustain.,* **2017**, *5*(21), 10073-10091. [http://dx.doi.org/10.1039/C7TA01142C]

[78]   Zhang, Y.; Feng, X.; Yuan, S.; Zhou, J.; Wang, B. Challenges and recent advances in MOF–polymer composite membranes for gas separation. *Inorg. Chem. Front.,* **2016**, *3*(7), 896-909. [http://dx.doi.org/10.1039/C6QI00042H]

[79]   Adatoz, E.; Avci, A.K.; Keskin, S. Opportunities and challenges of MOF-based membranes in gas separations. *Separ. Purif. Tech.,* **2015**, *152*, 207-237. [http://dx.doi.org/10.1016/j.seppur.2015.08.020]

[80]   Marti, A.M.; Venna, S.R.; Roth, E.A.; Culp, J.T.; Hopkinson, D.P. Simple fabrication method for mixed matrix membranes within situ MOF growth for gas separation. *ACS Appl. Mater. Interfaces,* **2018**, *10*(29), 24784-24790. [http://dx.doi.org/10.1021/acsami.8b06592] [PMID: 29952556]

[81]   Shahid, S.; Nijmeijer, K.; Nehache, S.; Vankelecom, I.; Deratani, A.; Quemener, D. MOF-mixed matrix membranes: Precise dispersion of MOF particles with better compatibility *via* a particle fusion approach for enhanced gas separation properties. *J. Membr. Sci.,* **2015**, *492*, 21-31.

[http://dx.doi.org/10.1016/j.memsci.2015.05.015]

[82] Li, J.R.; Kuppler, R.J.; Zhou, H.C. Selective gas adsorption and separation in metal–organic frameworks. *Chem. Soc. Rev.,* **2009**, *38*(5), 1477-1504.
[http://dx.doi.org/10.1039/b802426j] [PMID: 19384449]

[83] Dincă, M.; Long, J.R. Strong H$_{(2)}$ binding and selective gas adsorption within the microporous coordination solid Mg$_{(3)}$(O($_{(2)}$C-C($_{(10)}$H($_{(6)}$)-CO($_{(2)}$))($_{(3)}$. *J. Am. Chem. Soc.,* **2005**, *127*(26), 9376-9377.
[http://dx.doi.org/10.1021/ja0523082] [PMID: 15984858]

[84] Ma, S.; Wang, X.S.; Collier, C.D.; Manis, E.S.; Zhou, H.C. Ultramicroporous metal-organic framework based on 9,10-anthracenedicarboxylate for selective gas adsorption. *Inorg. Chem.,* **2007**, *46*(21), 8499-8501.
[http://dx.doi.org/10.1021/ic701507r] [PMID: 17854186]

[85] Navarro, J.A.R.; Barea, E.; Rodríguez-Diéguez, A.; Salas, J.M.; Ania, C.O.; Parra, J.B.; Masciocchi, N.; Galli, S.; Sironi, A. Guest-induced modification of a magnetically active ultramicroporous, gismondine-like, copper(II) coordination network. *J. Am. Chem. Soc.,* **2008**, *130*(12), 3978-3984.
[http://dx.doi.org/10.1021/ja078074z] [PMID: 18321099]

[86] Matsuda, R.; Kitaura, R.; Kitagawa, S.; Kubota, Y.; Belosludov, R.V.; Kobayashi, T.C.; Sakamoto, H.; Chiba, T.; Takata, M.; Kawazoe, Y.; Mita, Y. Highly controlled acetylene accommodation in a metal–organic microporous material. *Nature,* **2005**, *436*(7048), 238-241.
[http://dx.doi.org/10.1038/nature03852] [PMID: 16015325]

[87] Yang, R.T. *Adsorbents: fundamentals and applications*; John Wiley& Sons, **2003**.
[http://dx.doi.org/10.1002/047144409X]

[88] Dubbeldam, D.; Galvin, C.J.; Walton, K.S.; Ellis, D.E.; Snurr, R.Q. Separation and molecular-level segregation of complex alkane mixtures in metal-organic frameworks. *J. Am. Chem. Soc.,* **2008**, *130*(33), 10884-10885.
[http://dx.doi.org/10.1021/ja804039c] [PMID: 18651737]

[89] Chen, B.; Liang, C.; Yang, J.; Contreras, D.S.; Clancy, Y.L.; Lobkovsky, E.B.; Yaghi, O.M.; Dai, S. A microporous metal-organic framework for gas-chromatographic separation of alkanes. *Angew. Chem. Int. Ed.,* **2006**, *45*(9), 1390-1393.
[http://dx.doi.org/10.1002/anie.200502844] [PMID: 16425335]

[90] Xiang, S.C.; Zhang, Z.; Zhao, C.G.; Hong, K.; Zhao, X.; Ding, D.R.; Xie, M.H.; Wu, C.D.; Das, M.C.; Gill, R.; Thomas, K.M.; Chen, B. Rationally tuned micropores within enantiopure metal-organic frameworks for highly selective separation of acetylene and ethylene. *Nat. Commun.,* **2011**, *2*(1), 204.
[http://dx.doi.org/10.1038/ncomms1206] [PMID: 21343922]

[91] He, Y.; Krishna, R.; Chen, B. Metal–organic frameworks with potential for energy-efficient adsorptive separation of light hydrocarbons. *Energy Environ. Sci.,* **2012**, *5*(10), 9107.
[http://dx.doi.org/10.1039/c2ee22858k]

[92] Hu, T.L.; Wang, H.; Li, B.; Krishna, R.; Wu, H.; Zhou, W.; Zhao, Y.; Han, Y.; Wang, X.; Zhu, W.; Yao, Z.; Xiang, S.; Chen, B. Microporous metal–organic framework with dual functionalities for highly efficient removal of acetylene from ethylene/acetylene mixtures. *Nat. Commun.,* **2015**, *6*(1), 7328.
[http://dx.doi.org/10.1038/ncomms8328]

[93] Luo, F.; Yan, C.; Dang, L.; Krishna, R.; Zhou, W.; Wu, H.; Dong, X.; Han, Y.; Hu, T.L.; O'Keeffe, M.; Wang, L.; Luo, M.; Lin, R.B.; Chen, B. UTSA-74: a MOF-74 isomer with two accessiblebinding sites per metal center for highly selective gas separation. *J. Am. Chem. Soc.,* **2016**, *138*(17), 5678-5684.
[http://dx.doi.org/10.1021/jacs.6b02030] [PMID: 27113684]

[94] Lin, R.B.; Li, L.; Wu, H.; Arman, H.; Li, B.; Lin, R.G.; Zhou, W.; Chen, B. Optimized separation of acetylene from carbondioxide and ethylene in a microporous material. *J. Am. Chem. Soc.,* **2017**, *139*(23), 8022-8028.

[http://dx.doi.org/10.1021/jacs.7b03850] [PMID: 28574717]

[95]   Li, L.; Lin, R.B.; Krishna, R.; Wang, X.; Li, B.; Wu, H.; Li, J.; Zhou, W.; Chen, B. Flexible-robust metal-organic frameworkfor efficient removal of propyne from propylene. *J. Am. Chem. Soc.,* **2017,** *139*(23), 7733-7736.
[http://dx.doi.org/10.1021/jacs.7b04268] [PMID: 28580788]

[96]   Sun, C.Y.; Qin, C.; Wang, X.L.; Su, Z.M. Metal-organic frameworks as potential drug delivery systems. *Expert Opin. Drug Deliv.,* **2013,** *10*(1), 89-101.
[http://dx.doi.org/10.1517/17425247.2013.741583] [PMID: 23140545]

[97]   Horcajada, P.; Gref, R.; Baati, T.; Allan, P.K.; Maurin, G.; Couvreur, P.; Férey, G.; Morris, R.E.; Serre, C. Metal-organic frameworks in biomedicine. *Chem. Rev.,* **2012,** *112*(2), 1232-1268.
[http://dx.doi.org/10.1021/cr200256v] [PMID: 22168547]

[98]   An, J.; Geib, S.J.; Rosi, N.L. Cation-triggered drug release from a porous zinc-adeninate metal-organic framework. *J. Am. Chem. Soc.,* **2009,** *131*(24), 8376-8377.
[http://dx.doi.org/10.1021/ja902972w] [PMID: 19489551]

[99]   Nadar, S.S.; Vaidya, L.; Maurya, S.; Rathod, V.K. Polysaccharide based metal organic frameworks (polysaccharide–MOF): A review. *Coord. Chem. Rev.,* **2019,** *396*, 1-21.
[http://dx.doi.org/10.1016/j.ccr.2019.05.011]

[100] Horcajada, P.; Chalati, T.; Serre, C.; Gillet, B.; Sebrie, C.; Baati, T.; Eubank, J.F.; Heurtaux, D.; Clayette, P.; Kreuz, C.; Chang, J.S.; Hwang, Y.K.; Marsaud, V.; Bories, P.N.; Cynober, L.; Gil, S.; Férey, G.; Couvreur, P.; Gref, R. Porous metal–organic-framework nanoscale carriers as a potential platform for drug delivery and imaging. *Nat. Mater.,* **2010,** *9*(2), 172-178.
[http://dx.doi.org/10.1038/nmat2608] [PMID: 20010827]

[101] Taylor, K.M.L.; Rieter, W.J.; Lin, W. Manganese-based nanoscale metal-organic frameworks for magnetic resonance imaging. *J. Am. Chem. Soc.,* **2008,** *130*(44), 14358-14359.
[http://dx.doi.org/10.1021/ja803777x] [PMID: 18844356]

[102] Horcajada, P.; Márquez-Alvarez, C.; Rámila, A.; Pérez-Pariente, J.; Vallet-Regí, M. Controlled release of Ibuprofen from dealuminated faujasites. *Solid State Sci.,* **2006,** *8*(12), 1459-1465.
[http://dx.doi.org/10.1016/j.solidstatesciences.2006.07.016]

[103] Horcajada, P.; Serre, C.; Maurin, G.; Ramsahye, N.A.; Balas, F.; Vallet-Regí, M.; Sebban, M.; Taulelle, F.; Férey, G. Flexible porous metal-organic frameworks for a controlled drug delivery. *J. Am. Chem. Soc.,* **2008,** *130*(21), 6774-6780.
[http://dx.doi.org/10.1021/ja710973k] [PMID: 18454528]

[104] Orellana-Tavra, C.; Baxter, E.F.; Tian, T.; Bennett, T.D.; Slater, N.K.H.; Cheetham, A.K.; Fairen-Jimenez, D. Amorphous metal–organic frameworks for drug delivery. *Chem. Commun. (Camb.),* **2015,** *51*(73), 13878-13881.
[http://dx.doi.org/10.1039/C5CC05237H] [PMID: 26213904]

[105] Rieter, W.J.; Pott, K.M.; Taylor, K.M.L.; Lin, W. Nanoscale coordination polymers for platinum-based anticancer drug delivery. *J. Am. Chem. Soc.,* **2008,** *130*(35), 11584-11585.
[http://dx.doi.org/10.1021/ja803383k] [PMID: 18686947]

[106] Miller, S.R.; Heurtaux, D.; Baati, T.; Horcajada, P.; Grenèche, J.M.; Serre, C. Biodegradable therapeutic MOFs for the delivery of bioactive molecules. *Chem. Commun. (Camb.),* **2010,** *46*(25), 4526-4528.
[http://dx.doi.org/10.1039/c001181a] [PMID: 20467672]

[107] de Paiva, R.E.F.; Abbehausen, C.; Gomes, A.F.; Gozzo, F.C.; Lustri, W.R.; Formiga, A.L.B.; Corbi, P.P. Synthesis, spectroscopic characterization, DFT studies and antibacterial assays of a novel silver(I) complex with the anti-inflammatory nimesulide. *Polyhedron,* **2012,** *36*(1), 112-119.
[http://dx.doi.org/10.1016/j.poly.2012.02.002]

[108] Wang, H.N.; Meng, X.; Yang, G.S.; Wang, X.L.; Shao, K.Z.; Su, Z.M.; Wang, C.G. Stepwise

assembly of metal–organic framework based on a metal–organic polyhedron precursor for drug delivery. *Chem. Commun. (Camb.),* **2011**, *47*(25), 7128-7130.
[http://dx.doi.org/10.1039/c1cc11932j] [PMID: 21614372]

[109]   Sun, C.Y.; Qin, C.; Wang, C.G.; Su, Z.M.; Wang, S.; Wang, X.L.; Yang, G.S.; Shao, K.Z.; Lan, Y.Q.; Wang, E.B. Chiral nanoporous metal-organic frameworks with high porosity as materials for drug delivery. *Adv. Mater.,* **2011**, *23*(47), 5629-5632.
[http://dx.doi.org/10.1002/adma.201102538] [PMID: 22095878]

[110]   Sun, C.Y.; Qin, C.; Wang, X.L.; Yang, G.S.; Shao, K.Z.; Lan, Y.Q.; Su, Z.M.; Huang, P.; Wang, C.G.; Wang, E.B. Zeolitic imidazolate framework-8 as efficient pH-sensitive drug delivery vehicle. *Dalton Trans.,* **2012**, *41*(23), 6906-6909.
[http://dx.doi.org/10.1039/c2dt30357d] [PMID: 22580798]

CHAPTER 8

Understanding Synthesis and Characterization of Oxide Semiconductor Nanostructures through the Example of Nanostructured Nickel Doped Hematite

Sharmila Kumari Arodhiya[1,2], Jaspreet Kocher[1], Jiri Pechousek[3], Shashank Priya[4], Ashok Kumar[5,*] and Shyam Sundar Pattnaik[6]

[1] *Department of Physics, National Institute of Technology, Kurukshetra-136119, Haryana, India*

[2] *Department of Physics, Rajiv Gandhi Government College for Women, Bhiwani-127021, Haryana, India*

[3] *Regional Centre of Advanced Technologies and Materials, Department of Experimental Physics, Faculty of Science, Palacky University, Olomouc, Czech Republic*

[4] *Materials Research Institute, Penn State University, PA 16801, USA*

[5] *Department of Applied Sciences, National Institute of Technical Teachers Training and Research, Chandigarh-160019, India*

[6] *Media Engineering, National Institute of Technical Teachers Training and Research, Chandigarh-160019, India*

Abstract: Hematite is an n-type semiconductor, and its semiconducting properties can further be improved by nano-structuring and doping. In several optoelectronic devices, such as thermoelectric and solar cells, both n- and p-type semiconductors are required. The p-type hematite can be synthesized by doping cations, such as Ni^{2+}, Mg^{2+}, Cu^{2+}, and Mn^{2+}. Furthermore, hematite is a weak ferromagnetic material, and its magnetic properties vary with the size of nanoparticles, doping of cations as well as doping concentration. This chapter discusses various properties of nanostructured nickel-doped hematite. As nickel is a ferromagnetic divalent dopant with a high magnetic moment, its doping in hematite together with nano-structuring shows a large variation in both electrical and magnetic properties in nickel.

Keywords: Hematite, Magnetic behaviour, Nano-structuring, Nickel doped.

---

* **Corresponding author Ashok Kumar:** Department of Applied Sciences, National Institute of Technical Teachers Training and Research, Chandigarh-160019, India, Tel: +91-1722759772; Email: ashokku@nitttrchd.ac.in

INTRODUCTION

Hematite (Fe_2O_3) is one of the most abundant materials on the earth's surface. In it, the iron atom has four unpaired electrons in its 3d orbital; therefore, it exhibits a strong magnetic moment. Furthermore, the crystals of hematite show different magnetic states with variations in temperature. At room temperature, the spins are canted at a small angle with the c-axis of hematite, due to which a net magnetic moment arises. Therefore, it behaves as a weak ferromagnetic (WF) material. However, as the temperature increases, the randomness in the orientation of spins increases. As a consequence, above Neel temperature ~ 950 K, the net magnetic moment becomes zero, and, in turn, it behaves as a paramagnetic material. When the temperature goes down to ~265 K (Morin temperature of bulk hematite), the basal spin plane (111) gets reoriented and aligns in an exactly antiparallel configuration along the axis of the electric field gradient, giving a zero net magnetic moment again, and, in turn, it attains antiferromagnetic (AF) state [1, 2].

The net magnetization of the bulk material, in the absence of an applied external magnetic field, is less than the vector sum of the magnetic moment of the individual atoms. This is because the bulk material contains domains whose magnetic moments may not be aligned in the same direction, consequently leading to reduced magnetization. Furthermore, the magnetic behaviour of hematite depends on crystallinity, particle size, morphology, nature and concentration of the dopant [1, 3, 4].

Hematite is widely used in technical applications like pigmentation, ceramics, wastewater treatment, photocatalyst, solar cells, gas sensors, lithium-ion batteries, thermoelectric devices [5 - 14], etc. Moreover, solar cells and photocatalytic applications need a material with a high absorption coefficient in a wide spectral range of absorption. The hematite is an n-type semiconductor exhibiting both direct (2.7eV) and indirect (2eV) bandgaps [15, 16].

Furthermore, nanostructures provide considerably different magnetic properties in comparison to their bulk counterpart due to their high surface-to-volume ratio and different crystal structure. Among magnetic nanoparticles, mainly iron oxides, *e.g.*, tetragonal maghemite and hexagonal hematite, are used in biomedical applications, such as *in-vivo* analysis. These are non-toxic and naturally found in many biological systems. Nanoparticles of 10-50 nm size with nearly uniform size distribution are preferred for these applications. Metallic magnetic nanoparticles like Co, Fe and Ni are toxic to human bodies [17].

This chapter discusses the effect of nickel doping on the properties of nanostructured hematite. Due to its ferromagnetic nature, it significantly affects magnetic behaviour and also alters other properties of hematite, like electrical and optical properties.

SYNTHESIS PROCESS

The nickel-doped hematite nanoparticles have been synthesized using the facile wet chemical method. For this, the ferric nitrate nonahydrate and nickel nitrate hexahydrate precursors were dissolved separately in the ethanol. The oxalic acid (taken equal to nickel and iron gram equivalent weights) was taken as a chelating agent. Nickel nitrate hexahydrate was used according to [Ni/(Ni+Fe)], *i.e.*, 0, 1, 2 and 4 wt% of nickel content. Finally, all solutions were mixed via continuous stirring to get a homogeneous solution and dried subsequently at 70°C to make flakes, and after grinding, a fine powder was formed. The final product formed was in the form of oxalate, which was then calcined at 600°C to obtain nanoparticles of pure and nickel-doped hematite. The crystallite sizes calculated using the Scherer formula, $d = k\lambda/\beta\cos\theta$, were 44, 47, 43 and 40 nm for 0, 1, 2, and 4 wt% nickel contents, where 'd' is crystallite size, 'k' shape factor, 'λ'X- ray wavelength, 'β' full width half maximum of diffraction peak and 'θ' is the Bragg's angle [1, 18].

PROPERTIES

Structural Properties

The formation of nickel-doped hematite (JCPDS # 86-0550) and nickel ferrite (JCPDS # 74-1913) phase by the wet chemical method was verified by the XRD pattern, as shown in Fig. (**1a**). Hematite has the same structure as corundum, Al_2O_3, with hexagonal or rhombohedral unit cells. The cell parameters of pure hematite were for a and b= 5.021 Å, c=13.620 Å and the volume of the cell was 297.3 Å^3 [18]. The average crystallite size values measured with the help of the Scherer formula from Fig. (**1a**) were 44, 47, 43 and 40 nm for 0, 1, 2 and 4 wt % nickel, respectively. In the case of bulk hematite synthesized by thermal decomposition of Fe and Ni oxinates at 700 °C in the presence of air, it was found that nickel doping decreased the cell parameters, though the ionic radius of Ni^{2+}cation (0.69 Å) was larger than Fe^{3+}cation (0.64 Å) [2]. However, in the case of nanostructured hematite synthesized by the facile wet chemical method, the cell parameters and cell volume increased continuously with the increase in nickel content. The morphology of nanostructured hematite synthesized by the wet chemical method, as shown in Fig. (**1**), looks like the clusters of closely packed nanorods oriented in different directions.

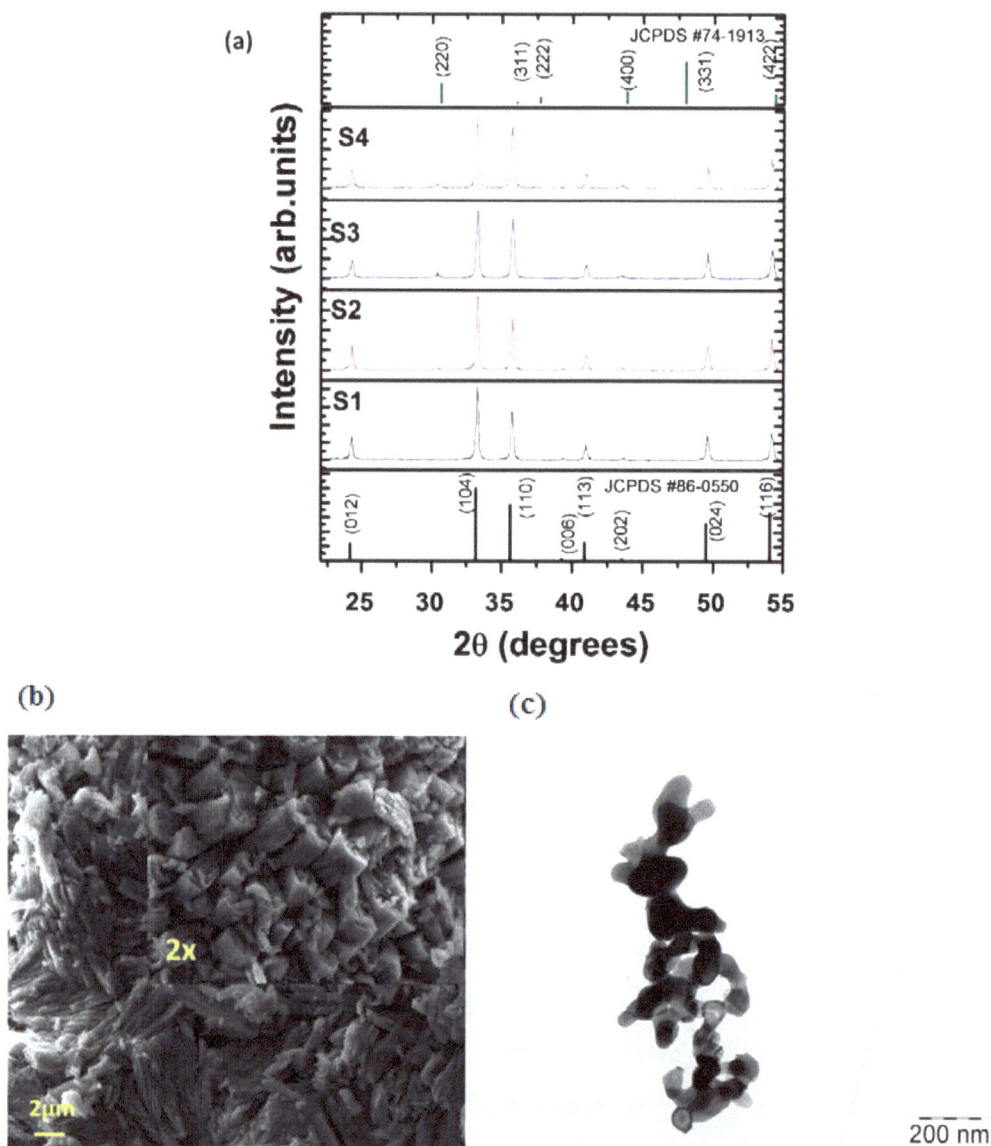

Fig. (1). The crystallography and morphology of nanostructured hematite synthesized by wet chemical method **(a)** XRD pattern of (S1) 0%, (S2) 1%, (S3) 2%, and (S4) 4% nickel content with JCPDS of hematite and nickel ferrite **(b)** SEM image **(c)** TEM image [1]. Reprinted by permission of the publisher (Taylor & Francis Ltd, http://www.tandfonline.com).

Magnetic Properties

The magnetic properties of nickel-doped hematite vary with nickel content and temperature. Since nickel is ferromagnetic material and possesses a high magnetic moment, the weak ferromagnetic nature of hematite becomes strong ferromagnetic due to nickel doping. The pure hematite exhibits weak ferromagnetic behaviouris due to the exchange interaction between trapped electrons in the oxygen vacancies.

The ferromagnetic behaviour increases as the Ni^{2+} replace the Fe^{3+} cation. Imran *et al.* synthesized the nickel-doped hematite nanoparticles by sol-gel method, which shows strong ferromagnetic properties at 8% nickel content [19]. Due to external magnetic field H, all the spins are oriented along the field axis; hence, the magnetization M increases continuously with an increase in the applied field until the saturation value M_S is reached. After reaching the saturation value M_S, the magnetization remains non-zero even after switching the external magnetic field H to zero. The M-H curve exhibits a hysteresis loop. This is because all the domains do not return to their initial state of direction. This remaining magnetization is called remnant magnetization M_R, which will be zero only when a coercive field H_C is applied opposite to the initial field direction.

The nanoparticles of nickel-doped hematite synthesized by the wet chemical method showed an increase in saturation magnetization with the increase in nickel content, as shown in Fig. (**2**). The parameters of the hysteresis loop are given in Table **1**.

Table 1. Parameters of the hysteresis loop at 5 K and 300 K, where M_{max+} (5 T) is a maximum magnetization at 5 T, M_{max-} is a maximum magnetization at -5 T, B_{C+} is a positive coercivity, B_{C-} is negative coercivity, M_{R+} is positive remanent magnetization and M_{R-} is negative remanent magnetization. Reprinted by permission of the publisher (Taylor & Francis Ltd, http://www.tandfonline.com).

Sample	T (K)	M_{max+} (5 T) (Am^2/kg)	M_{max-} (-5 T) (Am^2/kg)	B_{c+} (mT)	B_{C-} (mT)	M_{R+} (Am^2/kg)	M_{R-} (Am^2/kg)
1%_doped	5	2.85	-2.87	20.32	-30.31	0.66	-0.45
2%_doped	5	5.10	-5.13	22.15	-28.54	1.42	-1.16
4%_doped	5	9.39	-9.39	21.55	-27.56	3.10	-2.44
1%_doped	300	2.87	-2.87	19.48	-20.06	0.48	-0.47
2%_doped	300	4.85	-4.85	14.13	-14.12	3.10	-2.44
4%_doped	300	8.66	-8.67	14.66	-15.23	1.98	-1.95

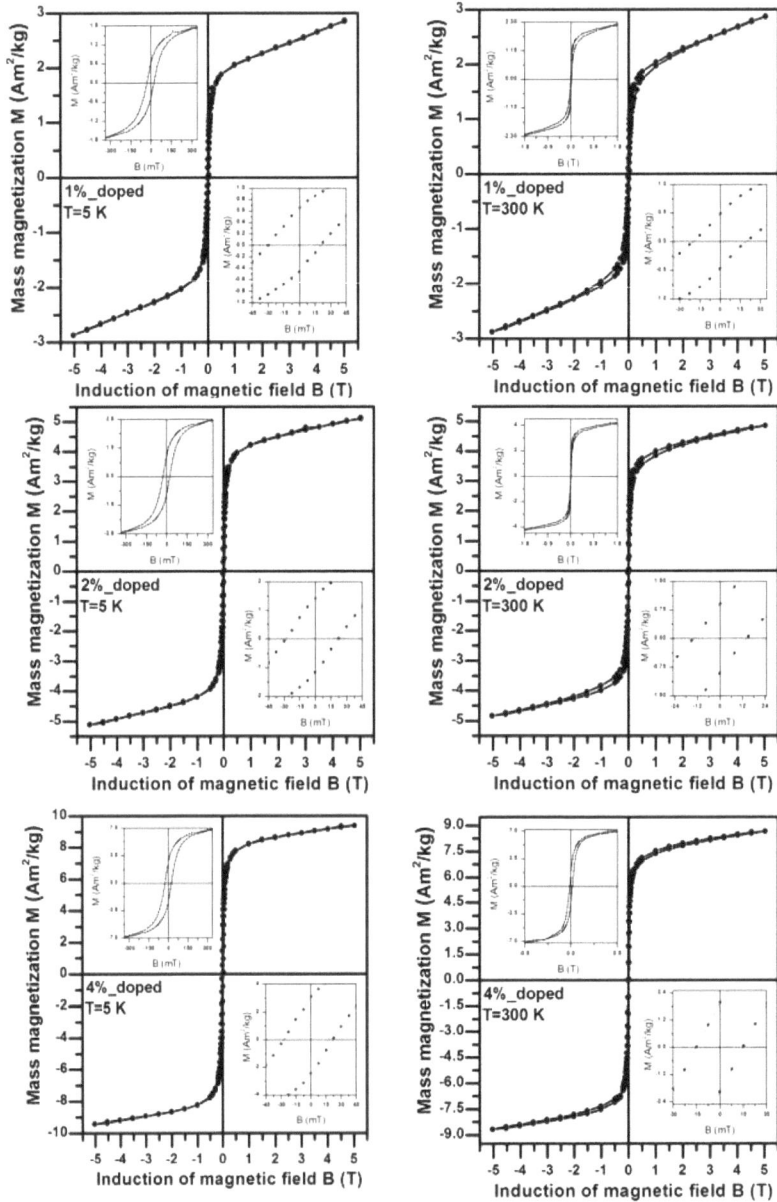

Fig. (2). Magnetic hysteresis loops of nanostructured nickel doped hematite synthesized by wet chemical method [1]. Reprinted by permission of the publisher (Taylor & Francis Ltd, http://www.tandfonline.com).

It was found that the hysteresis loop was not symmetric for doped hematite and shifted toward the left. This type of asymmetry was observed in the core-shell structure having different spin orientations. This was due to the exchange coupling between the differently oriented spins in the core and shell. Moreover, this effect was more prominent at low temperatures. Saturation magnetization also depends on the size of the nanoparticles, and it increases as the size of particles decreases because of more uncompensated spin at the surface [17]. The magnetic nanoparticles with a single domain do not display a hysteresis loop and behave as a superparamagnetic material. The hematite nanoparticles with a size of fewer than 20 nm exhibit superparamagnetic nature at room temperature [5].

The Mössbauer spectra of Ni-containing hematite as shown in Figs. (**3**) and (**4**), synthesized by the wet chemical method showed only one magnetically splitted component [1]. At or above 2 wt% nickel content, the second phase of nickel ferrite, *i.e.*, $NiFe_2O_4$, appeared. The hyperfine parameters, such as isomer shift which was $\delta = 0.37$ mm/s, quadrupole splitting, which was $\Delta E_Q = -0.21$ mm/s, and hyperfine magnetic field, which was $B_{hf} = 51.4$ T, for nickel doped hematite nanoparticles were calculated, which indicated pure weekly ferromagnetic state. These parameters were found to be the same as for pure hematite nanoparticles [20, 21]. The hyperfine parameters, such as δ, ΔE_Q, and B_{hf}, for bulk nickel-doped hematite were reported 0.62 ± 0.02, -0.20 ± 0.01, 53.6 ± 0.1, respectively [2, 22]. The low temperature (20 K) Mössbauer spectra of the Ni-doped hematite displayed a pure antiferromagnetic state, as shown in Fig. (**4**).

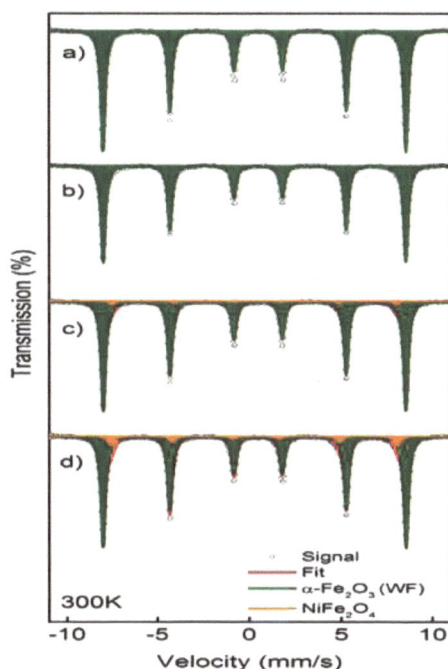

Fig. (3). Mössbauer spectra of nanostructured (**a**) 0%, (**b**) 1%, (**c**) 2%, and (**d**) 4% nickel doped hematite measured at 300 K [1]. Reprinted by permission of the publisher (Taylor & Francis Ltd, http://www.tandfonline.com).

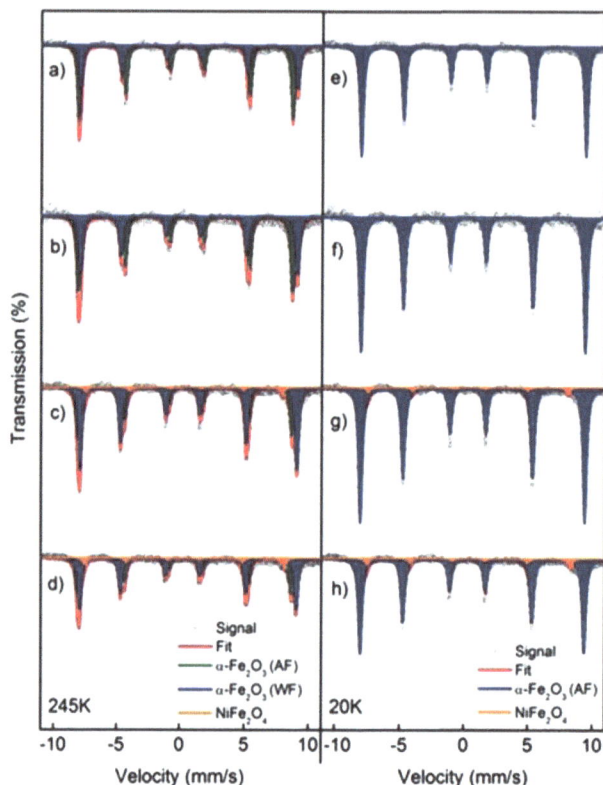

Fig. (4). Mössbauer spectra of nanostructured (a, e) 0%, (b, f) 1%, (c, g) 2%, and (d, h) 4% nickel doped hematite measured at 245 K and 20 K [1]. Reprinted by permission of the publisher (Taylor & Francis Ltd, http://www.tandfonline.com).

The Morin temperature at which the material undergoes a first-order spin reorientation transition [23] depends on the size of the crystallites, and it decreases with a decrease in size. The Morin transitions temperature measured by Mössbauer spectra for nanostructured 0, 1, 2, and 4 wt% nickel doped hematite was found to be 248, 244, 250 and 249 K, respectively. It is clear from the temperature *versus* relative area of the subspectrum curve of weak ferromagnetic and antiferromagnetic states of hematite (Fig. **5**) that the transition between these two states is not instant. Both states exist simultaneously over a certain temperature range known as the width of the Morin transition, and this width depends on the size distribution of nanoparticles. Thus, the Morin temperature is the temperature at which the sextets corresponding to antiferromagnetic and ferromagnetic states have equal spectral area. The width of the Morin transition increases with the incorporation of more nickel content.

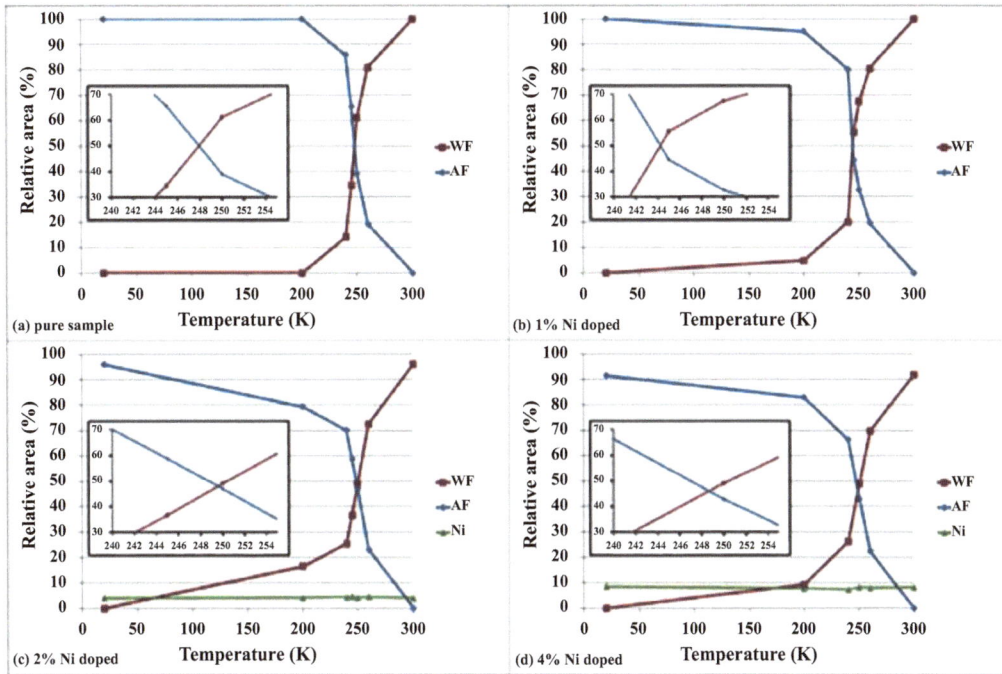

Fig. (5). The variation in the area of weak ferromagnetic and antiferromagnetic states of **(a)** 0%, **(b)** 1%, **(c)** 2%, and **(d)** 4% nickel-doped nanostructured hematite with temperature. The Ni represents the nickel ferrite phase in the samples [1]. Reprinted by permission of the publisher (Taylor & Francis Ltd, http://www.tandfonline.com).

The quadrupole splitting, ΔE_Q, of ferromagnetic and antiferromagnetic states slightly varies with temperature. The transition of the ΔE_Q value from – 0.21 mm/s to +0.40 mm/s reveals that the spins are reoriented by 90° at the Morin transition as shown in (Fig. **6**) [20, 21].

The effect of nickel doping on Morin temperature can be understood by considering the presence of vacancies and anisotropic lattice distortion due to doping and the presence of nickel ions. The change in Morin transition temperature is determined by the relative change in local ion anisotropy and magnetic dipolar anisotropy; hence, the three aforementioned aspects should affect the magnitude and temperature dependence of these anisotropies in a different way.

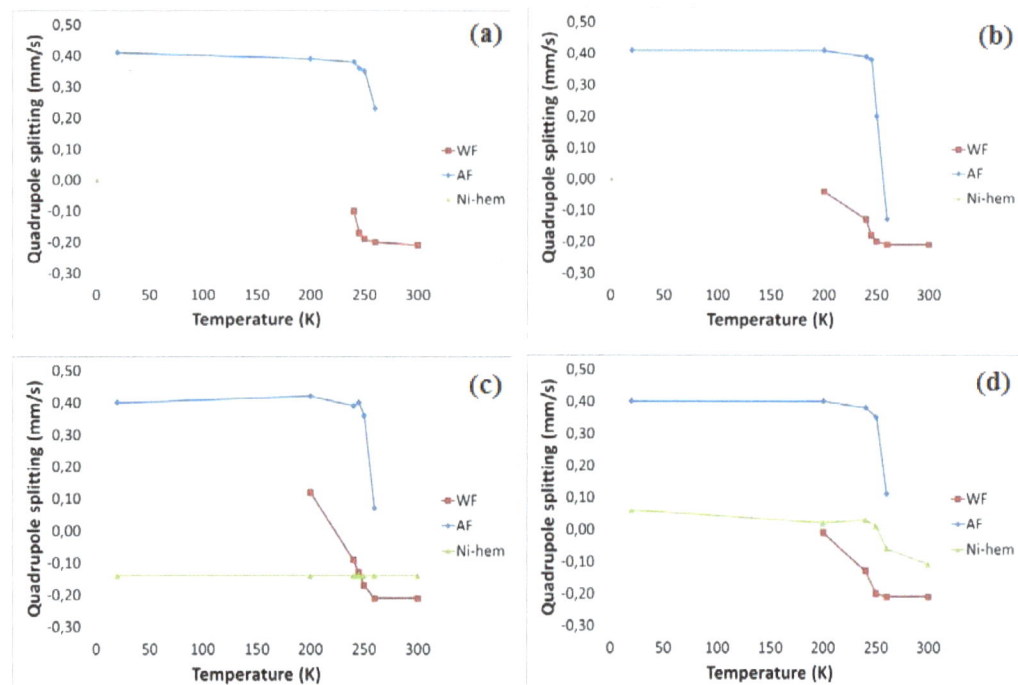

Fig. (6). Temperature dependence of quadrupole splitting of weak ferromagnetic and antiferromagnetic states of nanostructured (**a**) 0%, (**b**) 1%, (**c**) 2%, and (**d**) 4% nickel-doped hematite. Ni-hem represents the nickel ferrite phase in the samples [1]. Reprinted by permission of the publisher (Taylor & Francis Ltd, http://www.tandfonline.com).

The local ion anisotropy positively affects the antiferromagnetic behaviour of hematite, which decreases due to an increase in lattice defects created by doping, but dipolar anisotropy remains unaffected. Furthermore, nickel doping modifies the ferromagnetic character of the hematite, thus, resulting in reduced Morin temperature with nickel doping. In the case of nanostructured hematite synthesized by wet chemical method, at 1 wt% doping of Ni [1, 24], the Morin temperature decreases; however, it increases at 2 wt% (Fig. **7**) because of the appearance of the secondary phase, $NiFe_2O_4$. At 4 wt%, further decrease in Morin temperature may be ascribed to the predominance of size factor as the size of the nanocrystallites decreased with an increase in nickel content. However, the bulk hematite with nickel doping and synthesized by thermal decomposition [2] did not show much variation in Morin temperature (266, 264, and 262 K, at 5%, 10%, and 15% nickel content). The width of the Morin transition was 50 ± 5 K at 5% Ni content and showed a small increment with an increase in Ni content.

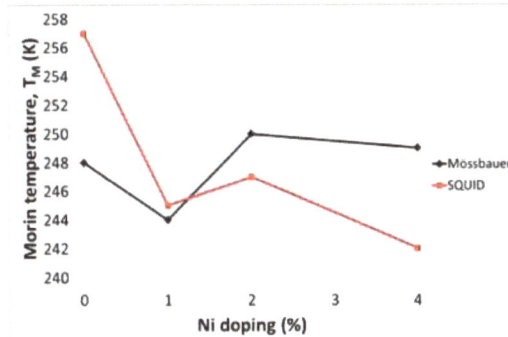

Fig. (7). Morin temperature of nanostructured nickel doped hematite measured by Mössbauer and SQUID [1]. Reprinted by permission of the publisher (Taylor & Francis Ltd, http://www.tandfonline.com).

In magnetization *versus* temperature curves, generally, two curves are obtained, one is zero field cooled (ZFC), and another is field cooled (FC) curve. In the case of superparamagnetic materials, in the ZFC curve, first, the magnetization increases with an increase in temperature and then decreases, while in the case of the FC curve, the magnetization decreases. Fig. (**8**) shows the temperature dependence of ZFC and FC curve of nanostructured hematite of size (a)~ 8 nm with separating temperature 103 K and blocking temperature 52 K and (b) ~ 100×30×10 nm with blocking temperature of 16K [25, 26]. The peak at the ZFC curve represents the average blocking temperature T_B of the sample. The magnetization of ferromagnetic nanoparticles with small size becomes thermally fluctuated and fluctuates more prominently at higher temperatures. At low temperatures, the magnetic moments are blocked due to low thermal energy. This temperature is called blocking temperature. The temperature at which ZFC and FC curve starts separating is called separating or irreversibility temperature, which corresponds to the blocking temperature of the biggest particle in the sample. The difference between separating temperature and blocking temperature represents the width of the size distribution. Fig. (**9**) shows the FC and ZFC magnetization curve and differential curve (dM / dT vs. T) of 0 (pure sample), 1, 2 and 4 wt% Ni doped hematite in the temperature range of 5-300 K at an applied field of 0.1 T. The plateau that appeared in the FC magnetization curve below the blocking temperature represents the existence of strong inter-particle interactions [27]. The T_B for nickel-doped hematite was 255 K, and the separating temperature was calculated as 275-280 K. The abrupt change in the magnetization curve of ZFC and FC indicated the Morin transition in the sample, and the main inflection point of this curve represented Morin temperature. The Morin temperatures of 0, 1, 2 and 4 wt% Ni doped hematite measured by a superconducting quantum interference device (SQUID) were calculated as 257, 245, 247 and 242 K, respectively.

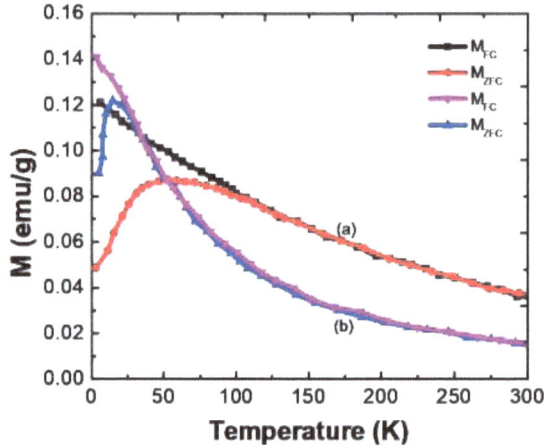

Fig. (8). Temperature dependence of ZFC and FC curve of hematite nanoparticles synthesized by hydrothermal method, showing superparamagnetic behaviour [25, 26].

Electronic Properties

Since the hematite is an n-type semiconductor having a direct bandgap of 2.7 eV and an indirect bandgap of 2.2 eV, its electronic properties can be varied via doping. Nickel is a divalent cation; therefore, nickel doping in hematite creates vacancies or holes due to the replacement of Fe^{3+} with Ni^{2+}, and hence a p-type semiconductor is developed. Maiti *et al.* [28] examined the electrical conductivity of iron-nickel mixed oxide sintered at 1050 to 1300 K. In the case of pure hematite, the conductivity varied linearly and increased with an increase in sintering temperature, but the variation was very complex for mixed oxide due to the dependence on nickel oxide content. With the addition of nickel oxide to hematite, the majority of charge carriers change from electrons to holes, *i.e.*, the hematite converted from an n-type to a p-type semiconductor. Furthermore, the positive value of the Seebeck coefficient confirms the holes as the majority charge carriers in the nickel-containing hematite, as shown in Fig. (**10**). Seebeck coefficient is negative for n-type semiconductors.

At lower concentrations, nickel doping acts as an acceptor impurity and therefore increases the hole density. The resistivity values were calculated as 57.96, 27.90, 97.34 and 94.98 Ωm for 0, 1, 2 and 4 wt% nickel-doped hematite [18]. The resistivity decreased at 1 wt% nickel content, and at higher concentrations, it was found to decrease because of precipitation of the secondary phase of nickel ferrite, which has a high resistivity value.

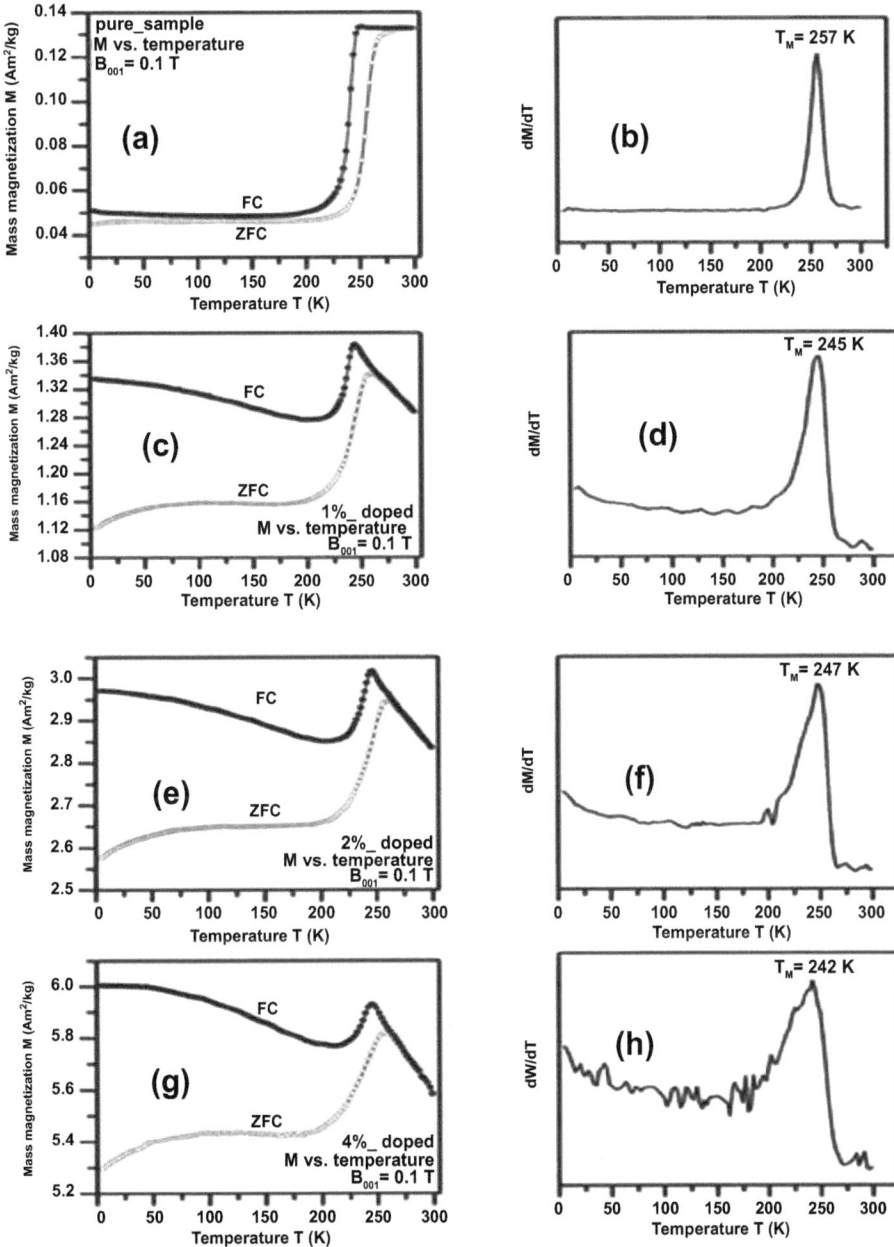

Fig. (9). (**a, c, e** and **g**) Temperature dependence of ZFC and FC curve, (**b, d, f** and **h**) temperature derivative of magnetization *versus* temperature for 0%, 1%, 2% and 4%, respectively, nickel doped hematite synthesized by wet chemical method, showing Morin transition [1]. Reprinted by permission of the publisher (Taylor & Francis Ltd, http://www.tandfonline.com).

Optical Properties

UV-Vis. Diffuse Reflectance Spectra

Fig. (**11**) shows the absorbance spectra of pure and nickel-doped hematite nanoparticles. The direct and indirect bandgap energies are calculated by using the theory of Kubelka–Munk (insets show Tauc plots). The reflectance and absorbance coefficients are related by the equation, $K/S = (1-R_d)^2 /2R_d = F(R)$, (where K, S and R stand for absorption coefficient, scattering coefficient and absolute diffuse reflectance, respectively, and F(R) is Kubelka-Munk function). As the above relation shows the F(R) varies directly with the molar absorption coefficient and, therefore, F(R) can be put in place of α in the Tauc plot equation $(h\nu\alpha)^{1/n} = A(h\nu - E_g)$ (where h denotes Planck's constant, ν frequency of vibration, α absorption coefficient, E_g bandgap and A proportionality constant). The value of the exponent n is ½ for direct transitions and 2 for indirect transitions.

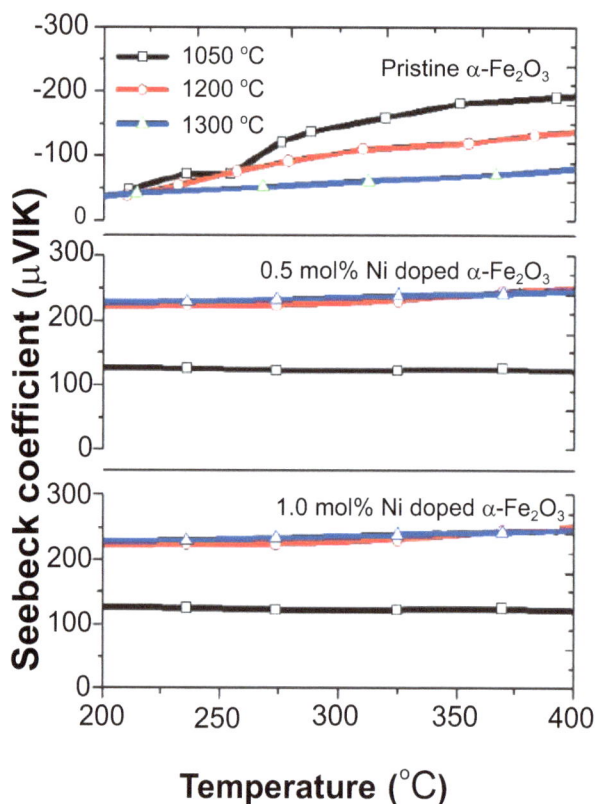

Fig. (10). Seebeck coefficient of nickel-containing hematite sintered at (a) 1050 K (b) 1200 K (c) 1300 K [28].

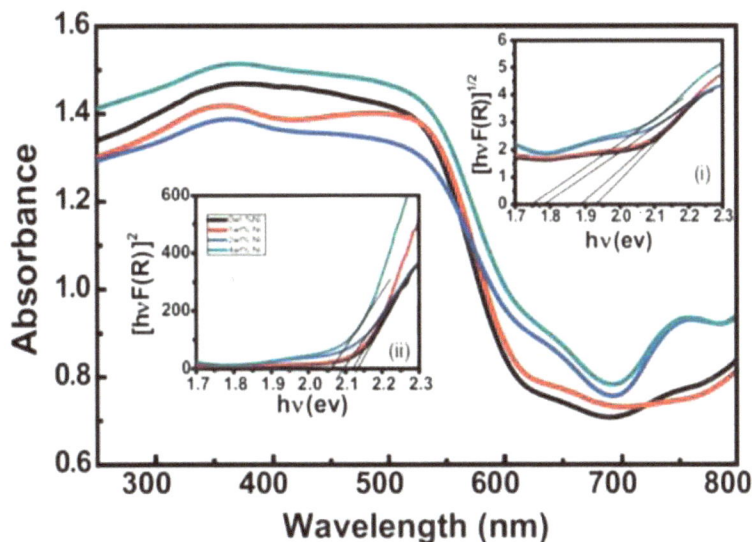

Fig. (11). The absorbance spectra of pure and nickel-doped hematite nanoparticles and insets show the estimated Tauc plots for direct (i) and indirect (ii) bandgap [18]. Reprinted by permission of the publisher (John Wiley and Sons, https://onlinelibrary.wiley.com/).

Several researchers have reported the values for indirect bandgap 1.38 to 2.09 eV and direct bandgap 1.95 to 2.35 eV. The width of the bandgap depends on many factors, such as the synthesis method, phase and average crystallite size. It was observed that the hematite possesses either direct, indirect or both types of band transitions [29 - 36]. For 0, 1, 2, and 4 wt% nickel doped nanostructured hematite, the indirect bandgap values were found to be 1.94, 1.86, 1.78 and 1.75 eV, respectively, and the direct bandgap values were found to be 2.14, 2.12, 2.09 and 2.06 eV, respectively. The reduction in bandgap due to nickel doping was due to the 3d electrons of nickel lying near the Fermi level and below the O2p energy state. Therefore, the shifting in the lowest unoccupied molecular orbital was reported to be comparatively greater than the highest occupied molecular orbital of Fe [37].

Photoluminescence Spectra

Fig. (**12**) shows the photoluminescence (PL) behaviour of pure and nickel-doped hematite nanoparticles synthesized by the wet chemical method. The photoluminescence spectra of hematite nanoparticles exhibited a sharp peak at a wavelength of ~ 578 nm and showed a small red shift with nickel doping. The bandgap corresponding to this wavelength was found to be the same, *i.e.*, 2.14 eV, as for direct bandgap value in the UV-Vis spectrum, showing that the major transitions in hematite were of O2p-Fe3d type. Furthermore, it was observed that the intensity varies inversely with the addition of nickel content, which is an indication of enhanced carrier density with the addition of nickel content. The existence of a nickel ferrite phase in the sample is expected to absorb the emitted radiation [18].

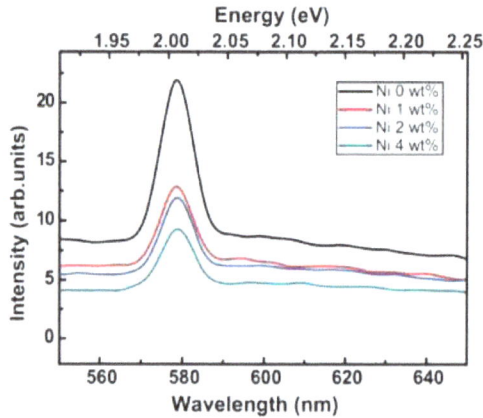

Fig. (12). Photoluminescence (PL) behaviour of pure and nickel-doped hematite nanoparticles [18]. Reprinted by permission of the publisher (John Wiley and Sons, https://onlinelibrary.wiley.com/).

Dielectric Properties

The dielectric constant, also known as relative permittivity, measures the ability of a material to store electrical energy in the electric field and can be expressed as the ratio of the permittivity of the material to the permittivity of free space. If all other factors remain constant, the stored electric flux increases with an increase in the dielectric constant. This enables the material to hold more charge for a comparatively long time and is therefore used in high-capacity capacitors. In equation1,ε represents the dielectric constant or relative permittivity, C stands for capacitance, A area and d separation between the plates of the capacitor and equation 2 represents a tangent loss.

$$\varepsilon = \frac{Cd}{\varepsilon_{\circ}A} \qquad (1)$$

$$\tan \delta = \frac{1}{2\pi f \varepsilon_{\circ} \, \varepsilon \rho} \qquad (2)$$

The *'f'* stands for frequency, and 'ρ' stands for resistivity. Tangent loss includes loss in the medium in the form of heat due to damping of the vibrating dipole moments and loss in the conductivity of that material. For lossless material, it will be zero. From equation 2, it is clear that tangent loss is frequency-dependent and decreases with a decrease in frequency due to space charge carriers of that material. The electrons approach the grain boundaries by moving through the grains leading to space charge polarization. The space charge polarization and turning direction polarization contribute only at low frequency, while electron displacement polarization contributes at low as well as high frequency. Furthermore, at higher frequencies, the resonance effect, *i.e.*, frequency of jumping ions and applied electric field, coincides and plays an important role in increasing the dielectric constant [38]. Imran *et al.* [19] reported that the dielectric constant of hematite nanoparticles increases as the nickel content increases. It was found to increase from 45 to 75 and then 98 with the increase in nickel content from 2% to 4% and 8%, respectively. This increment in dielectric constant is ascribed to the increase in crystallite size with an increase in nickel content.

CONCLUSION

Nickel doping and nano-structuring in hematite produce enhanced magnetization. Furthermore, the Ni doping in bulk hematite did not show much variation in Morin temperature. However, the nanostructured hematite showed a significant change in Morin temperature with the incorporation of Ni content. The Morin temperature was found to reduce from 257 K to 245 K at 1% Ni content. Furthermore, the semiconducting behaviour of hematite changed from n-type to p-type due to nickel doping. The dielectric constant increased with an increase in nickel content.

ACKNOWLEDGEMENTS

The author, Ashok Kumar, acknowledges the financial support of the Council of Scientific and Industrial Research (CSIR), New Delhi (India) (F. No. 22(0778)/18/EMR-II).

REFERENCES

[1]   Arodhiya, S.K.; Placke, A.; Kocher, J.; Kumar, A.; Pechousek, J.; Malina, O.; Machala, L. Nickel-induced magnetic behaviour of nano-structured α-Fe₂O₃, synthesised by facile wet chemical route. *Philos. Mag.,* **2018**, *98*(26), 2425-2439.
      [http://dx.doi.org/10.1080/14786435.2018.1490036]

[2]   Saragovi, C.; Arpe, J.; Sileo, E.; Zysler, R.; Sanchez, L.C.; Barrero, C.A. Changes in the structural and magnetic properties of Ni?substituted hematite prepared from metal oxinates. *Phys. Chem. Miner.,* **2004**, *31*(9), 625-632.
      [http://dx.doi.org/10.1007/s00269-004-0422-y]

[3]   Raming, T.P.; Winnubst, A.J.A.; van Kats, C.M.; Philipse, A.P. The synthesis and magnetic properties of nanosized hematite (α-Fe₂O₃) particles. *J. Colloid Interface Sci.,* **2002**, *249*(2), 346-350.
      [http://dx.doi.org/10.1006/jcis.2001.8194] [PMID: 16290607]

[4]   Bødker, F.; Mørup, S. Size dependence of the properties of hematite nanoparticles. *Europhys. Lett.,* **2000**, *52*(2), 217-223.
[http://dx.doi.org/10.1209/epl/i2000-00426-2]

[5]   Teja, A.S.; Koh, P.Y. Synthesis, properties, and applications of magnetic iron oxide nanoparticles. *Prog. Cryst. Growth Charact. Mater.,* **2009**, *55*(1-2), 22-45.
[http://dx.doi.org/10.1016/j.pcrysgrow.2008.08.003]

[6]   Cheng, X-L.; Jiang, J-S.; Jin, C-Y.; Lin, C-C.; Zeng, Y.; Zhang, Q-H. Cauliflower-like α-Fe$_2$O$_3$ microstructures: Toluene–water interface-assisted synthesis, characterization, and applications in wastewater treatment and visible-light photocatalysis. *Chem. Eng. J.,* **2014**, *236*, 139-148.
[http://dx.doi.org/10.1016/j.cej.2013.09.089]

[7]   Zhu, L.P.; Bing, N.C.; Wang, L.L.; Jin, H.Y.; Liao, G.H.; Wang, L.J. Self-assembled 3D porous flowerlike α-Fe$_2$O$_3$ hierarchical nanostructures: Synthesis, growth mechanism, and their application in photocatalysis. *Dalton Trans.,* **2012**, *41*(10), 2959-2965.
[http://dx.doi.org/10.1039/c2dt11822j] [PMID: 22277922]

[8]   Hosseini, S.G.; Ayoman, E. Synthesis of α-Fe$_2$O$_3$ nanoparticles by dry high-energy ball-milling method and investigation of their catalytic activity. *J. Therm. Anal. Calorim.,* **2017**, *128*(2), 915-924.
[http://dx.doi.org/10.1007/s10973-016-5969-6]

[9]   Jiao, Y.; Liu, Y.; Qu, F.; Wu, X. Dendritic α-Fe$_2$O$_3$ hierarchical architectures for visible light driven photocatalysts. *CrystEngComm,* **2014**, *16*(4), 575-580.
[http://dx.doi.org/10.1039/C3CE41994K]

[10]   Hou, Y.; Wang, D.; Yang, X.H.; Fang, W.Q.; Zhang, B.; Wang, H.F.; Lu, G.Z.; Hu, P.; Zhao, H. J.; Yang, H.G. Rational screening low cost-counter electrodes for dye sensitized solar cell. *Nat. Commun.,* **2013**, *4*(1-8), 1584.

[11]   Shinde, S.S.; Bansode, R.A.; Bhosale, C.H.; Rajpure, K.Y. Physical properties of hematite a-Fe$_2$O$_3$ thin film: Applications to photoelectrochemical solar cells. *J. Semicond.,* **2011**, 32, 013001.

[12]   Wu, C.Z.; Yin, P.; Zhu, X.; OuYang, C.Z.; Xie, Y. Synthesis of hematite (α-Fe$_2$O$_3$) nanorods: Diameter-size and shape effects on their applications magnetism, lithium-ion batteries, and gas sensors. *J. Phys. Chem. B,* **2006**, *110*, 17806-17812.
[http://dx.doi.org/10.1021/jp0633906] [PMID: 16956266]

[13]   Niu, H.; Zhang, S.; Ma, Q.; Qin, S.; Wan, L.; Xu, J.; Miao, S. Dye sensitized solar cells based on flower shaped α-Fe$_2$O$_3$ as a photoanode and reduced graphene oxide-polyanilene composite as a counter electrode. *RSC Advances,* **2013**, *3*, 17228-17235.
[http://dx.doi.org/10.1039/c3ra42214c]

[14]   Hwang, H.K.; Seo, J.W.; Seo, W.S.; Lim, Y.S.; Park, K. Thermoelectric properties of P-doped and V-doped Fe$_2$O$_3$ for renewable energy conversion. *Int. J. Energy Res.,* **2014**, *38*(2), 241-248.
[http://dx.doi.org/10.1002/er.3052]

[15]   Mochizuki, S. Electrical conductivity of α-Fe$_2$O$_3$. *Phys. Status Solidi, A Appl. Res.,* **1977**, *41*(2), 591-594.
[http://dx.doi.org/10.1002/pssa.2210410232]

[16]   Glasscock, J.A.; Barnes, P.R.F.; Plumb, I.C.; Bendavid, A.; Martin, P.J. Structural, optical and electrical properties of undoped polycrystalline hematite thin films produced using filtered arc deposition. *Thin Solid Films,* **2008**, *516*(8), 1716-1724.
[http://dx.doi.org/10.1016/j.tsf.2007.05.020]

[17]   Issa, B.; Obaidat, I.; Albiss, B.; Haik, Y. Magnetic nanoparticles: surface effects and properties related to biomedicine applications. *Int. J. Mol. Sci.,* **2013**, *14*(11), 21266-21305.
[http://dx.doi.org/10.3390/ijms141121266] [PMID: 24232575]

[18]   Kocher, J.; Kumar, A.; Kumar, A.; Priya, S.; Kumar, J. Nickel-induced structural, optical, magnetic, and electrical behavior of α-Fe$_2$O$_3$. *Phys. Status Solidi, B Basic Res.,* **2014**, *251*(8), 1552-1557.

[http://dx.doi.org/10.1002/pssb.201451183]

[19]   Imran, M.; Riaz, S.; Nayani, N.Z.; Naseem, S. Study of magnetic and dielectric behaviour of Ni doped α-Fe₂O₃ nanopowder. **2016**.

[20]   Tuček, J.; Tuček, P.; Čuda, J.; Filip, J.; Pechoušek, J.; Machala, L.; Zbořil, R. Iron(III) oxide polymorphs and their manifestations in in-field ⁵⁷FeMössbauer spectra. *AIP Conf. Proc.,* **2012**, *1489*, 56-74.
[http://dx.doi.org/10.1063/1.4759474]

[21]   Tuček, J.; Machala, L.; Frydrych, J.; Pechoušek, J.; Zbořil, R. *Mossbauer spectroscopy in study of nanocrystalline iron oxides from thermal processes in Mössbauer Spectroscopy: Applications in Chemistry, Biology, and Nanotechnology*; Sharma, V.K.; Klingelhofer, G.; Nishida, T., Eds.; John Wiley and Sons: New Jersey, **2013**, pp. 351-392.

[22]   Barrero, C.A.; Arpe, J.; Sileo, E.; Sánchez, L.C.; Zysler, R.; Saragovi, C. Ni- and Zn- doped hematite obatained by combustion of mixed metal oxinates. In: *Physica B: Condensed. Matter*; , **2004**; 354, pp. (1–4), 27-34.

[23]   Lu, H.M.; Meng, X.K. M. andMengX. K., "Morin temperature and Neel temperature of hematite nanocrystals". *J. Phys. Chem. C,* **2010**, *114*(49), 21291-21295.
[http://dx.doi.org/10.1021/jp108703b]

[24]   Verma, K.C.; Goyal, N.; Singh, M.; Singh, M.; Kotnala, R.K. Hematite α-Fe₂O₃ induced magnetic and electrical behavior of NiFe₂O₄ and CoFe₂O₄ ferrite nanoparticles. *Results Phys.,* **2019**, *13*, 102212.
[http://dx.doi.org/10.1016/j.rinp.2019.102212]

[25]   Tadic, M.; Panjan, M.; Damnjanovic, V.; Milosevic, I. Magnetic properties of hematite (α-Fe₂O₃) nanoparticles prepared by hydrothermal synthesis method. *Appl. Surf. Sci.,* **2014**, *320*, 183-187.
[http://dx.doi.org/10.1016/j.apsusc.2014.08.193]

[26]   Xu, Y.Y.; Wang, L.; Wu, T.; Wang, R.M. Magnetic properties of α-Fe₂O₃ nanopallets. *Rare Met.,* **2019**, *38*(1), 14-19.
[http://dx.doi.org/10.1007/s12598-017-0938-1]

[27]   Xu, Y.Y.; Rui, X.F.; Fu, Y.Y.; Zhang, H. Magnetic properties of α-Fe₂O₃ nanowires. *Chem. Phys. Lett.,* **2005**, *410*(1-3), 36-38.
[http://dx.doi.org/10.1016/j.cplett.2005.04.090]

[28]   Maiti, H.S.; Rajendran, S.; Rao, V.S. Electrical conductivity and thermoelectric power studies of iron (III)-nickel (II) mixed oxides. *Phys. Status Solidi, A Appl. Res.,* **1984**, *84*(2), 631-638.
[http://dx.doi.org/10.1002/pssa.2210840236]

[29]   Dghoughi, L.; Elidrissi, B.; Bernède, C.; Addou, M.; Lamrani, M.A.; Regragui, M.; Erguig, H. Physico-chemical, optical and electrochemical properties of iron oxide thin films prepared by spray pyrolysis. *Appl. Surf. Sci.,* **2006**, *253*(4), 1823-1829.
[http://dx.doi.org/10.1016/j.apsusc.2006.03.021]

[30]   Özer, N.; Tepehan, F. Optical and electrochemical characteristics of sol–gel deposited iron oxide films. *Sol. Energy Mater. Sol. Cells,* **1999**, *56*(2), 141-152.
[http://dx.doi.org/10.1016/S0927-0248(98)00152-4]

[31]   Akl, A.A. Optical properties of crystalline and non-crystalline iron oxide thin films deposited by spray pyrolysis. *Appl. Surf. Sci.,* **2004**, *233*(1-4), 307-319.
[http://dx.doi.org/10.1016/j.apsusc.2004.03.263]

[32]   Zotti, G.; Schiavon, G.; Zecchin, S.; Casellato, U. Electrodeposition of amorphous α-Fe₂O₃ film by reduction of iron perchlorate in acetonitrile. *J. Electrochem. Soc.,* **1998**, *145*(2), 385-389.
[http://dx.doi.org/10.1149/1.1838273]

[33]   Miller, E.L.; Paluselli, D.; Marsen, B.; Rocheleau, R.E. Low-temperature reactively sputtered iron oxide for thin film devices. *Thin Solid Films,* **2004**, *466*(1-2), 307-313.
[http://dx.doi.org/10.1016/j.tsf.2004.02.093]

[34]   Gartner, M.; Crisan, M.; Jitianu, A.; Scurtu, R.; Gavrila, R.; Oprea, I.; Zaharescu, M. Spectroellipsometric characterization of multilayer so-gel α-Fe_2O_3 films. *J. Sol-Gel Sci. Technol.,* **2003**, *26*(1/3), 745-748.
[http://dx.doi.org/10.1023/A:1020706423230]

[35]   Vayssieres, L.; Lindquist, S-E.; Hagfeldt, A.; Photoelectrochemical studies oriented nanorod thin film of hematite. *J. Electrochem. Soc.,* **2000**, *147*, 2456-2461.
[http://dx.doi.org/10.1149/1.1393553]

[36]   Gahlawat, S.; Singh, J.; Yadav, A.K.; Ingole, P.P. Exploring Burstein–Moss type effects in nickel doped hematite dendrite nanostructures for enhanced photo-electrochemical water splitting. *Phys. Chem. Chem. Phys.,* **2019**, *21*(36), 20463-20477.
[http://dx.doi.org/10.1039/C9CP04132J] [PMID: 31502609]

[37]   Pozun, Z.D.; Henkelman, G. Hybrid density functional theory band structure engineering in hematite. *J. Chem. Phys.,* **2011**, *134*(22), 224706.
[http://dx.doi.org/10.1063/1.3598947] [PMID: 21682532]

[38]   Rathi, P.L.; Gowrishankar, R.; Deepa, S. Dielectric and magnetic properties of oxides of tin and iron and their composites- a comparative study. *AIP Conf. Proc.,* **2020**, *2265*, 030157.
[http://dx.doi.org/10.1063/5.0016761]

CHAPTER 9

Nanostructures for Cosmetics and Medicine

Baby[1] and **Saurabh Gupta**[1,*]

[1] *Department of Microbiology, Mata Gujri College, Fatehgarh Sahib-140 407, Punjab, India*

Abstract: The application of various nanoparticles and nanotechnology in cosmetics and pharmaceuticals is an interesting area of research and development. The use of nanotechnology has also emerged as an important tool for gene manipulations, diagnosis of several diseases along with improvement in treatment efficacy. This chapter has emphasized the use of nano-materials in cosmetics and pharmaceuticals globally with associated legislation in different countries. More than 100 different products have been listed and discussed with their uses in different fields along with associated concerns.

Keywords: Nanotechnology, Nano-carriers, Nano-cosmetics, Nano-pharmaceuticals.

INTRODUCTION

Nanotechnology is recognized as the most promising technology of the twenty-first century, and it is expected to be a major boon to science, where atoms are manipulated at the nanometer scale ($1nm=10^{-9}m$). Nanotechnology creates new legislative structures; they demonstrate the various behaviours and features of previously recognized materials [1]. Nanotechnology is made up of two words: technology and the Greek number "nano," which signifies dwarf in Greek. As a result, nanotechnology is defined as the science and technology used to improve or utilize particles with a diameter of 1 to 100 nanometers [2]. The Egyptians, Greeks, and Romans all recorded the practice of nanotechnology throughout the 4000 BC era, with the concept of hair dye processing using nanotechnology. Nanotechnology has risen to prominence in a number of scientific domains over the previous decade. Meanwhile, from 1959, nanotechnology developed in several sectors, such as engineering, physics, chemistry, biology, and science, and approximately 40 years later, it has infiltrated the domains of cosmetics, health products, and dermal necessities, as shown in Fig. **(1)** [3].

---

[*] **Corresponding author Saurabh Gupta:** Department of Microbiology, Mata Gujri College, Fatehgarh Sahib-140 407, Punjab, India; Email: sau27282@gmail.com

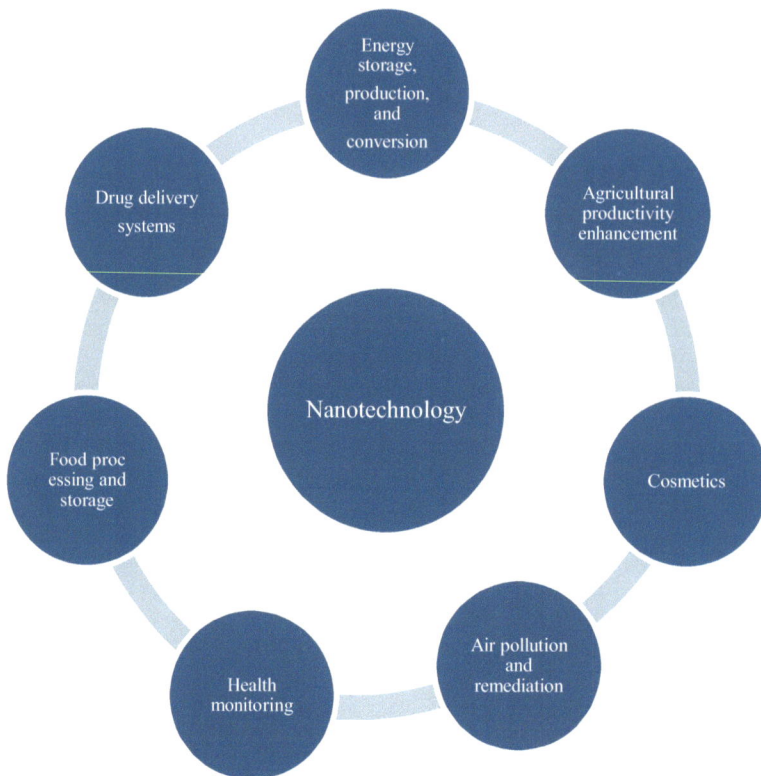

Fig. (1). Applications of nanotechnology in different fields.

Nanotechnology is progressively introduced into our everyday existence and is employed as a raw material in cosmetics and pharmaceutical goods, as well as in the packaging industry [4]. Nanotechnology also emerges as a vector in gene manipulation, diagnosis and anticancer treatments. The technology acts as an adjuvant in the formulations of antimicrobials and vaccines. This plays a significant part in the progress and designing of new tools and analytical instruments [5]. The *National Science Foundation* evaluates that nanotechnology reached the global economy by about $ 1 trillion in 2015, demanding around two million employees. In the United States of America, the European Union and Japan, developments in the field of nanotechnology are on the highest level, investing about a billion dollars per year and symbolizing more than half of the world's reserves. Countries like Russia, China, India, and Brazil, on the other hand, have made significant investments in nanotechnology in recent years. Between 2001 and 2006, the Brazilian government invested $ 140 million in nanotechnology research networks and projects [6].

THE ROLE OF NANOMATERIALS IN COSMETICS

Nanotechnology is one of the most rapidly advancing scientific topics nowadays. Nanotechnology products have become common in our daily lives. Supervisory agencies, on the other hand, have yet to agree on a unified definition for nanomaterials and nanotechnology. As a result, each country has its own definitions and legislation for regulating nanomaterials-containing items. As these materials are relatively new, there is no long-term research on their impact on human health and the environment [7]. Subsequently, countries regulate the amount of nanomaterials in cosmetics, allowing consumers to choose between products that include or do not contain nano-materials. As a result, the primary goal of this research was to identify the most commonly used nanomaterials in cosmetics and determine whether these formulations are in violation of laws in force in the United States, the European Union, and Brazil, thereby determining whether cosmetics on the market are compliant with these three economic powers' laws. This research is unique, and it will contribute to a broader discussion of existing rules governing the use of nanotechnology in cosmetics [6]. In 1995, *Lancôme,* the luxury division of *L'Oréal,* was the first company to introduce a nanotechnology-based cosmetic, with the launch of a face cream composed of nano-capsules of pure Vitamin E to combat skin aging [8]. In the meantime, companies, such as Christian Dior, Anna Pegova, Procter & Gamble, Estee Lauder, Dermazone Solution, Johnsons & Johnsons, Skinceuticals, Shiseido, Garnier, Chanel, and Revlon, developed products in this track [6]. Other international companies started to finance research to develop nano-cosmetics. The first company to develop and market a nano-cosmetic in Brazil was *O Boticário.* Nano-serum, an anti-aging lotion for the eyes, forehead, and around the mouth, was developed by the company. The nanostructure work includes active ingredients, such as vitamins A, C, and K, as well as a whitening substance. The technology developed in collaboration with the French laboratory Comucel and cost $ 14 million is part of the active range, which was introduced in 2005 [6]. VitActive Nano-peeling Renovator Microdermabrasion, a functional nanotechnology-based anti-aging cosmetic, was introduced in 2007. During these days, lift serum anti-aging and anti-aging 65+ advanced systems were introduced. At the same time, Natura released "Brumas de Leite," a body hydration product containing particles that are approximately 150 nanometers in size. It also introduced the "Refreshing Body Spray" to the male public in the same year [8].

Nano-cosmetic purposes products, which are anticipated for use on the face and body skin, with anti-aging and light protection properties, are capable of penetrating the deep layers of the skin and potentiating the active ingredients [6]. Fronza and their collaborators defined nano-cosmetic as "comparing with the traditional products, a cosmetic preparation that contains actives or other nanostructured elements has improved performance attributes" [9]. Shampoos, conditioners, toothpaste, anti-wrinkle creams, anti-cellulite creams, whitening skin, hydrating, face powders, aftershave lotions, deodorants, soaps, sunscreens, foundations, perfumes, and nail polishes all include nanoparticles; these are nano-cosmetics prepared in the cosmetic industries by using nanotechnology [6, 7] for better dispersion and coordinated release of active substances, which are used as vehicles. Emulsions with small, homogeneous droplets in size range of 20 to 500 nm are known as nano-emulsions [10]. Nano-emulsions can be classified into two types based on their morphology, as presented in Fig. (**2**).

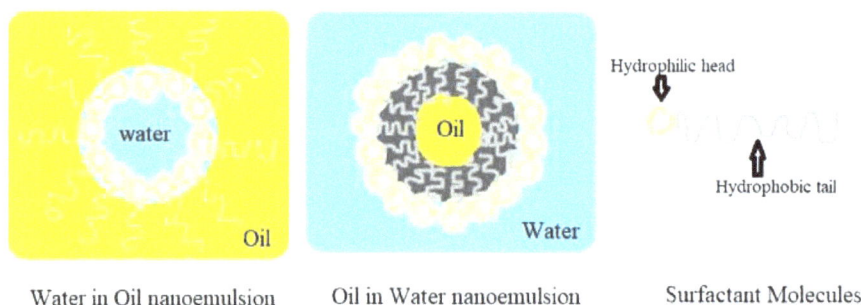

Fig. (2). Diagrammatic representation of water-in-oil (W/O) and oil-in-water (O/W) nano-emulsion consisting of surfactant micelles.

A 'water-based' or oil-in-water (O/W) emulsion has water as the continuous phase and oil as the dispersed phase [11, 12], whereas the 'oil-based' or water-in-oil (W/O) emulsion yields in inversed condition, on the other hand, micro-emulsions require surfactant concentrations of 20% or higher. As a result, nano-emulsions appear more fluid (at low oil concentrations), displaying appealing physical properties and skin feel, especially in the absence or the small use of thickeners [13].

During the 70s, there were some formulations used as sunscreen, which contained chemicals capable of penetrating the epidermis but only in the stratum corneum. It is critical to select the active substance's carrier with caution. During the '80s, α-hydroxyl acids appeared, with a little higher penetration potential. Liposomes, which are microscopic particles made up of fat and water, penetrating the deeper layers of skin but not the basal layers, were discovered in 1990 and have materialized in nano-cosmetics [8]. As a result, the beauty sector employs nanotechnology due to a number of advantages associated with its application, particularly in terms of increased active penetration capacity in skin layers. However, with a larger and more effective development of this technology in the near future, more clearly seen practical benefits and the safety of products given with this charm will be observed. The usage of nanoparticles may pose hazards, such as toxicity and lack of biocompatibility of the materials utilized. The advantages of utilizing nano-cosmetics include protecting substances from chemical or enzymatic degradation, limiting their release, especially in the case of irritants at high concentrations, and extending the time cosmetic remains active, or medications stay in the stratum corneum [6, 7]. Due to their great entrapment efficiency and good sensory qualities, cosmetics are more stable than traditional cosmetics. In the preparation of anti-wrinkle creams, moisturizing creams, skin-whitening creams, hair-repairing shampoos, conditioners, and hair serums, nano-materials are widely used [14, 15]. Everything in this universe has both positive and negative elements, as is the rule of nature. Some of the positive and negative aspects related to nano-cosmeceuticals are discussed in Fig. (**3**) [16 - 18].

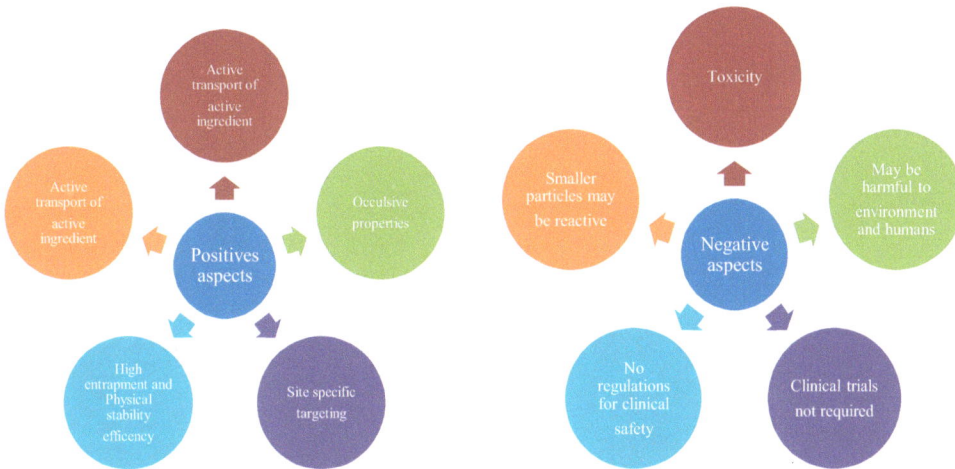

Fig. (3). Pictorial presentation of positive and negative aspects of nano-cosmeceutical.

Positive aspects enhance the use of nanotechnology, but its drawback is that it can affect human health, is toxic to the tissues and cells, and damages DNA, RNA, and lipids in the cells. Nano-cosmeceuticals may also be harmful to the environment. As no clinical trials are required for nano-cosmeceutical approval, there is concern about toxicity after use [19].

In the past years, a tremendous increase in the production and application of a wide variety of cosmeceutical products that contain nano-materials has been reported, thus resulting in an enhanced exposure of the workers and customers to the nanoparticle-based products through different routes (Fig. **4**). Despite their enormous potential value, little is understood about the short- and long-term health consequences on the environment and organisms. There may be limitations due to health risks, product functionality, and environmental considerations.

Concerns have been raised about the potential risks of nanoparticles penetrating the skin after their application on the skin.

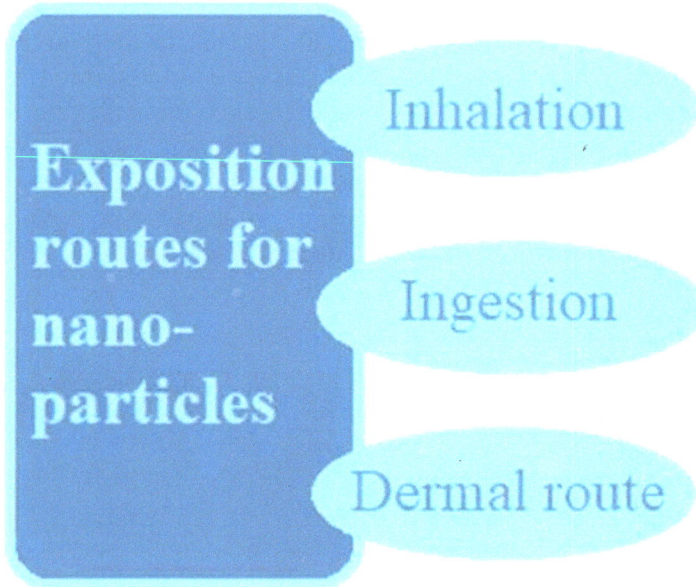

Fig. (4). Exposition of routes for nano-particles.

Major Classes in Nano-cosmeceuticals

Cosmetics is expected to be the fastest-growing sector of the personal care market. Nano-cosmeceuticals are found in abundance in nail, hair, lip, and skincare products. Fig. (5) [20] depicts the major kinds of nano-cosmeceuticals. Cosmeceuticals for skincare products improve the texture and function of the skin by encouraging collagen formation and counteracting the damaging effects of free radicals. These cosmeceuticals help the skin to be healthier by keeping the keratin structure in good shape. Zinc oxide and titanium dioxide nanoparticles are the most efficient minerals in sunscreens, as they protect the skin by entering deep into the layers of the skin and making the product less oily, odourless, and clear [21]. Hair nano-cosmeceutical shampoos, conditioners, hair growth accelerators, colouring, and style products are among the products available. The hair follicle is targeted, the shaft is targeted, and the active component quantity is enhanced. The purpose of conditioning nano-cosmeceuticals agents is to add softness, sheen, silkiness, and gloss and improve hair disentangling [22].

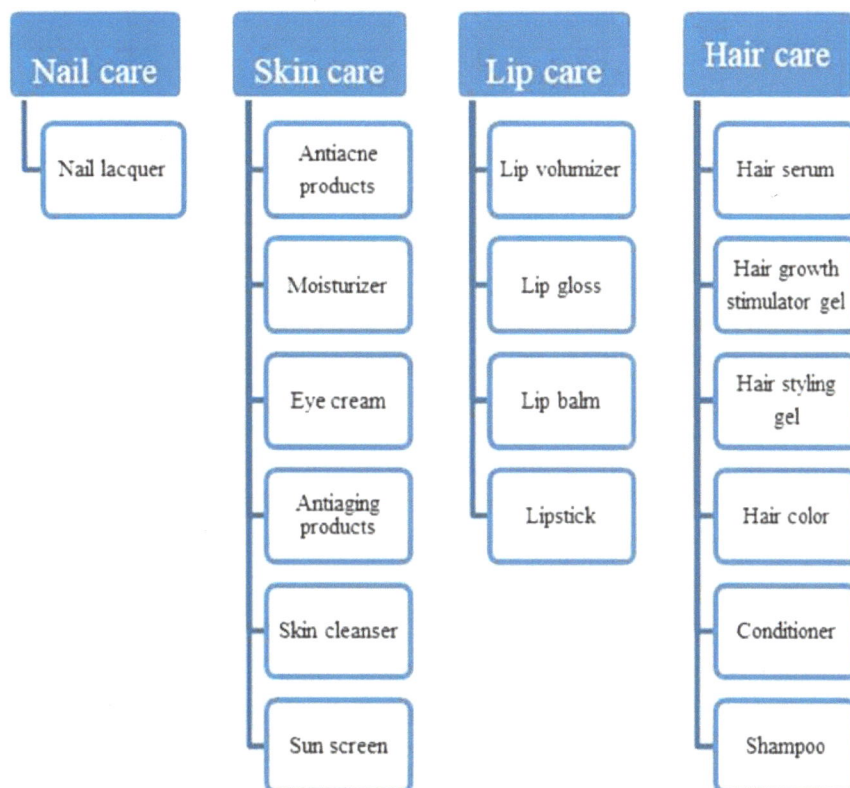

Fig. (5). Major classes in nano-cosmeceuticals.

Lipstick, lip balm, lip gloss, and lip volumizer are some of the lip care items available in nano-cosmeceuticals. Lip gloss and lipstick can contain a variety of nanoparticles that soften the lips by preventing transepidermal water loss and preventing pigment migration from the lips, allowing the colour to last longer. Lip volumizer with liposomes boosts lip volume, moisturizes and defines the lips, and smoothes out wrinkles in the lip contour [23]. Nail care solutions based on nano-cosmeceuticals are more effective than traditional ones. Nanotechnology-based nail paints include advantages, such as increased toughness, quick drying, durability, chip resistance, and ease of application due to flexibility. In nail paints, new tactics, such as amalgamating silver and metal oxide nanoparticles, have shown antifungal qualities for the treatment of fungal infections in toenails [24].

THE ROLE OF NANO-MATERIALS IN MEDICINES

In the twentieth century, nanotechnology and its applications in the area of medicine and pharmaceuticals evolved as nano-materials have size-tunable properties like electrical conductivity, chemical reactivity, magnetic ability, and optical, thermal and mechanical properties. Nano-materials, due to their size-dependent smart properties, are used for a broad range of applications in pharma and nano-devices. Photonics applications in nano-electronics and nano-engineering work by changing individual atoms and molecules into nanostructures having good bio-compatibility with biological systems [25]. In today's era, various diseases like cancer, diabetes, Parkinson's disease, Alzheimer's disease, cardiovascular diseases, multiple sclerosis, and different kinds of inflammatory or infectious diseases lead to a higher number of complex illnesses, resulting in a major problem for humankind. Nano-medicines are produced using nanotechnology, which is used for the treatment of various kinds of diseases. The use of molecular nanotechnology in nano-medicines has also strengthened in several ways [26]. These are used as nano-tags and labels, which improve the sensitivity and flexibility of testing. With the development of nano-devices, such as gold nanoparticles, which can be tagged with short segments of DNA and used to detect genetic sequences in a sample, gene sequencing has become more proficient. Nanotechnology helps in the repair of damaged tissues. Artificially stimulated cells are employed in tissue engineering, which could revolutionize organ transplantation and artificial implant placement. Carbon nanotubes can be used to create innovative biosensors with unusual properties (CNTs). These biosensors could be employed in astrobiology, shedding light on the origins of life. Nanotechnology is also being employed to build cancer diagnostic sensors. Although CNTs are inert, they can be functionalized with a probe molecule at the tip. In the field of stem cell research, nanotechnology has made an excellent contribution.

Magnetic nanoparticles (MNPs) have been utilized to isolate and group stem cells with great effectiveness. Quantum dots have been employed for molecular imaging and tracing of stem cells, as well as the transport of genes and medicines into stem cells. This will hopefully lead to stem cell-based therapeutics for the prevention, diagnosis and treatment of human diseases [27]. Nanotechnology has applications in basic science as well as translational medicines. Nano-carriers are mixed with biological molecules and used to modulate the stem cells. Nano-devices can be used for intracellular access, intelligent delivery and sensing of biomolecules. These technologies have a great impact on stem cell microenvironment and tissue engineering studies and have great potential for biomedical applications [28]. Different kinds of nano-systems and nano-materials are used for pharmaceutical preparations. The nano-systems used for pharmaceutical preparations are represented in Fig. (**6**).

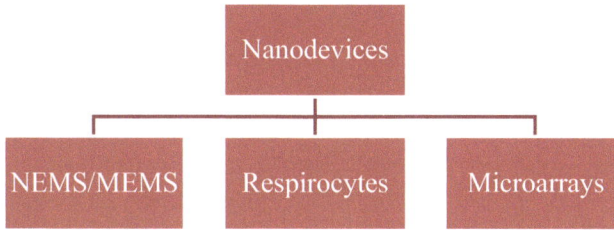

Fig. (6). Schematic diagram of various types of pharmaceutical nano-systems.

NANO-CARRIERS FOR COSMECEUTICALS AND MEDICINE

For the conveyance of nano-cosmeceuticals and nano-pharmaceuticals, carrier technology is used, which allows for the fast delivery of active substances. Several innovative nano-carriers for the delivery of cosmeceuticals and medicines have been developed, as depicted in Fig. (7) [29, 30]. Liposomes, niosomes, dendrimers, gold nanoparticles, nano-emulsion, nanostructured lipid carriers, polymersomes, cubosomes, *etc.*, are used as nano-carriers in cosmetics and pharma nanotechnology. Liposomes are vesicular structures with an aqueous centre and a hydrophobic lipid bilayer surrounding them. Niosomes are bilayer-structured vesicles produced from self-assembled hydrated nonionic surfactants, with or without lipids. They can be multilamellar or unilamellar vesicles that contain an aqueous solute solution and lipophilic components [31].

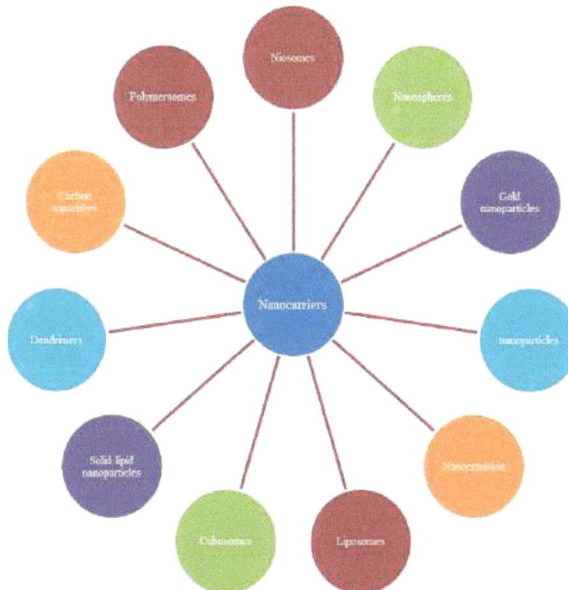

Fig. (7). Various nano-carriers used in cosmeceuticals and pharmaceutical industries.

Nano-spheres are spherical particles having a core-shell structure that range in diameter from 10 to 200 nanometers. Nano-spheres are generally used for the entrapment of drugs [32]. Polymersomes are artificial vesicles made up of copolymer blocks that enclose an aqueous chamber in the centre. They are extremely adaptable and biologically stable. Drugs, proteins, peptides, enzymes, DNA, and RNA fragments are among the sensitive substances that polymersomes can encapsulate and preserve [33, 34]. Cubosomes are advanced nanostructured particles that are discrete, submicron-sized, and self-assembled liquid crystalline particles of surfactants with a specific water ratio that gives them unique features. When aqueous lipid and surfactant systems are mixed with water and nanostructures in a specific ratio, they produce self-assembled structures. Solid lipid nanoparticles (SLN) were introduced in the early 1990s as an alternative to traditional lipoidal carriers, such as emulsions and liposomes. Solid lipid nanoparticles range in size from 50 to 1000 nm [35]. The size of nano-gold or gold particles ranges from 5-400 nm, and they have different shapes like nano-spheres, nano-shell, nano-rod, nano-cube, *etc.* Dendrimers are highly branched, uni-molecular, globular, and multivalent nanoparticles whose production theoretically yields mono-disperse components. Dendrimers are derived from the Greek words 'dendron,' meaning tree and 'meros' meaning portion [36]. While carbon nanotubes are utilized to present one of the most outstanding inventions in the field of nanotechnology. CNTs are seamless cylindrical hollow fibres made from rolled graphene with sp2 hybridization and walls created by graphene as a hexagonal carbon lattice. The size varies from 0.7 to 50 nm in diameter and 10's microns in length [37, 38]. Various industries exploit these nano-carriers to produce products using nanotechnology. In the field of cosmetics and medicines, nanotechnology plays a significant role. Various products related to the cosmetic and medicine industries, their manufacturers, and their uses are given below in Table **1** [39].

Table 1. List of marketed formulations of novel nano-carriers.

Marketed formulations of nano-carriers		
Liposomes		
Product name	**Marketed by**	**Uses**
Dermosome	Microfluidics	Moisturizer
Decorte Moisture Liposome Face Cream	Decorte	Moisturizer
Advanced Night Repair Protective Recovery Complex	Est´ee Lauder	Skin repair
Clinicians Complex Liposome Face and Neck Lotion	Clinicians Complex	Nourishes skin and prevents photoaging

(Table 1) cont.....

Marketed formulations of nano-carriers		
Niosomes		
Identik Shampooing Floral Repair	Identik	Hair repair shampoo
Base Cream of Mayu Niosome	Laon Cosmetics	Moisturizing and Whitening
Niosome+ Perfected Age Treatment	Lancome	Removes wrinkles
Solid lipid nanoparticles		
Parfum Bottlle of Allure	Chanel	Perfume
Body Cream of Allure	Chanel	Body moisturizer
Parfum Spray of Allure Eau	Chanel	Perfume
Soosion Facial Lifting Cream	SLN Technology Soosion	Anti-wrinkle cream
NLC		
Q10's Nanorepair Intensive Serum	Dr. Rimpler	Fights signs of aging, anti-wrinkle serum
Q10's Nanorepair Cutanova-Cream	Dr. Rimpler	Promotes restructuring of skin and aging, Smoothing of fine lines
Q10's Nanovital Cutanova Cream	Dr. Rimpler	Used for antiaging treatment with UV-protection
Extra Moist Softener of Iope Supervital	Amore Pacific	Moisturizes dry and rough skin
Nanoemulsion		
Nanocream	Sinerga	Wet wipes
Bepanthol-Protect Facial Cream	Bayer HealthCare	Ultra Moisturizing, antiaging, and antipollution
Moiture Cream of Nanovital Vitanics Crystal	Vitacos Cosmetics	Skin moisturizing, elastic, and lightening effects
Vitacos Vita-Herb Nona-Vital Skin Toner	Vitacos Cosmetics	Moisturizer
Gold nanoparticles		
Nano Gold Energizing Eye Serum of Chantecaille	Chantecaille	Prevents aging, promotes collagen, reduces inflammation, and cell growth repair
Nano Ultra Silk Serum of Orogold 24K	Orogold	Restores loss of moisture, improves wrinkles and fine lines, and maintains healthy skin
O3+ 24K Gold Gel Cream	O3+	Makes skin glow and shine
Nanospheres		
Fresh As A Daisy Body Lotion	Kara Vita	Moisturizing body lotion
Nanosphere Plus	Dermaswiss	Anti-aging and anti-wrinkle

Marketed formulations of nano-carriers		
Cell Act DNA Filler Intense Cream	CellAct Switzerland	Reduces firms skin and wrinkles

CONCLUSION

The progress of nanotechnology in the world has been mainly regarded and sustained on the foundation that this new and emerging technology has huge potential to help humanity by addressing communal challenges, such as the endowment of drinking water, healthcare, personal care, agriculture, *etc.*, and concurrently achieve economic gains through growth in the nanotech-based industrial sector. This aspect surrounds the convenience of techniques and materials required for the preparation of nano-materials. Nano-emulsions are used for a wide range of nano-materials. In conclusion, nanotechnology for cosmeceutical and pharmaceutical applications is a promising technology of the future. Nanomaterials gain attention due to size-tunable smart properties. The major focus of nanotechnology-based industries is on the development, awareness and commercialization of smart nano-products for markets and consumers. In addition, this technology focuses on the regulatory framework at different levels to address the risks and safety parameters. Nanotechnology has a bright future due to its integration with other technologies, thus, resulting in complex and inventive hybrid technologies. Nanotechnology and biology-based technologies are linked. Nanotechnology is already being utilized to change genetic data, and biological components are being employed to create nano-materials. Medicine, cosmetics, information technology, cognitive science, and biotechnology are all being transformed by nanotechnology to engineer matter at the atomic level. Nanotechnology can be useful for every aspect of human life, such as medicine, regenerative medicine, stem cell research, nutraceuticals and cosmetics.

REFERENCES

[1]   Cadioli, L.; Salla, L.D. *R. C. Exatas e Tecnol.,* **2006**, *1*, 98.

[2]   Maynard, A. D. Nanotechnology: a research strategy for addressing risk. **2006**.

[3]   Bangale, M.S.; Mitkare, S.S.; Gattani, S.G.; Sakarkar, D.M. Recent nanotechnological aspects in cosmetics and dermatological preparations. *Int. J. Pharm. Pharm. Sci.,* **2012**, *4*(2), 88-97.

[4]   Kimbrell, G.A. Nanotechnology and nanomaterial personal care products: necessary oversight and recommendations. Global regulatory issues for the cosmetics industry. William Andrew Publishing, **2007**; pp. 117-153.
[http://dx.doi.org/10.1016/B978-081551567-8.50012-7]

[5]   Knopp, D.; Tang, D.; Niessner, R. Review: Bioanalytical applications of biomolecule-functionalized nanometer-sized doped silica particles. *Anal. Chim. Acta,* **2009**, *647*(1), 14-30.
[http://dx.doi.org/10.1016/j.aca.2009.05.037] [PMID: 19576381]

[6]   Baril, M. B.; Franco, G. F.; Viana, R. S.; Zanin, S. M. W. Nanotechnology applied to cosmetics. Academic Vision. **2012**, *13*(1).

[7]   Neves, K. Nanotechnology in cosmetics. Cosmetics & Toiletries, **2008**, *20*, 22.

[8]   Kim, J.; Park, S.; Lee, J.E.; Jin, S.M.; Lee, J.H.; Lee, I.S.; Yang, I.; Kim, J.S.; Kim, S.K.; Cho, M.H.; Hyeon, T. Designed fabrication of multifunctional magnetic gold nanoshells and their application to magnetic resonance imaging and photothermal therapy. *Angew. Chem. Int. Ed.,* **2006**, *45*(46), 7754-7758.
[http://dx.doi.org/10.1002/anie.200602471] [PMID: 17072921]

[9]   Fronza, T.; Guterres, S.; Pohlmann, A.; Teixeira, H. Nanocosmetics: towards the establishment of regulatory frameworks. In: *Porto Alegre: UFRGS Graphics*; , **2007**.

[10]   Capek, I. Degradation of kinetically-stable o/w emulsions. *Adv. Colloid Interface Sci.,* **2004**, *107*(2-3), 125-155.
[http://dx.doi.org/10.1016/S0001-8686(03)00115-5] [PMID: 15026289]

[11]   Mason, T.G.; Wilking, J.N.; Meleson, K.; Chang, C.B.; Graves, S.M. Nanoemulsions: formation, structure, and physical properties. *J. Phys. Condens. Matter,* **2006**, *18*(41), R635-R666.
[http://dx.doi.org/10.1088/0953-8984/18/41/R01]

[12]   Singh, Y.; Meher, J.G.; Raval, K.; Khan, F.A.; Chaurasia, M.; Jain, N.K.; Chourasia, M.K. Nanoemulsion: Concepts, development and applications in drug delivery. *J. Control. Release,* **2017**, *252*, 28-49.
[http://dx.doi.org/10.1016/j.jconrel.2017.03.008] [PMID: 28279798]

[13]   Maruno, M.; Rocha-Filho, P.A. O/W nanoemulsion after 15 years of preparation: a suitable vehicle for pharmaceutical and cosmetic applications. *J. Dispers. Sci. Technol.,* **2009**, *31*(1), 17-22.
[http://dx.doi.org/10.1080/01932690903123775]

[14]   Mu, L.; Sprando, R.L. Application of nanotechnology in cosmetics. *Pharm. Res.,* **2010**, *27*(8), 1746-1749.
[http://dx.doi.org/10.1007/s11095-010-0139-1] [PMID: 20407919]

[15]   Nohynek, G.J.; Lademann, J.; Ribaud, C.; Roberts, M.S. Grey goo on the skin? Nanotechnology, cosmetic and sunscreen safety. *Crit. Rev. Toxicol.,* **2007**, *37*(3), 251-277.
[http://dx.doi.org/10.1080/10408440601177780] [PMID: 17453934]

[16]   Dahiya, A.; Romano, J.F. Cosmeceuticals: a review of their use for aging and photoaged skin. *Cosmetic Dermatology-cedar Knolls,* **2006**, *19*(7), 479-484.

[17]   Starzyk, E.; Frydrych, A.; Solyga, A. Nanotechnology: does it have a future in cosmetics?. *SÖFW-Journal,* **2008**, *134*(6).

[18]   Antonio, J.R.; Antônio, C.R.; Cardeal, I.L.S.; Ballavenuto, J.M.A.; Oliveira, J.R. Nanotechnology in Dermatology. *An. Bras. Dermatol.,* **2014**, *89*(1), 126-136.
[http://dx.doi.org/10.1590/abd1806-4841.20142228] [PMID: 24626657]

[19]   Mukta, S.; Adam, F. Cosmeceuticals in day-to-day clinical practice. *J. Drugs. Dermatol.,* **2010**, *9*(5), pp. 62-6.

[20]   Lohani, A.; Verma, A.; Joshi, H.; Yadav, N.; Karki, N. Nanotechnology-Based Cosmeceuticals. *ISRN Dermatol.,* **2014**, 1-14.
[http://dx.doi.org/10.1155/2014/843687] [PMID: 24963412]

[21]   Smijs, T.; Pavel, S. Titanium dioxide and zinc oxide nanoparticles in sunscreens: focus on their safety and effectiveness. *Nanotechnol. Sci. Appl.,* **2011**, *4*, 95-112.
[http://dx.doi.org/10.2147/NSA.S19419] [PMID: 24198489]

[22]   Rosen, J.; Landriscina, A.; Friedman, A. Nanotechnology-based cosmetics for hair care. *Cosmetics,* **2015**, *2*(3), 211-224.
[http://dx.doi.org/10.3390/cosmetics2030211]

[23]   Sundari, P.T.; Anushree, H. Novel delivery systems: current trend in cosmetic industry. *Eur. J. Pharm. Med. Res.,* **2017**, *4*(8), 617-627.

[24]   Bethany, H. Zapping nanoparticles into nail polish. *Laser Ablation Method Makes Cosmetic and Biomedical Coatings in a Flash,* **2017**, *95*(12), 9.

[25]   Boisseau, P.; Loubaton, B. Nanomedicine, nanotechnology in medicine. *C. R. Phys.,* **2011**, *12*(7), 620-636.
[http://dx.doi.org/10.1016/j.crhy.2011.06.001]

[26]   Wang, Z.; Ruan, J.; Cui, D. Advances and prospect of nanotechnology in stem cells. *Nanoscale Res. Lett.,* **2009**, *4*(7), 593-605.
[http://dx.doi.org/10.1007/s11671-009-9292-z] [PMID: 20596412]

[27]   Ricardo, P.N.; Lino, F. Stem cell research meets nanotechnology. Revista Da Sociedade Portuguesa D Bioquimica. *CanalBQ,* **2010**, *7*, 38-46.

[28]   Deb, K.D.; Griffith, M.; Muinck, E.D.; Rafat, M. Nanotechnology in stem cells research: advances and applications. *Front. Biosci.,* **2012**, *17*(1), 1747-1760.
[http://dx.doi.org/10.2741/4016] [PMID: 22201833]

[29]   Nasir, A. Nanotechnology and dermatology: Part II—risks of nanotechnology. *Clin. Dermatol.,* **2010**, *28*(5), 581-588.
[http://dx.doi.org/10.1016/j.clindermatol.2009.06.006] [PMID: 20797523]

[30]   Fox, C. (1998). Cosmetic and pharmaceutical vehicles: skin care, hair care, makeup and sunscreens. Cosmetics and toiletries, 113(1), 45-56.

[31]   Duarah, S.A.; Pujari, K.U.; Durai, R.D.; Narayanan, V.H. Nanotechnology-based cosmeceuticals: a review. *International Journal of Applied Pharmaceutics.,* **2016**, *8*(1), 8-12.

[32]   Singh, A.; Garg, G.; Sharma, P.K. Nanospheres: a novel approach for targeted drug delivery system. *Int. J. Pharm. Sci. Rev. Res.,* **2010**, *5*(3), 84-88.

[33]   Kim, S.H.; Shum, H.C.; Kim, J.W.; Cho, J.C.; Weitz, D.A. Multiple polymersomes for programmed release of multiple components. *J. Am. Chem. Soc.,* **2011**, *133*(38), 15165-15171.
[http://dx.doi.org/10.1021/ja205687k] [PMID: 21838246]

[34]   Discher, D.E.; Eisenberg, A. Polymer Vesicles. *Science,* **2002**, *297*(5583), 967-973.
[http://dx.doi.org/10.1126/science.1074972] [PMID: 12169723]

[35]   Puri, D.; Bhandari, A.; Sharma, P.; Choudhary, D. Lipid nanoparticles (SLN, NLC): A novel approach for cosmetic and dermal pharmaceutical. *J. Glob. Pharma Technol.,* **2010**, *2*(9), 1-15.

[36]   Fruchon, S.; Poupot, R. Pro-inflammatory versus anti-inflammatory effects of dendrimers: The two faces of immuno-modulatory nanoparticles. *Nanomaterials (Basel),* **2017**, *7*(9), 251.
[http://dx.doi.org/10.3390/nano7090251] [PMID: 28862693]

[37]   Ibrahim, K. S. Carbon nanotubes-properties and applications: a review. *Carbon letters,* **2013**, *14*(3), 131-144.

[38]   Kaushik, B.K.; Majumder, M.K. Carbon nanotube: Properties and applications. *Carbon Nanotube Based VLSI Interconnects*; Springer, **2015**, pp. 17-37.

[39]   Akbarzadeh, A.; Rezaei-Sadabady, R.; Davaran, S.; Joo, S.W.; Zarghami, N.; Hanifehpour, Y.; Samiei, M.; Kouhi, M.; Nejati-Koshki, K. Liposome: classification, preparation, and applications. *Nanoscale Res. Lett.,* **2013**, *8*(1), 102.
[http://dx.doi.org/10.1186/1556-276X-8-102] [PMID: 23432972]

CHAPTER 10

Nano-cosmetics and Nano-medicines

Balwinder Kaur[1,*], **Subhash Chand**[2] and **Rajesh Kumar**[3]

[1] *Department of Chemistry, Punjabi University Patiala-147002, Punjab, India*

[2] *Department of Chemistry, L.R.D.A.V. College, Jagraon-142 026, Punjab, India*

[3] *Department of Chemistry, Government Degree College, Sugh-Bhatoli-1760 022, Kangra, Himachal Pradesh, India*

Abstract: In today's fast-moving scenario, nanotechnology has already spread its wings to nanocosmetics and nanomedicines due to the wide range of physical and chemical properties associated with nanoparticles. Different types of nanoparticles, like nanoliposomes, fullerenes, solid lipid nanoparticles *etc.*, have made their entrance into the nanocosmetic industry. However, the safety concern of nanoparticles has forced the cosmetic industry to limit their applications. The pharmaceutical industry has explored the benefits of nanotechnology; it has developed dendrimers, micelles, drug conjugates, metallic nanoparticles *etc.* The brief explanation of these nanoparticles provides a salient glimpse of why they are used in nano pharmaceutical and medicinal chemistry.

• Metallic nanoparticles: Used for drug delivery, cancer treatment, and also in biosensors.

• Nano-liposomes: Bio-compatible and possess entrapment efficiency.

• Nano-emulsions: Used for controlled delivery of bioactive materials.

Keywords: Fullerenes, Metallic nanoparticles, Nanocosmetics, Nanomedicines, Nanoliposomes, Solid lipid nanoparticles.

INTRODUCTION: NANO-COSMETICS

Introduction to Cosmetics

Cosmetics are a class of skincare and personal care products that are used to modify and improve a person's appearance. A large number of cosmetics are available in the market; each is designed for different purposes and possesses di-

* **Corresponding author Balwinder Kaur**: Department of Chemistry, Punjabi University Patiala-147002, Punjab, India; Email: balwindertaheem1987@gmail.com

Karamjit Singh Dhaliwal (Ed.)

verse characteristics. Cosmetics are constituted of a mixture of chemical compounds. Cosmetics designed for skincare can be used to perform the following functions:

a. Cleanse
b. Exfoliate
c. Protect the skin
d. Replenishing
e. Toners, moisturizers, and balms

Cosmetics designed for personal care are as follows:

a. Shampoo
b. Body wash, which can be used to cleanse the body

Cosmetics designed to enhance one's appearance (makeup) are as follows:

a. To conceal blemishes
b. Enhance one's natural features (such as the eyebrows and eyelashes)
c. Add color to a person's face

Extreme forms of makeup are used for performances, fashion shows and by people in costume to change the appearance of the face entirely to resemble a different person, creature or object. Cosmetics can also be designed to add fragrance to the body. Fig. (**1**) shows the pictorial presentation of the classification of cosmetics based on the area of usage and side effects on human health based on the information provided by the Food and Drug Administration (FDA).

Opportunities in Cosmetics and Problems

Around the globe, the cosmetic industry is the fastest-growing industry in 2023, garnering $ 429.9 billion [1]. Out of the list of more than 12000 chemicals, which are used in cosmetics, only 205 chemicals are considered to be safer to use [2]. Cosmetic products are toxic in terms of neuro and reproductive functionalities, carcinogenic and considered endocrine disruptors [3]. Rapid innovation and inclination in the cosmetic industry have arisen due to various toxic chemicals used as additives to enhance functionalities. Table **1** shows some toxic chemicals, which are widely used in almost every cosmetic product, whether skincare or personal care, with the transmission route and hazardous effects on human health.

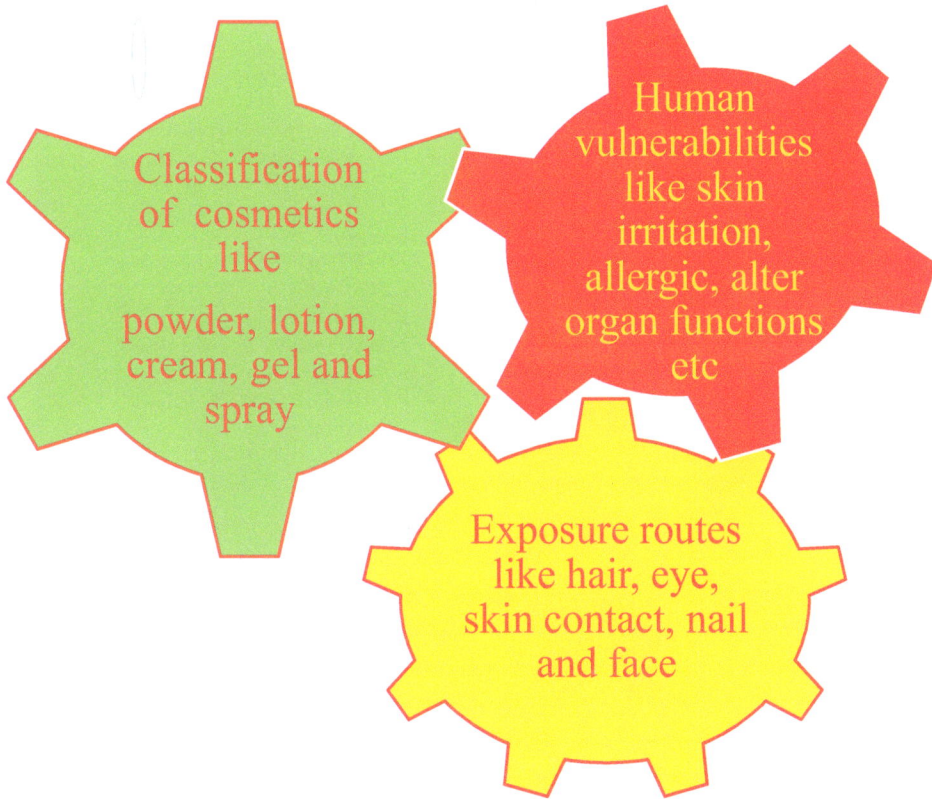

Fig. (1). Classification of cosmetics and side-effects on human health.

Table 1. Commonly used toxic chemicals in the cosmetic industry.

Sr. No.	Chemical	Formula (Chemical or Molecular)	Transmission route	Exposure	Human Vulnerabilities	References
1.	1,4- dioxane	$C_4H_8O_2$	1. Inhalation 2. Skin contact 3. Ingestion	Personal care products like mouthwash, shampoos, toothpaste	Carcinogenic, breathing issues, irritation to eyes and throat.	[4, 5]

(Table 1) cont.....

Sr. No.	Chemical	Formula (Chemical or Molecular)	Transmission route	Exposure	Human Vulnerabilities	References
2.	Formaldehyde and Para-formaldehyde (Polymer of formaldehyde)	Formaldehyde and $(CH_2O)_n$ Paraformaldehyde (Polymer of formaldehyde)	1. Inhalation 2. Skin contact 3. Ingestion	Personal care products like a preservative in liquid soaps, shampoos, and lotions.	Skin rashes, altered lung function, breathing issues, irritation to eyes and throat, myeloid leukemia	[5, 6]
3.	Imidazolidinyl urea (Formaldehyde releasing due to degradation) and diazolidinyl urea	$C_{11}H_{16}N_8O_8$	1. Inhalation 2. Dermal 3. Ingestion 4. Eye absorption	Anti-microbial preservatives in personal care products like eye, face, nails and hair make-up products.	Carcinogenic, dizziness, headache, joint pain.	[7]

Need for Nanoparticles in Cosmetics

An innovative science accompanies the design, production, characterization and utility of designed (engineered) structures by carefully manipulating sizes and shapes at nanoscale dimensions and gives rise to "Nanotechnology". It is the futuristic technology in the world of cosmetics and medicines. The cosmetic world/manufacturer employs a nanoscale version of bulk materials to provide the best UV protector material with a long-lasting effect and deeper skin penetration formula. As one knows, in today's scenario, nanotechnology is a widely accepted technology due to the ease of manipulation of particles' size and shape. As compared to large-scale particles, the most attractive properties of materials at the nanoscale are as follows:

a. Color
b. Solubility
c. Transparency
d. Thermal reactivity

The above properties are very influential in personal care as well as the cosmetics industry [8].

Front Running Brands of Naño-cosmetics

Various major cosmetic industries are using nanomaterials like ZnO and TiO_2 in various skincare and personal care products. Estee Lauder introduced nanoparticles in the industry in 2006. L'Oreal, the highest brand cosmetic industry, has invested a giant amount worth 600 million dollars. L'Oreal is the 6th ranked patent holder in the U.S in nanotechnology/nano cosmetics [9]. Similarly, Coloresuince, Doctor's Dermatologic Formula (DDF) and Freeze 24/7 are the other companies that are using nanoparticles in various cosmetic products [10]. Following are some types of nanomaterials used in cosmetics:

Nano-Liposomes

Liposomes at the nanoscale are called nano-liposomes. Commonly, nano-liposomes and liposomes have indistinguishable chemical, thermodynamic and structural properties. Nano-liposomes offer the following properties:

a. Enhanced surface area.
b. The ability to enhance solubility.
c. Improved controlled release and increased bioavailability.
d. Provide meticulousness in the targeting of the encapsulated material compared with liposomes.

The aqueous volume is enclosed by concentric lipid bi-layered vesicles. The bi-layer is composed of natural or synthetic phospholipids (The chief element: bilayers of the phospholipid molecules applied in the structure of lipid vesicles). The content of liposomes is released by coupling the bi-layer with the cell membrane of the targeted site for the successful delivery of vitamins. They are biocompatible and biodegradable. They are highly recommendable carriers for fragrance delivery in deodorants and lipsticks. Moreover, nano-liposomes enhance the absorption of active ingredients with continuous absorption into cells, thus, making nano-liposomes suitable for cosmetic industries [11]. Also, in dairy products, nano-liposomes encapsulate the enzymes and keep them away from the surrounding food environment and provide a protective situation for enzymes that would otherwise barricade activity. Otherwise, early proteolysis could be a result of adding a free enzyme to milk, which causes low yields and undesirable curd consistency. Nano-liposomal encapsulation has been revealed to stabilize the encapsulated components against a range of chemical changes.

Nano-emulsion

Emulsion (oil-in-water or water-in-oil) is the term famous for the dispersed solute in the dispersion medium. Similarly, nano-emulsions are nanoscale droplets dispersed in some dispersion medium and possess optical, stable, rheological, and ingredient-desirable properties of delivery, which are superior to traditional ones. Nano-emulsions are formed in a non-equilibrium state to afford droplets of remarkably small size in the range of 20-200 nm. The use of nanoparticles in nano-emulsion has increased the life of cosmetic products and higher stability in cosmetic delivery applications [12]. A nano-emulsion, especially oil-in-water containing hydro glycolic extract of *Opuntia ficus-indica* (L.) Mill was made, and its thermal stability and moisturizing property were characterised by its effectiveness in containing 1 percent of the extract of *Opuntia ficus-indica* (L.) Mill. Sufficient stability was shown for a minimum of 2 months and was characterized by its thermal stability and moisturizing efficacy. In addition, the formulation was able to increase the water content of the stratum corneum, showing its moisturizing efficacy. The hyaluronan (hyaluronic acid, HA)-glycerol a-monostearate (GMS) based nanoemulsion resulted in an interesting colloidal transdermal carrier suitable for applications in skincare and cosmetic products.

Nano-capsules

Polymeric nanoparticles are named nano-capsules when they contain a polymeric wall composed of non-ionic surfactants, macromolecules, phospholipids and an oil core. Nano-capsules can either be used as a protective ingredient or directly applied to the skin as final cosmetic products. A very famous nano-capsule is poly-1-Lactic acid, which is suitable for the sustained release of perfume by entrapping fragrance molecules in polymeric nano-carrier. The capsule alters the penetration of UV light into the skin and filters the octyl methoxycinnamate [13]. Poly (isobutylcyanoacrylate) nanocapsules used for oral delivery of peptides and proteins and siRNAs encapsulated in nanocapsules can be used to target estrogen receptor alpha (ERα), which is responsible for tumor cell growth.

Solid Lipid Nanoparticles (SLNs)

These are the alternative carrier systems to liposomes and emulsions. SLNs are considered core-shell structures, as SLNs are comprised of a phospholipid-coated solid hydrophobic core matrix (containing the hydrophobic tails of the phospholipid section). SLNs are developed on principle based on adhesive forces between two liquids. Normally, the interfacial tension between two liquids is less than their surface tension because of the weaker adhesive forces compared to that with gas. Molecules at the interface constitute surface-free energy of interfacial tension while they undergo agitation and form a spherical system to minimize the surface-free energy. SLNs are very similar to emulsions, where solid lipids are used as a substitute for the oil phase and melted and mixed with the aqueous phase. Adding a surfactant reduces the interfacial tension between the two liquids, thereby reducing the surface energy, and stable SLNs are formed. For example, Poloxamer 407 and Tween-80 are widely used surfactants to increase the efficacy of SLNs during drug delivery. Solid lipid nanoparticles are colloidal dispersions with modified properties of other nanoparticles, such as microemulsions, suspensions, liposomes, and polymeric nanoparticles. The major problems encountered with nanoparticles can be successively avoided using SLNs, and finally, a chemically stable and physiologically suitable drug delivery system can be achieved with lesser limitations. Moreover, they can help to increase the water content on the skin (skin hydration formula). They also demonstrate enhancement in skin penetration as an efficient carrier, biocompatibility, and stability agent for distribution. In moisturizing creams, SLNs can be used in sunscreen and act as potential UV blockers. They play an important role in cosmetic delivery applications as they screen encapsulated ingredients from degradation over a long period of time and also improve the penetration power of active ingredients into *stratum corneum* due to their smaller size [14]. In *in vivo* study, it is seen that solid lipid nanoparticles are the blocker of 3,4,5-trimethoxy benzoylchitin [15].

Nano-silver and Nano-gold

Nano-silver and nano-gold are famous on the medicinal front. Various cosmetic manufacturers claim that nano-silver used in deodorant offers antibacterial protection up to 24/7.

Similarly, nano-gold is used as a bacterial disinfectant and used in mouthwash and as an additive in toothpaste [16]. Nano-silver is also used as a preservative in cosmetics, and it has been reported that nano-silver remained stable, without exhibiting sedimentation, for longer than 1 year. The ability of nano-silver to prevent the spread of infection is due to the release of silver (Ag^+) ions from the surface of this material that is capable of destroying compounds that contain sulfur and phosphorus, such as the DNA and proteins present within bacteria, fungi or viruses. Nano-silver has become increasingly popular in such items due to its ability to destroy pathogens in superior potency at lower concentrations as compared to bulk silver in larger quantities. Gold is an antioxidant. It has anti-inflammatory properties. The metals (gold and silver) can calm acne inflammation, reduce skin redness and protect against free radicals that lead to wrinkles and sun damage. It vitalizes the fiber tissues under the skin and accelerates blood circulation.

Buck Minister Fullerence

The iconic nanomaterial is "C-60", which is structurally named Buckminsterfullerene. Their approximate diameter is 1 nm. Since it is iconic, it is used in very expensive face creams. It is a potent scavenger of free radicals [17]. Buck minster fullerene is commonly referred to as "fullerene" in skincare products. Its main thrilling property is that it can act as a powerful antioxidant that can scavenge free radicals because free radicals are responsible for aging, cancer, acne and rosacea. The double bonds in fullerene can easily react with free radicals, drench them like a "radical sponge" and damage them.

Warnings for Nano-cosmetics: How and Why?

The toxicity of nanoparticles is arising due to [18]:

a. Smaller size
b. Shape
c. Large surface area to volume ratio
d. Skin penetration capability

No doubt, the opportunity for the nanoparticles is given by smaller size/size alteration properties, which can manipulate their physiochemical properties. The smaller size particles easily access to bloodstream *via* adsorption into skin cells and are transported to various organs, leading to malfunctioning like (a) inhibited cell growth, (b) DNA strand breakage, (c) effect an inflammatory response to the abdominal wall.

Smaller nanoparticles with a high surface-to-volume ratio are much more reactive because of the large number of atoms lying over the surface rather than volume. Like nanoscale dimensional, SiO_2 may undergo explosion if suspended in air under the contact of some ignition source. Furthermore, the flexed skin invites the deep penetration of nanoparticles. The skin penetration formula of nanoparticles is enhanced by physical and chemical means like inducing/altering the vesicular system. The broken skin also causes the penetration of nanoparticles *via* wounds, acne and eczema.

Environmental Risks of Nanoparticles

Everything which is excess in nature causes risk(s) to the well-being of the whole ecosystem, like a gradual release of nanoparticles into the air, soil and water by various means. For example, after the usage of nanoparticles for potential activities and during the engineering of nanoparticles, some byproducts are released, which are hazardous. The very famous TiO_2 nanoparticles are used in personal care products, and they are also anti-bacterial in nature. These potential applications can also induce harsh effects. Although these materials are anti-bacterial in nature, if they release in excess or inappropriate amount, they can terminate many beneficial bacteria, which are useful for sewage and wastewater treatment [19].

NANOMEDICINES

Introduction

Nano-medicine is the medicinal branch of chemistry and nanotechnology that makes use of nanoparticles to cure ill health in humankind. Nano-medicine has targeted deadly diseases like cancer, Parkinson's disease (PD), Alzheimer's disease (AD), heart illness, and many more. The genetic sequence in DNA is detected by the invention of gold nanoparticles. Nanoparticles help in the diagnosis, treatment and tracking of illness because they are used in tags and labels in biological systems. Artificial implantation is gaining attention due to the nanotechnology front because nanoparticles are used for repairing and recovering damaged tissues from yielding artificially stimulated cells.

Medicinal use of Nanomaterials

Three factors are taken into consideration for coupling nanoparticles with drug delivery:

a. Encapsulation of drugs.
b. Delivery of drug to the targeted part.
c. Successful release of drug.

Following is a brief explanation of the medicinal use of nanoparticles:

Drug Delivery

Nanoparticles are used to deliver drugs at a specific site. A specific drug dose is delivered at a specific site to avoid its side effects. Especially, dendrimers are used for effective drug delivery, along with the reduction in the cost of drugs and pain to patients. In other words, drug delivery by nanoparticles is cost-effective as well as limits the use of excess drug consumption. Gold nanoparticles have found pounced application in drug delivery applications. Following are some examples of nanoparticles that are used in drug delivery:

• Abraxane (Albumin bound paclitaxel) nanoparticles are used for drug delivery in breast cancer and lung cancer [20].
• Long-chained iron oxide is used in drug delivery systems to prevent tumor growth [21].
• PEG nanoparticles carrying a payload of antibiotics are used to target bacterial infection.
• Biodegradable nanoparticles coated with plasminogen activators are used to degrade blood clots [22].
• Mini-cell nanoparticles are used for early-phase clinical trials for drug delivery to untreated cancer cells [23].

Protein and Peptide Delivery

Proteins and peptides are biopharmaceuticals and are used to target various diseases and disorders in the human body. The biodegradable polystyrene micro-particles capped by myelin sheath peptides induce autoimmunity in mice and prevent the occurrence of diseases.

Treatment of Cancer

Fluorescent quantum dots on coupling with MRI produce exceptional images of tumor sites. Nanoparticles (or quantum dots) are advantageous over organic dyes for imaging purposes in terms of cost-effectiveness. Not only imaging, multifunctional nanoparticles or quantum dots can detect and treat a tumor in cancer treatment [24]. Gold and cadmium selenide nanoparticles are used for the treatment of cancer cells. Gold nanoparticles have been studied to kill tumor cells in mice, as reported by Prof. Jennifer at Rice University, U.S.A [25]. Here, peptides are bound over nanoparticles and target cancerous cells by irradiating with an infrared laser, which heats the gold nanoparticles sufficiently and kills cancer cells. Cadmium selenide quantum dots are UV light absorbents, which are injected at the site of cancerous cells and further exposed to UV light. Since these nanoparticles are fluorescent and easily seen by a glowing tumor, this is called cancer photodynamic therapy.

Nanotechnology in the Treatment of Neurodegenerative Disorders

Another important application of nanotechnology after drug delivery is the treatment of neurodegenerative disorders (illnesses that involve the death of certain parts of the brain) like Parkinson's disease, Alzimer's disease and many more. Nanotechnology is being researched parallel with neurophysiology, neuropathology and cell biology to develop certain tools to protect Central Nervous System (CNS) directly or indirectly [26].

Alzheimer's Disease

Alzheimer's disease is caused by the abnormal build-up of proteins in and around brain cells. One of the proteins involved is called amyloid, which deposits and forms plaques around brain cells. The other protein is called tau, which deposits and forms tangles within brain cells.

More than 35 million of the worldwide population is suffering from AD. Two approaches can opt for the detection of AD markers:

a. In the first approach, the total amyloid-β (Aβ) or tau protein in the plasma or cerebrospinal fluid (CSF) is measured.
b. In the second approach, the pathogenic markers like Aβ-derived diffusible ligands and phosphorylated and cleaved tau protein are measured.

Nanotechnology has started to put forward a noteworthy bluster in neurology and neurodegeneration. The engineering of nanoparticles has made it possible for brain-specific endothelial cells (EC). The nanoparticles are specific in circulating Aβ, which may improve the AD condition by inducing a "sink effect". A recently developed method called the bio-barcode technique utilizes magnetic iron oxide core microparticles and gold nanoparticles (attached to a large number of *"barcode"* DNA oligonucleotide strands), which are suspended in solution and conjugated with ADDL-specific antibodies. The term *"barcode"* DNA is used for the exclusive marker exact to the target protein. Moreover, nanoparticle-based immune sensors have the capability to detect circulating amyloid-β (Aβ)$_{1-40}$ and Aβ_{1-42} and improve AD conditions [27].

Parkinson's Disease

A person suffers from Parkinson's disease (PD) when neurons in the brain that controls movement become impaired.

PD is the second major neurodegenerative disease after Alzheimer's disease (AD). It affects the central nervous system (CNS) as well as neuroinflammatory responses and leads to malfunctioning in body motions. Nanotechnology, an emerging tool, has the potential to improve these circumstances as follows:

a. By introducing novel carrier-based platforms that will target the selective release of drug payload on-demand.
b. Controlled release kinetics and amplified attain *via* modulating or bypassing the blood-brain barrier.

Above are the ways to decrease the side effects of PD, an intracranial nano-enabled scaffold device (NESD) is being engineered for site-specific delivery of dopamine. Nanotechnology plays an important role in *in vitro* diagnostics, novel biomarker identification and bioimaging modalities for the diagnosis of PD. Tashkhourian *et al.* reported a modified carbon paste electrode for the simultaneous quantification of dopamine and ascorbic acid. The electrodes were modified to incorporate silver nanoparticles and carbon nanotubes. Zinc oxide (ZnO) nanowire could selectively detect dopamine along with urate and ascorbate by a method called differential pulse voltammetry. The addition of nanowires leads to high electrical conductivity and enhancement of the sensitivity of electrochemical biosensors, leading to a reduction in the detection limit to 1 nanomolar for dopamine and uric acid in the serum of PD patients. α-Synuclein is a very important neuronal protein associated with PD. An *et al.* developed highly ordered microfabricated arrays using gold-doped TiO_2 nanotubes for photo-electrochemical detection of α-Synuclein. The breath test is one such advancement in PD diagnosis. Specific volatile organic compounds like styrene, butylated hydroxytoluene and hexadecane have been found in the breath of PD patients. For sensing the above specific compounds, organically functionalized nanoparticles like carbon nanotubes and gold nanoparticles were used by Tisch and co-workers. The sensors were effective in distinguishing the breath prints of PD from healthy states with an accuracy of 78%, which is also helpful in identifying Alzheimer's disease and differentiating it from PD. Moreover, peptides and peptidic nanoparticles are being considered to cure PD.

Tuberculosis (TB) Treatment

Tuberculosis is a deadly infectious disease caused by the bacteria *Mycobacterium tuberculosis*. *Mycobacterium* primarily attacks the lungs, but it can also attack the kidney, lymphatic system, central nervous system (meningitis), circulatory system (military tuberculosis), genitourinary system and joints and bones.

The serious problem with the existing tuberculosis chemotherapy is that when the drug is taken intravenously or administered orally, it is distributed throughout the body *via* systemic blood circulation, and a majority of molecules do not reach their targets. For the treatment of TB, a long course of medicines needs to be used, which puts a burden on the patient. It also poses some side effects and risks. At the pediatric stage, there is no availability of first-line drugs. Nanotechnology is not only working in the field of nano-medicines but also trying to design such routes which could overcome the high dose of drugs, create shortcut courses of treatment and also break the interaction of injected drugs with antiretroviral therapies.

Application of Nanotechnology in Dentistry

Nanotechnology aims at engineering and utilizing particles of size 1 nm to 100 nm. In operational dentistry, nanofillers or nano-filled composite resins employ SiO_2 (with spherical morphology) with sizes 5-40 nm. Additionally, micro-hybrids with nanofillers are best chosen in dentistry. Nanofillers mimic natural hard tissues (enamels) over optical and mechanical aspects. Nanofillers possess wear resistance as well as luster resistance. Nano-filled composite resins are non-agglomerated discrete nanoparticles. Another nanofillers used is aluminosilicate powder with a mean size of 80 ran and1:4 M ratio of alumina to silica [28].

Application in Surgery

The application of nanoparticles in surgery was reported by Rice University. In this application, gold-coated nanoparticles were allowed to dribble along the seam between two fused pieces of chicken meat by flesh welding and further applied during surgeries when arteries were being cut during organ transplants [29].

Application in Visualization

The movement of drug distribution and its metabolism in the body is monitored by fluorescent dyes or quantum dots. Quantum dots can couple with proteins that go through cell membranes.

Application to Resist Anti-biotic Resistance

Sometimes various proteins interact with anti-biotics and resist their activities. To decrease anti-biotic resistance, anti-biotic drugs are coupled with specific nanoparticles, which enhance their activities [30].

CONCLUSION AND SCOPE

Due to nanoscale effects, nanotechnology has found beneficial applications in cosmetics and drugs. Materials and devices at nano-scale dimensions pose unique physiochemical and biological properties in comparison to their larger-sized counterparts. Nanotechnology on interlinking with pharmaceuticals has yielded interesting facts like ZnO nanoparticles used in sunscreen lotions. Moreover, it also increases the anti-biotic properties of ciprofloxacin drugs. Fluorescent quantum dots are used to tag affected cells in the body. C_{60} nanoparticles are used in expensive personal care products without penetrating the skin. Nanomaterials are used in drug delivery and peptide-protein delivery. It is a futuristic technology by means of the regeneration of cells, modifying nutraceutical sectors to a large extent.

ACKNOWLEDGMENTS

One of the authors, Mrs. Balwinder Kaur, is thankful to University Grants Commission (UGC), New Delhi, India, for providing financial support under the scheme Maulana Azad National Fellowships (MANF). The author, Dr. Subhash Chand, would like to thank Lajpat Rai DAV College Jagraon, Ludhiana, Punjab, India. The author Dr. Rajesh Kumar is thankful to the Department of Chemistry, Government Degree College Sugh-Bhatoli, Kangra, Himachal Pradesh, India.

REFERENCES

[1]   Allied Market Research. https://www.alliedmarketresearch.com/cosmetics-

[2]   Anne Houtman, S.K.A.J.I. Toxic bottles? On the trail of chemicals in our every day lives. American Environmental Science for a Changing World. Kate Ahr Paker, **2013**; p. 54.

[3]   Wang, Z.; Dinh, D.; Scott, W.C.; Williams, E.S.; Ciarlo, M.; DeLeo, P.; Brooks, B.W. Critical review and probabilistic health hazard assessment of cleaning product ingredients in all-purpose cleaners, dish care products, and laundry care products. *Environ. Int.,* **2019**, *125*, 399-417.
[http://dx.doi.org/10.1016/j.envint.2019.01.079] [PMID: 30743146]

[4]   Juhász, M.L.W.; Marmur, E.S. A review of selected chemical additives in cosmetic products. *Dermatol. Ther.,* **2014**, *27*(6), 317-322.
[http://dx.doi.org/10.1111/dth.12146] [PMID: 25052592]

[5]   Halla, N.; Fernandes, I.; Heleno, S.; Costa, P.; Boucherit-Otmani, Z.; Boucherit, K.; Rodrigues, A.; Ferreira, I.; Barreiro, M. Cosmetics preservation: a review on present strategies. *Molecules,* **2018**,

23(7), 1571.
[http://dx.doi.org/10.3390/molecules23071571] [PMID: 29958439]

[6] Zhang, L.; Freeman, L.E.B.; Nakamura, J.; Hecht, S.S.; Vandenberg, J.J.; Smith, M.T.; Sonawane, B.R. Formaldehyde and leukemia: epidemiology, potential mechanisms, and implications for risk assessment. *Environ. Mol. Mutagen.,* **2010**, *51*(3), 181-191.
[PMID: 19790261]

[7] Bilal, M.; Iqbal, H.M.N. An insight into toxicity and human-health-related adverse consequences of cosmeceuticals — A review. *Sci. Total Environ.,* **2019**, *670*, 555-568.
[http://dx.doi.org/10.1016/j.scitotenv.2019.03.261] [PMID: 30909033]

[8] Friends of the Earth Report. Nanomaterials, Sunscreens and Cosmetics: Small Ingredients Big Risks. Available from: http://www.foe.org

[9] Nano Science Institute. Scientific Committee Rules on the Safety of Nanocosmetics. Available from: http://www.nanoscienceinstitute. com/NanoCosmetics.htm

[10] Schueller, R.; Romanowski, P. Emerging Technologies and the Future of Cosmetic Science. Available from: http://www.specialchem4cosmetics.com/services/articles

[11] Jain, S.; Sapee, R.; Jain, N.K. Proultraflexible lipid vesicles for effective transdermal delivery of norgesterol. USA *Proceedings of 25th conference of C.R.S.,* **1998**, pp. 32-35.

[12] Sonneville-Aubrun, O.; Simonnet, J.T.; L'Alloret, F. Nanoemulsions: a new vehicle for skincare products. *Adv. Colloid Interface Sci.,* **2004**, *108-109*, 145-149.
[http://dx.doi.org/10.1016/j.cis.2003.10.026] [PMID: 15072937]

[13] Hwang, S.L.; Kim, J.C. *In vivo* hair growth promotion effects of cosmetic preparations containing hinokitiol-loaded poly(ε -caprolacton) nanocapsules. *J. Microencapsul.,* **2008**, *25*(5), 351-356.
[http://dx.doi.org/10.1080/02652040802000557] [PMID: 18465297]

[14] Müller, R.H.; Radtke, M.; Wissing, S.A. Solid lipid nanoparticles (SLN) and nanostructured lipid carriers (NLC) in cosmetic and dermatological preparations. *Adv. Drug Deliv. Rev.,* **2002**, *54* Suppl. 1, S131-S155.
[http://dx.doi.org/10.1016/S0169-409X(02)00118-7] [PMID: 12460720]

[15] Song, C.; Liu, S. A new healthy sunscreen system for human: Solid lipid nannoparticles as carrier for 3,4,5-trimethoxybenzoylchitin and the improvement by adding Vitamin E. *Int. J. Biol. Macromol.,* **2005**, *36*(1-2), 116-119.
[http://dx.doi.org/10.1016/j.ijbiomac.2005.05.003] [PMID: 16005509]

[16] Raj, S.; Jose, S.; Sumod, U.S.; Sabitha, M. Nanotechnology in cosmetics: Opportunities and challenges. *J. Pharm. Bioall. Sci.,* **2012**, *4*, 186-193.

[17] Bakry, R.; Vallant, R.M.; Najam-ul-Haq, M.; Rainer, M.; Szabo, Z.; Huck, C.W.; Bonn, G.K. Medicinal applications of fullerenes. *Int. J. Nanomedicine,* **2007**, *2*(4), 639-649.
[PMID: 18203430]

[18] Katz, L.M.; Dewan, K.; Bronaugh, R.L. Nanotechnology in cosmetics. *Food Chem. Toxicol.,* **2015**, *85*, 127-137.
[http://dx.doi.org/10.1016/j.fct.2015.06.020] [PMID: 26159063]

[19] Cimitile, M. Nanoparticles in Sunscreen Damage Microbes. *Environ. Health News,* **2009**, *24*, 2009.

[20] Hollmer, M. Carbon nanoparticles charge up old cancer treatment to powerful effect. *Fierce. Pharma.,* **2012**.

[21] Peiris, P.M.; Bauer, L.; Toy, R.; Tran, E.; Pansky, J.; Doolittle, E.; Schmidt, E.; Hayden, E.; Mayer, A.; Keri, R.A.; Griswold, M.A.; Karathanasis, E. Enhanced delivery of chemotherapy to tumors using a multicomponent nanochain with radio-frequency-tunable drug release. *ACS Nano,* **2012**, *6*(5), 4157-4168.
[http://dx.doi.org/10.1021/nn300652p] [PMID: 22486623]

[22]   Reddy, S.J. The Recent Advances in the Nanotechnology and Its Applications-A Review. *Nanotechnology,* **2020**, *50,* 24-30.

[23]   Suzanne, E. Bacterial 'minicells' deliver cancer drugs straight to the target. *Fierce. Drug Deliv.,* **2012**.

[24]   Nie, S.; Xing, Y.; Kim, G.J.; Simons, J.W. Nanotechnology applications in cancer. *Annu. Rev. Biomed. Eng.,* **2007**, *9*(1), 257-288.
        [http://dx.doi.org/10.1146/annurev.bioeng.9.060906.152025] [PMID: 17439359]

[25]   Loo, C.; Lin, A.; Hirsch, L.; Lee, M.H.; Barton, J.; Halas, N.; West, J.; Drezek, R. Nanoshell-enabled photonics-based imaging and therapy of cancer. *Technol. Cancer Res. Treat.,* **2004**, *3*(1), 33-40.
        [http://dx.doi.org/10.1177/153303460400300104] [PMID: 14750891]

[26]   Wong, H.L.; Wu, X.Y.; Bendayan, R. Nanotechnological advances for the delivery of CNS therapeutics. *Adv. Drug Deliv. Rev.,* **2012**, *64*(7), 686-700.
        [http://dx.doi.org/10.1016/j.addr.2011.10.007] [PMID: 22100125]

[27]   Brambilla, D.; Le Droumaguet, B.; Nicolas, J.; Hashemi, S.H.; Wu, L-P.; Moghimi, S.M.; Couvreur, P.; Andrieux, K. Nanotechnologies for Alzheimer's disease: diagnosis, therapy, and safety issues. *Nanomedicine,* **2011**, *7*(5), 521-540.
        [http://dx.doi.org/10.1016/j.nano.2011.03.008]

[28]   Sivaramakrishnan, S.M.; Neelakantan, P. Nanotechnology in Dentistry - What does the Future Hold in Store? *Dentistry,* **2014**, *4,* 1-3.

[29]   Gobin, A.M.; O'Neal, D.P.; Watkins, D.M.; Halas, N.J.; Drezek, R.A.; West, J.L. Near infrared laser-tissue welding using nanoshells as an exogenous absorber. *Lasers Surg. Med.,* **2005**, *37*(2), 123-129.
        [http://dx.doi.org/10.1002/lsm.20206] [PMID: 16047329]

[30]   Banoee, M.; Seif, S.; Nazari, Z.E.; Jafari-Fesharaki, P.; Shahverdi, H.R.; Moballegh, A.; Moghaddam, K.M.; Shahverdi, A.R. ZnO nanoparticles enhanced antibacterial activity of ciprofloxacin against Staphylococcus aureus and Escherichia coli. *J. Biomed. Mater. Res. B Appl. Biomater.,* **2010**, *93B*(2), 557-561.
        [http://dx.doi.org/10.1002/jbm.b.31615] [PMID: 20235250]

CHAPTER 11

Nano Ferrites for Biomedical Applications

Gulshan Dhillon[1], **Mansi Chitkara**[1] and **Inderjeet Singh Sandhu**[1,*]

[1] *Chitkara University Institute of Engineering and Technology, Chitkara University, Rajpura, Punjab, India*

Abstract: Superparamagnetic iron oxide nanoparticles have attracted attention due to their compatibility with various biomedical applications. The quantum confinement and increased surface area to volume ratio of the nanostructures alter their magnetic properties. There are several bottom-up techniques to synthesize superparamagnetic iron oxide nanoparticles; however, they offer certain limitations, like the existence of a secondary phase. The reaction parameters can be controlled to form pure-phase nanoparticles to increase their scope of applications in the field of medicine. Moreover, different applications demand different surface coatings of iron oxide nanoparticles.

Keywords: Drug delivery, Hyperthermia, Iron oxide, *In vitro*, Nanostructures, Superparamagnetism.

INTRODUCTION

Nanotechnology has emerged in recent years to the extent that it is now possible to synthesize, characterize, and modify its functional properties for biomedical applications [1 - 3]. Nanostructures and systems with tunable chemical and physical characteristics are preferably synthesized by inorganic nanoparticles [1, 4]. For some years, small iron oxide particles have been used for *in vitro* diagnostics [5]. There have been several research projects on various forms of magnetic iron oxide nanostructures from which magnetite, because of its biocompatibility, is the most promising candidate [6]. Magnetite (Fe_3O_4) is a magnetic type of iron oxide with a cubic inverse spinel structure and oxygen tightly packed into the FCC structure. The interstitial tetrahedral and octahedral positions are filled by the Fe cations [7]. Electrons appear to jump between Fe^{2+} and Fe^{3+} ions in octahedral positions at room temperature, making magnetite an important class of semi-metallic materials [8].

* **Corresponding author Inderjeet Singh Sandhu**: Chitkara University Institute of Engineering and Technology, Chitkara University, Punjab, India; Email: is.sandhu@chitkara.edu.in

Karamjit Singh Dhaliwal (Ed.)

Nanostructures are not physically or chemically distinguished by either an atom or a mass twin [9]. Quantum confinement effects and magnetic nanostructures' broad surface-to-volume ratio change some magnetical properties unexpectedly and reveal superparamagnetic behavior as every component acts as a single magnetic field [10]. Biomedical applications, such as drug delivery, hyperthermia, and MRI, based on the unique physical, chemical, thermal and mechanical features, are promising for superparamagnetic nanoparticles [11 - 14]. The particles should have combined characteristics of high magnetic saturation, biocompatibility, and surface interaction. However, it is possible, by adding different bioactive molecules, to change the surface of these particles by forming a few atomic layers that are suitable for more functions [15]. In the identification of such anti-cancer therapies, the aggregation of magnetic ions, *e.g.*, in tumour tissue, is crucial to the development and assessment of binding and drug carriers. The ideal surface properties of magnetic nanoparticles have high potential in many *in vitro* and *in vivo* applications [16]. The performance of the particles is based on different factors like:

a. Strong sensitivity to magnet enrichment for efficient purposes [17],
b. Particles of a single magnetic domain, that is, particles with uniform magnetization, with super-paramagnetic and high saturation values, will be of a size in the range of 6–15 nm (Particles below critical particle size (~15nm) [18], are easily separated by extravasations and clearance [19]),
c. Superparamagnetic conduct [16],
d. Produced unique bio-medical uses for surface chemistry [1], *etc.*

SYNTHESIS OF MAGNETIC IRON OXIDE NANOPARTICLES

For researchers, it is often a struggle to synthesize iron oxide nanoparticles of a particular shape and structure. Three key pathways, chemical, physical, and biological, can be used to synthesize iron oxide nanoparticles. These techniques have been studied with a view to synthesizing nanoparticles that are more stable, soluble and size-controlled [20]. Co-precipitation of Fe^{3+} and Fe^{2+} aqueous salt solutions by adding a base can be used to prepare iron oxide nanoparticles (Fe_3O_4 or γ-Fe_2O_3). Synthesized nanoparticles can be regulated by their size, structure, and shape based on the nature of the salt used, pH, and the medium's ionic strength [21].

Magnetite is typically formed in 2:1 molar ratios by adding a base to an aqueous mixture of Fe^{3+} and Fe^{2+} chlorides. The synthesized nanoparticles have black colour; the chemical reaction for the magnetite nanoparticles' co-precipitation route is presented below:

$$Fe^{2+} + Fe^{3+} + 8OH^- \rightarrow Fe_3O_4 + 4H_2O \qquad (1)$$

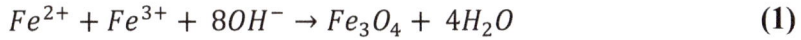

According to the thermodynamics of the above reaction, the total precipitation of magnetite is expected between a pH of 9 and 14. To prevent the oxidation of synthesized nanoparticles, the molar ratio of Fe^{3+}:Fe^{2+} as 2:1 should be preserved. Fig. (**1a**) shows magnetite formation at room temperature nearly immediately, while Fig. (**1b**) shows maghemite formation by chemisorption of magnetite oxygen. Fig. (**1c**) shows the final colloidal magnetite formed by peptizing an alkaline source.

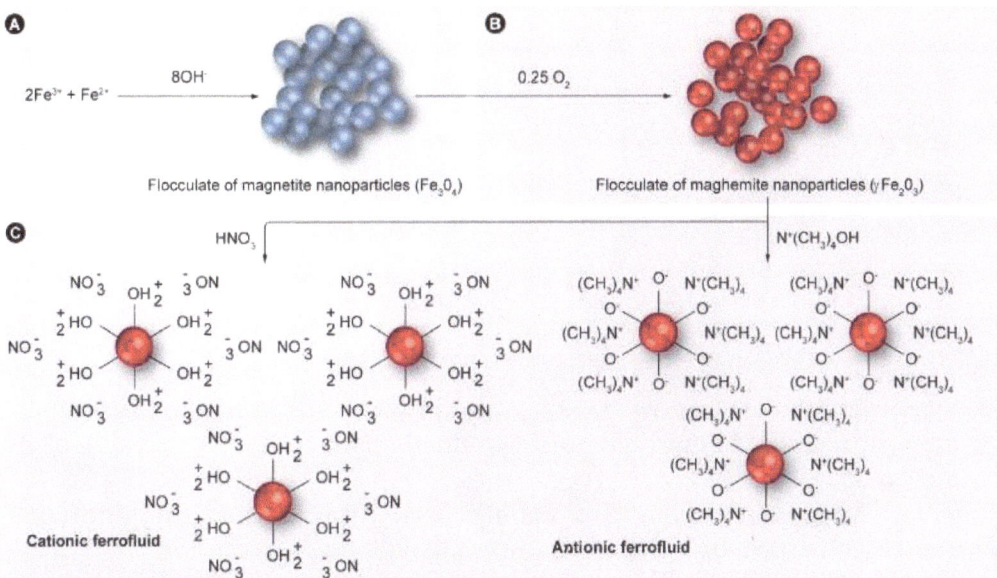

Fig. (**1**). (**a**) Flocculation of magnetite nanoparticles, (**b**) Flocculation of maghemite nanoparticles, (**c**) Colloidal magnetite dispersions obtained after peptization [22].

The physical and chemical properties of synthesized magnetic nanoparticles are seriously compromised by oxidation. They are also coated with organic/inorganic molecules during the precipitation process to avoid the oxidation of synthesized Fe_3O_4 nanoparticles in the air.

IRON OXIDE NANOPARTICLES COATING AND FUNCTIONALI-ZATION

Core-shell nanoparticles appear to bind different medicines with nanoparticles of iron oxide. In contrast to the surface layer working, stability enhancement, biocompatibility, and biodiversity [23], the center is the nanoparticles. Thus, the core is made of iron oxide nanoparticles. The shells of the iron oxide nanoparticles can be polymers (synthetic/natural), organic surfactants, inorganic compounds, and bioactive molecules (Fig. **2**).

Nanoparticles functionalization - surface coating

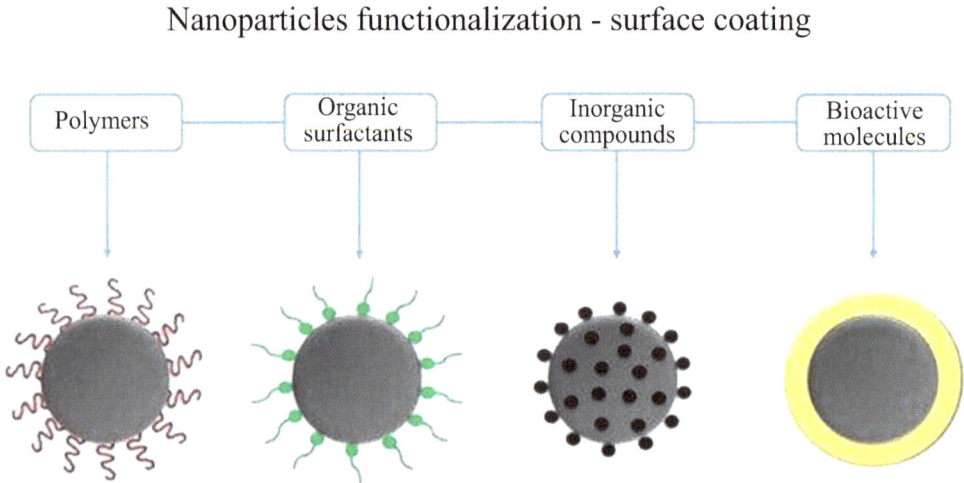

Fig. (2). Schematic illustration of the main shells for functionalization of iron oxide nanoparticles (IONPs). Grey circles represent the core of IONPs [22].

- *Synthetic and Natural Polymers*: Polymers prevent oxidation and offer stability to nanoparticles, and hence are most commonly used for surface coating of iron oxide nanoparticles [24].
- *Organic Surfactants*: They are mainly used in functionalizing iron oxide nanoparticles, especially when synthesized in organic solutions [25].
- *Inorganic Compounds*: These compounds have the ability to increase the antioxidant properties of iron oxide nanoparticles [26]. The most experimented inorganic compounds for iron oxide nanoparticles are silica, carbon, metals, oxides, and sulfides.
- *Bioactive Molecules*: Bioactive structures, including lipids, peptides, and proteins, are included in this category [20, 27]. Biomolecules can maintain the stability and magnetic properties of iron oxide nanoparticles.

BINDING DRUGS TO IRON OXIDE NANOPARTICLES

In contrast to micrometric equivalents, the reactive zone and the ability to resolve biological obstacles, magnetic nanoparticles are best used to transport drugs. Fig. (**3**) demonstrates the relationship between various medicines directly with nanoparticles of iron oxide or with nanosystems of the core-shell. Adsorption, nuclear encapsulation, polymer-matrix delivery, and covalent addition to the surface can be used to accomplish this binding [28]. This association aims to boost the pharmacological properties of iron oxide nanostructures. Furthermore, the carriers of antifungal, antibiotic, and anticancer agents have been made from magnetic nanoparticles.

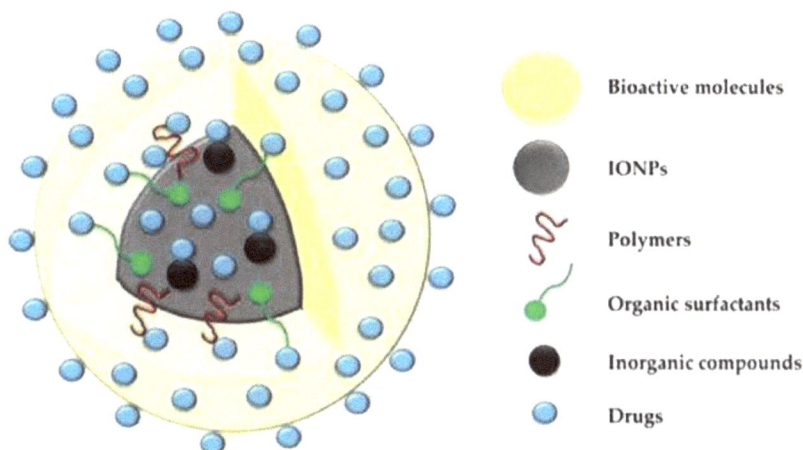

Fig. (3). Schematic illustration of drugs directly bound to iron oxide nanoparticles (IONPs) or core-shell nanosystems [22].

CHARACTERIZATION

The magnetism of the iron oxide nanoparticles is determined by their scale, form, composition, and chemical phase. By using various techniques, as shown in the following Fig. (**4**), the properties and chemical composition of nanostructures can be determined. Electron microscopy helps to determine the nano-structure particle size. This method calculates the overall particle size and defines the size distribution, while x-ray diffraction helps to determine the crystalline structure of nanoparticles. In addition, the Scherrer formula allows for determining the crystal dimensions from the x-ray diffraction pattern line broadening. Mossbauer spectroscopy is a means of accessing the structure of the crystal. This provides information on the relaxation timescale of Néel, an essential complement of superparamagnetic particles. With magnetometry, the superparamagnetic activity of the nanostructures is confirmed. This technique also shows the mean diameter and magnetization of the lenses.

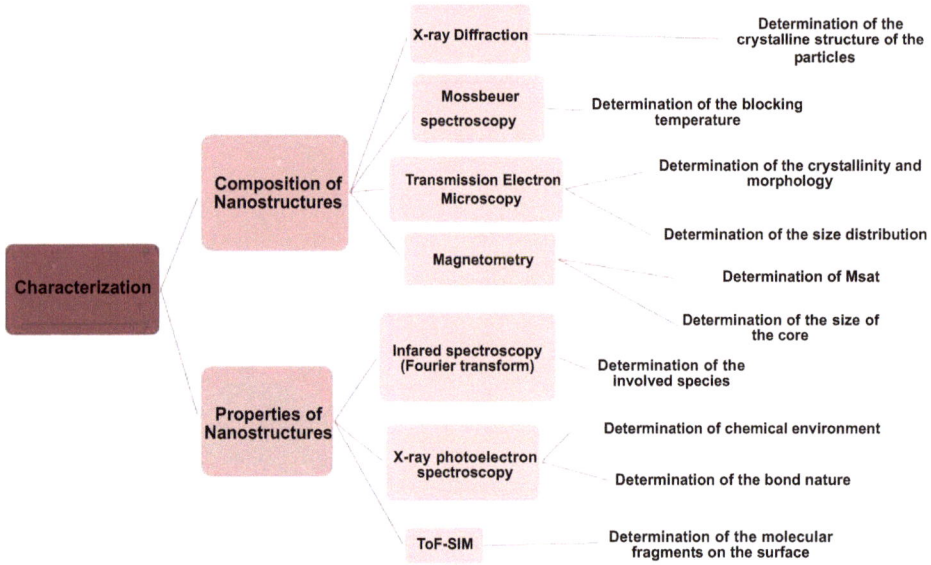

Fig. (4). Flowchart shows the various characterization techniques used to identify the composition and properties of nanostructures.

APPLICATIONS

Superparamagnetic nanoparticles are biocompatible and can also be recycled by cells using the usual biochemical mechanisms for iron metabolism [29]. This is a biodegradable iron oxide mineral nucleus. The surface covering of the molecules and functional groups also allows for a chemical interconnection. This encourages the utilization of nanoparticles to treat a specific illness or organ [30].

- *Tissue Repair:* Iron oxide nanoparticles help in tissue repair either by welding or soldering. In welding, two tissue surfaces are sufficiently heated to join, while in soldering, protein or synthetic polymer-coated nanoparticles are placed between the surfaces of two tissues to join. Nanoparticles that absorb sunlight more strongly are also helpful in tissue-repairing procedures. The nanoparticles are coated on the surfaces of two tissues at the site requiring joining. This method offers techniques to minimize tissue damage with the use of the least hazardous wavelength of light sources.
- *Targeted Drug Delivery:* Chemotherapy is usually practiced for the transportation of anticancer drugs, which have various negative effects on the human body. Hence, targeted drug delivery serves as an alternative to this toxic treatment. Magnetic nanoparticles are generally coated with a biocompatible layer to functionalize the nanoparticles.

- *Magnetic Hyperthermia:* It is a therapeutic technique, where heat therapy is used to tear down cancer tissues or cells. The temperature of the infected or diseased area is raised to 41-46 °C to destroy the cancerous cells without damaging the healthy cells. Magnetic fluid hyperthermia is based on the principle of converting electromagnetic energy into heat. The magnetic nanoparticles are distributed around the target site, and an alternating magnetic field is applied. This alternating magnetic field supplies energy, which helps the magnetic moments in the particles to overcome the reorientation energy barrier.

CONCLUSION

The scope of application of superparamagnetic iron oxide nanoparticles is very wide in biomedicine. Various attempts have been made to modify the surface behaviour of the nanostructures to increase their magnetic behaviour and decrease the toxic effects on the human body. However, since iron changes its oxidation state very rapidly, it is important to stabilize the nanostructures by adopting an efficient synthesis technique. The various futuristic biomedical applications of nanostructures are expected to revolutionize the medical sector.

ACKNOWLEDGEMENT

Declared none.

REFERENCES

[1]   Moghimi, S.M.; Hunter, A.C.H.; Murray, J.C. Long-circulating and target-specific nanoparticles: theory to practice. *Pharmacol. Rev.,* **2001**, *53*(2), 283-318.
       [PMID: 11356986]

[2]   Curtis, A.; Wilkinson, C. Nanotechniques and approaches in biotechnology. *Trends Biotechnol.,* **2001**, *19*(3), 97-101.
       [http://dx.doi.org/10.1016/S0167-7799(00)01536-5] [PMID: 11179802]

[3]   Wilkinson, J.M. Nanotechnology applications in medicine. *Med. Device Technol.,* **2003**, *14*(5), 29-31.
       [PMID: 12852120]

[4]   Panyam, J.; Labhasetwar, V. Biodegradable nanoparticles for drug and gene delivery to cells and tissue. *Adv. Drug Deliv. Rev.,* **2003**, *55*(3), 329-347.
       [http://dx.doi.org/10.1016/S0169-409X(02)00228-4] [PMID: 12628320]

[5]   Gilchrist, R.K.; Medal, R.; Shorey, W.D.; Hanselman, R.C.; Parrott, J.C.; Taylor, C.B. Selective inductive heating of lymph nodes. *Ann. Surg.,* **1957**, *146*(4), 596-606.
       [http://dx.doi.org/10.1097/00000658-195710000-00007] [PMID: 13470751]

[6]   Schwertmann, U.; Cornell, R.M. *Iron oxides in the laboratory: preparation and characterization;* VCH: Weinheim, Cambridge, **1991**.

[7]   Cornelis, K.; Hurlburt, C.S. *Manual of Mineralogy;* Wiley: New York, **1977**.

[8]   Kwei, G.H.; Von Dreele, R.B.; Williams, A.; Goldstone, J.A.; Lawson, A.C., II; Warburton, W.K. Structure and valence from complementary anomalous X-ray and neutron powder diffraction. *J. Mol. Struct.,* **1990**, *223*, 383-406.

[http://dx.doi.org/10.1016/0022-2860(90)80482-Y]

[9]   Babes, L.; Denizot, B.; Tanguy, G.; Le Jeune, J.J.; Jallet, P. Synthesis of iron oxide nanoparticles used as MRI contrast agents: a parametricstudy. *J. Colloid Interface Sci.,* **1999**, *212*(2), 474-482.
[http://dx.doi.org/10.1006/jcis.1998.6053] [PMID: 10092379]

[10]   Goya, G.F.; Berquó, T.S.; Fonseca, F.C.; Morales, M.P. Static and dynamic magnetic properties of spherical magnetite nanoparticles. *J. Appl. Phys.,* **2003**, *94*(5), 3520-3528.
[http://dx.doi.org/10.1063/1.1599959]

[11]   Arbab, A.S.; Bashaw, L.A.; Miller, B.R.; Jordan, E.K.; Lewis, B.K.; Kalish, H.; Frank, J.A. Characterization of biophysical and metabolic properties of cells labeled with superparamagnetic iron oxide nanoparticles and transfection agent for cellular MR imaging. *Radiology,* **2003**, *229*(3), 838-846.
[http://dx.doi.org/10.1148/radiol.2293021215] [PMID: 14657318]

[12]   Reimer, P.; Weissleder, R. [Development and experimental use of receptor-specific MR contrast media]. *Radiologe,* **1996**, *36*(2), 153-163.
[http://dx.doi.org/10.1007/s001170050053] [PMID: 8867433]

[13]   Pankhurst, Q.A.; Connolly, J.; Jones, S.K.; Dobson, J. Applications of magnetic nanoparticles in biomedicine. *J. Phys. D Appl. Phys.,* **2003**, *36*(13), R167-R181.
[http://dx.doi.org/10.1088/0022-3727/36/13/201]

[14]   Ha¨ feli U, Schu¨ tt W, Teller J, Zborowski M, editors. Scientific and clinical applications of magnetic carriers. New York: Plenum Press; **1997**.

[15]   Berry, C.C.; Curtis, A.S.G. Functionalisation of magnetic nanoparticles for applications in biomedicine. *J. Phys. D Appl. Phys.,* **2003**, *36*(13), R198-R206.
[http://dx.doi.org/10.1088/0022-3727/36/13/203]

[16]   Tartaj, P.; Morales, M.P.; Veintemillas-Verdaguer, S. Gonza´ lez- Carren˜ o T, Serna CJ. The preparation of magneticnanopartic les for applications in biomedicine. *J. Phys. D Appl. Phys.,* **2003**, *36*, R182-R197.
[http://dx.doi.org/10.1088/0022-3727/36/13/202]

[17]   Jordan, A.; Scholz, R.; Maier-Hauff, K.; Johannsen, M.; Wust, P.; Nadobny, J.; Schirra, H.; Schmidt, H.; Deger, S.; Loening, S.; Lanksch, W.; Felix, R. Presentation of a new magnetic field therapy system for the treatment of human solid tumors with magnetic fluid hyperthermia. *J. Magn. Magn. Mater.,* **2001**, *225*(1-2), 118-126.
[http://dx.doi.org/10.1016/S0304-8853(00)01239-7]

[18]   Chatterjee, J.; Haik, Y.; Chen, C.J. Size dependent magnetic properties of iron oxide nanoparticles. *J. Magn. Magn. Mater.,* **2003**, *257*(1), 113-118.
[http://dx.doi.org/10.1016/S0304-8853(02)01066-1]

[19]   Pratsinis, S.E.; Vemury, S. Particle formation in gases: A review. *Powder Technol.,* **1996**, *88*(3), 267-273.
[http://dx.doi.org/10.1016/S0032-5910(96)03130-0]

[20]   Wu, W.; He, Q.; Jiang, C. Magnetic iron oxide nanoparticles: synthesis and surface functionalization strategies. *Nanoscale Res. Lett.,* **2008**, *3*(11), 397-415.
[http://dx.doi.org/10.1007/s11671-008-9174-9] [PMID: 21749733]

[21]   Hadjipanayis, G.C.; Siegel, R.W. Nanophase materials: synthesis, properties and applications. NATO ASI Series, Applied Sciences. Kluwer: Dordrecht, **1993**. Vol. E260.

[22]   Laurent, S.; Bridot, J.L.; Elst, L.V.; Muller, R.N. Magnetic iron oxide nanoparticles for biomedical applications. *Future Med. Chem.,* **2010**, *2*(3), 427-449.
[http://dx.doi.org/10.4155/fmc.09.164] [PMID: 21426176]

[23]   Wu, W.; Chen, B.; Cheng, J.; Wang, J.; Xu, W.; Liu, L.; Xia, G.; Wei, H.; Wang, X.; Yang, M.; Yang, L.; Zhang, Y.; Xu, C.; Li, J. Biocompatibility of Fe_3O_4/DNR magnetic nanoparticles in the treatment of hematologic malignancies. *Int. J. Nanomedicine,* **2010**, *5*, 1079-1084.

[PMID: 21170355]

[24] Arias, LaísSalomão, JulianoPelimPessan, Ana Paula Miranda Vieira, Taynara Maria Toito de Lima, Alberto Carlos BotazzoDelbem, and Douglas Roberto Monteiro. Iron oxide nanoparticles for biomedical applications: a perspective on synthesis, drugs, antimicrobial activity, and toxicity. *Antibiotics (Basel),* **2018,** *7*(2), 46.

[25] Couto, D.; Freitas, M.; Carvalho, F.; Fernandes, E. Iron oxide nanoparticles: An insight into their biomedical applications. *Curr. Med. Chem.,* **2015,** *22*(15), 1808-1828.
[http://dx.doi.org/10.2174/0929867322666150311151403] [PMID: 25760089]

[26] Xu, J.K.; Zhang, F.F.; Sun, J.J.; Sheng, J.; Wang, F.; Sun, M. Bio and nanomaterials based on Fe$_3$O$_4$. *Molecules,* **2014,** *19*(12), 21506-21528.
[http://dx.doi.org/10.3390/molecules191221506] [PMID: 25532846]

[27] Drmota, A.; Drofenik, M.; Koselj, J.; Žnidaršič, A. Microemulsion method for synthesis of magnetic oxide nanoparticles. *Microemulsions—An Introduction to Properties and Applications,* 1st ed; Najjar, R., Ed.; InTech: Rijeka, Croatia, **2012,** pp. 191-215.
[http://dx.doi.org/10.5772/36154]

[28] Rădulescu, M.; Andronescu, E.; Holban, A.; Vasile, B.; Iordache, F.; Mogoantă, L.; Mogoşanu, G.; Grumezescu, A.; Georgescu, M.; Chifiriuc, M. Antimicrobial nanostructured bioactive coating based on Fe$_3$O$_4$ and patchouli oil for wound dressing. *Metals (Basel),* **2016,** *6*(5), 103.
[http://dx.doi.org/10.3390/met6050103]

[29] Hans, M.L.; Lowman, A.M. Nanoparticles for Drug Delivery.*Nanomaterials Handbook,* 1st ed; Gogotsi, Y., Ed.; Taylor & Francis: Boca Raton, FL, USA, **2006,** pp. 637-664.
[http://dx.doi.org/10.1201/9781420004014.ch23]

[30] Sun, C.; Lee, J.S.H.; Zhang, M. Magnetic nanoparticles in MR imaging and drug delivery. *Adv. Drug Deliv. Rev.,* **2008,** *60*(11), 1252-1265.
[http://dx.doi.org/10.1016/j.addr.2008.03.018] [PMID: 18558452]

CHAPTER 12

Rare-earth Induced Nano-crystallization Study of Borate Glass System

Dinesh Kumar[1], S. M. Rao[2] and Supreet Pal Singh[1],*

[1] *Department of Physics, Punjabi University, Patiala-147002, Punjab, India*

[2] *Institute of Physics, Academia Sinica, Taipei-11529, Taiwan*

Abstract: The study of the physical and structural properties of rare-earth doped borate glasses is discussed in this chapter. The glasses have been synthesized using the conventional melt quenching technique. X-ray diffractometer has been used to determine the amorphous nature of the prepared glass samples. Upon closer inspection of the XRD patterns, it is observed that the peak width changes, which indicates that the increase in the concentration of rare earth induces a localized devitrification of the prepared glasses around the rare earth. To investigate the structural dependence on the chemical composition of the manufactured glass, various parameters were determined. The molecular vibrations and rotations related to covalent bonds found in the glasses were studied using the infrared spectrum.

Keywords: Borate glass, Fourier transform infrared spectroscopy, Nano-crystallization, Rare-earth, X-ray diffraction.

INTRODUCTION

In solids, the absence of long-range order is one of the main characteristics of the amorphous materials, while short-range order is exhibited by crystalline materials. The word glass tells about the state of the material. It can be defined as an inorganic product of fusion, which has been cooled to a rigid condition without crystallizing. Thus, glass is an amorphous solid that possesses glass transition temperature and is obtained by supercooling of melt without any crystallization [1]. Generally, when a liquid is cooled slowly from the high temperature to its melting temperature, T_m, the liquid attains a crystalline phase. To form the glass, the cooling must be very fast, so that it bypasses the crystalline phase. It is safe to say that we live in a world, where glass is an important part of the daily life schedule. Therefore, it is very important to study the properties of glass to explore

* **Corresponding author Supreet Pal Singh:** Department of Physics, Punjabi University, Patiala-147002, Punjab, India; Email: supreet_phy@pbi.ac.in

Karamjit Singh Dhaliwal (Ed.)

its applications in different areas. The widely used glasses are the oxide glasses and based on the former used, the glasses are of different types: borate, silicate, germanate, phosphate, *etc.* [2].

Borate glasses have become increasingly prevalent in the industrial, medical, and environmental monitoring aspects of modern society. The research interest in borate glasses is due to their promising properties, like high transparency, low melting point, effective atomic number, high thermal stability, and good solubility of rare earth (RE) ions [3 - 5]. However, the hygroscopic characteristic of these glasses has a detrimental effect on their properties, making them unstable. Modifiers, such as alkali and alkaline earth metals, have been used to mitigate thisproblem. These dopants have been found to not only improve moisture resistance but also function as activators in enhancing the glasses' emission qualities [6 - 8]. Of late, sodium oxide (Na_2O) and strontium oxide (SrO) are considered to be especially good modifiers for improving the features of borate glasses, such as moisture resistance and structural stability, to a substantial measure [9].

The rare-earth ions (RE^{3+}) possess a wide number of applications in the field of laser technology, solid-state lighting system, computer technology and optical communication systems. The applications of the rare-earth can be further enhanced, if they are doped in borate glass, which acts as a suitable host material. The change in the chemical composition of the glass matrix can modify the properties of RE^{3+} doped glasses, which are dependent on the environment around rare-earth ions [10]. It is interesting to analyze the effect of different modifiers on the physical and structural properties of the borate glasses.

In view of this, a study has been undertaken to analyze the physical and structural properties of the sodium strontium borate glass system doped with the different concentrations of dysprosium oxide (Dy_2O_3), erbium oxide (Er_2O_3), and neodymium oxide (Nd_2O_3). The synthesis method used for the glass preparation is the melt quenching, which is fast, easy and can be used to prepare glasses in bulk quantities. Important physical parameters have been calculated using suitable techniques, and structural properties have been analyzed using X-ray diffraction (XRD) and Fourier transform infrared spectroscopy (FTIR).

EXPERIMENTAL PROCEDURE

The choice of method for the preparation of the samples for any study is very important. There are various methods, which can be employed for the synthesis of glasses. In the present work, the glasses have been synthesized using the conventional melt quenching technique. Stoichiometric amounts of analytical reagent (AR) grade B_2O_3, Na_2O and SrO in the form of H_3BO_3, Na_2CO_3 and $SrCO_3$ were taken as initial glass constituents. The sodium strontium borate glasses were doped with the different concentrations of Dy_2O_3, Er_2O_3, and Nd_2O_3, respectively. The constituents were weighed on an electronic balance with the sensitivity of 1mg, mixed, and grinded in an agate mortar for 1 hour. The mixture was then transferred to a silica crucible and placed in an electronic muffle furnace. The temperature was initially maintained at 120°C for 2-3 hours to remove the moisture. After that, the temperature was increased slowly to 1000°C and maintained there for 1 hour to attain a bubble-free melt by stirring frequently. The melt was cast into a prepared graphite mould after it reached the desired viscosity. The mould was then put in an annealing furnace,set at 350°C for two hours to reduce internal stress, after which the glass sample was left in the furnace to cool. The prepared samples were transparent, uniform, and free of bubbles. The nomenclature and the chemical composition of the glass samples are detailed out in Tables **1**, **2**, and **3**.

Table 1. Chemical composition (in mol %) and glass codes nomenclature of dysprosium oxide doped sodium strontium borate glass samples (DyNaSrB).

Glass codes	B_2O_3	Na_2O	SrO	Dy_2O_3
Dy0	70.0	20.0	10.0	0.0
Dy1	69.8	20.0	10.0	0.2
Dy2	69.6	20.0	10.0	0.4
Dy3	69.4	20.0	10.0	0.6
Dy4	69.2	20.0	10.0	0.8
Dy5	69.0	20.0	10.0	1.0

Table 2. Chemical composition (in mol %) and glass codes nomenclature of erbium oxide doped sodium strontium borate glass samples (ErNaSrB).

Glass codes	B_2O_3	Na_2O	SrO	Er_2O_3
Er0	70.0	20.0	10.0	0.0
Er1	69.8	20.0	10.0	0.2
Er2	69.6	20.0	10.0	0.4
Er3	69.4	20.0	10.0	0.6
Er4	69.2	20.0	10.0	0.8
Er5	69.0	20.0	10.0	1.0

Table 3. Chemical composition (in mol %) and glass codes nomenclature of neodymium oxide doped sodium strontium borate glass samples (NdNaSrB).

Glass codes	B_2O_3	Na_2O	SrO	Nd_2O_3
Nd0	70.0	20.0	10.0	0.0
Nd1	69.8	20.0	10.0	0.2
Nd2	69.6	20.0	10.0	0.4
Nd3	69.4	20.0	10.0	0.6
Nd4	69.2	20.0	10.0	0.8
Nd5	69.0	20.0	10.0	1.0

ANALYSIS OF DyNaSrB GLASS SERIES

X-ray Diffraction

XRD technique is employed to confirm the phase of the prepared glass system. Fig. (1) shows the XRD pattern of sodium strontium borate glasses doped with different concentrations of Dy_2O_3. There is no sharp crystalline peak, which confirms the amorphous nature of the prepared samples. The X-ray scattering pattern generally shows maxima and minima when there are some definite interatomic distances in the sample material. Two broad peaks around $20-30°$ and $40-50°$ are due to $NaBO_3$ and $SrBO_3$ structures, respectively [11 - 13]. The peak position as well as the width of the peak increase as the Dy^{3+} content increases. The peaks broaden because theparticle size is reduced, and the small shift in peak position can be attributed to two factors: (1) the decrease in lattice parameters when particle size decreases; (2) the inhomogeneous stress distribution in the glass matrix. It is possible that some dysprosium ions are taking the place of strontium ions. The melting point of borate glasses is 700-800°C, while Dy_2O_3 is approximately 1500°C. When a small amount of Dy_2O_3 is introduced to the melt, the Dy_2O_3 particles act as nucleation sites, causing the glass to freeze locally. The number of particles per unit volume and the number of nucleating centres increase as Dy_2O_3 concentration increases. The solidified glass particle size appears to be smaller due to the reduced dispersion and the full width at half maximum (FWHM) of XRD patterns increases as a result of this. The particle size, calculated using the Scherrer equation to fit these graphs, is 0.60 nm at the lowest concentration and 0.57 nm at the maximum concentration. This is in line with the previous argument.

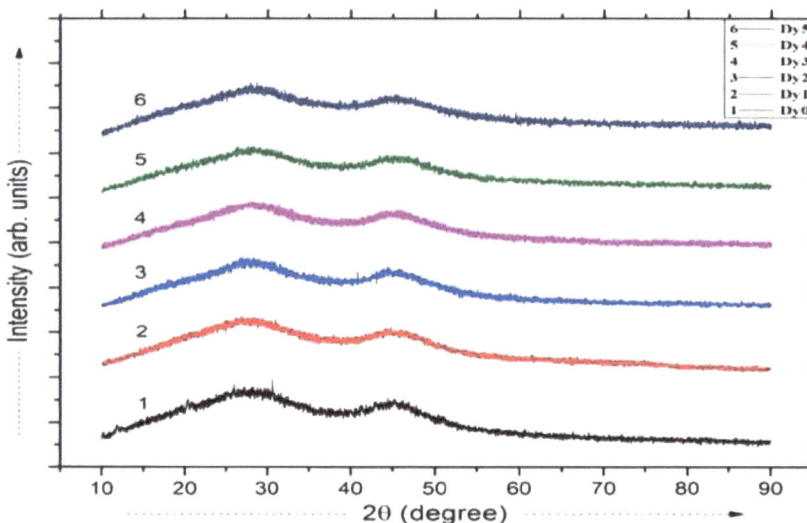

Fig. (1). XRD pattern of DyNaSrB glass system.

Physical Parameters

The different calculated physical parameters of the prepared DyNaSrB glasses are listed in Table **4**. From the table, it is observed that the experimentally measured density 'ρ_{exp}' and theoretically calculated density 'ρ_{cal}' along with the average molecular weight increase with the rising concentration of Dy_2O_3 because the mass of Dy_2O_3 is higher than that of B_2O_3. The increase in density is due to the Dy^{+3}ions-induced nano-crystallization as confirmed by the XRD pattern. The increase in density can also be linked to change in coordination and geometrical configuration [14, 15]. The molar volume, V_m, increases initially with the increasing Dy_2O_3 concentration up to 0.8 mol%, and then reduces for 1.0 mol% Dy_2O_3. The formation of non-bridging oxygens (NBOs) and the breaking of the covalent bonds between boron and oxygen can be linked to this decrease in molar volume [16]. Fig. (**2**) depicts the variation of molar volume and density, ρ_{exp}, as a function of Dy_2O_3 concentration.

Table 4. Physical and optical parameters of DyNaSrB glass system.

Parameters	Glass codes					
	Dy0	Dy1	Dy2	Dy3	Dy4	Dy5
$\rho_{exp}(gcm^{-3})$	2.740	2.774	2.798	2.800	2.811	2.854
$\rho_{cal}(gcm^{-3})$	2.647	2.656	2.667	2.678	2.688	2.699
M(g)	71.490	72.097	72.704	73.311	73.718	74.524

(Table 4) cont.....

Parameters	Glass codes					
$V_m (cm^3)$	26.088	25.991	25.981	26.181	26.290	26.115
O (g-atm/l)	91.984	92.342	92.363	91.664	91.516	91.911
N ($\times 10^{22}$ ionscm^3)	-	0.463	0.927	1.380	1.832	2.306
Vmb	43.481	43.032	42.731	42.779	42.679	42.121
$<d_{B-B}> \times 10^{-8} cm$	4.164	4.150	4.140	4.142	4.138	4.120
$r_p (Å)$ -	-	2.418	1.919	1.680	1.529	1.416
$r_i (Å)$	-	5.996	4.760	4.169	3.793	3.513
$F \times 10^{16}$-	-	0.513	0.815	1.063	1.283	1.496

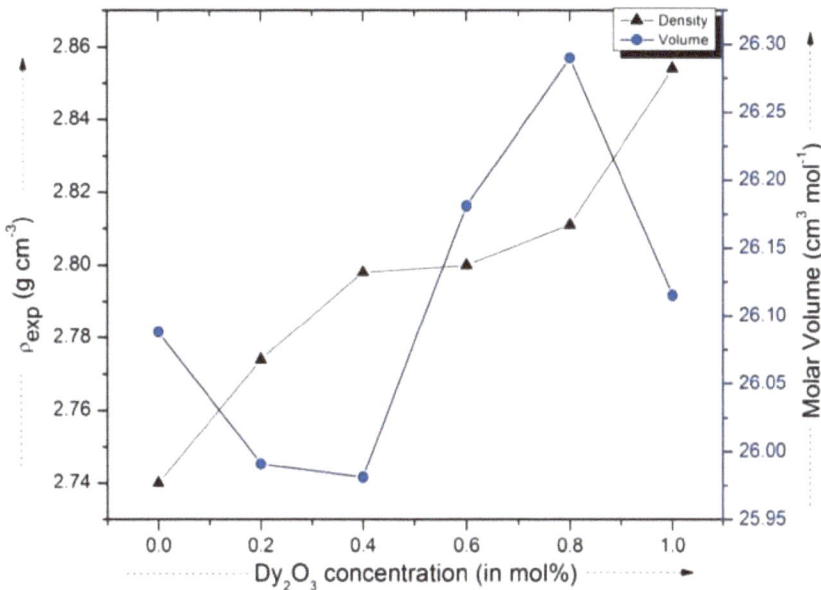

Fig. (2). Variation of ρ_{exp} and molar volume of DyNaSrB glasses.

Additionally, as shown in Table **4**, the polaron radius (r_p) decreases as the concentration of Dy_2O_3 increases, most likely due to a rise in the value of Dy^{3+} ion concentration (N). The average RE-oxygen distance (r_i) decreases as the concentration of dysprosium in the glass matrix increases, resulting in a significant rise in field strength due to the strong interaction between Dy^{3+} and B ions. The formation of this linkis caused by the displacement of the Dy^{3+} ion by newly generated oxygen during the conversion of BO_3 to BO_4 units. In DyNaSrB glass systems, a decrease in the average boron-boron distance, $<d_{B-B}>$, is observed whilethe inter-nuclear distance decreases as well, resulting in a more compact boron network [17, 18].

Fourier Transform Infrared Spectroscopy (FT-IR)

FT-IR spectrum has been taken to study the structural changes, which occurred in the glass network due to the addition of the rare-earth oxide. Fig. (**3**) shows the infrared spectra of the DyNaSrB glasses. Because of the bending and str*etc.*hing vibrations of glass particles, various bands corresponding to different wavenumbers can be seen and it is also noticeable from the Fig. (**3**) that IR spectrum of glass samples accommodates a number of shoulder bands. The first band in the region 1200-1600 cm^{-1} describes the asymmetric B-O str*etc.*hing of trigonal BO_3 units. The second band seen in the range 800-1200 cm^{-1} is due to B-O str*etc.*hing of tetrahedral BO_4 units. The third band in the range 800-600 cm^{-1} corresponds to the bending of different B-O-B linkages in the glass network [19, 20]. The vibration of sodium in their oxygen linkage causes a band at approximately 580 cm^{-1}. The meta centre shifts in the direction of lower/higher wavenumber. The intensity of the absorption band increases as the concentration of dysprosium oxide increases.

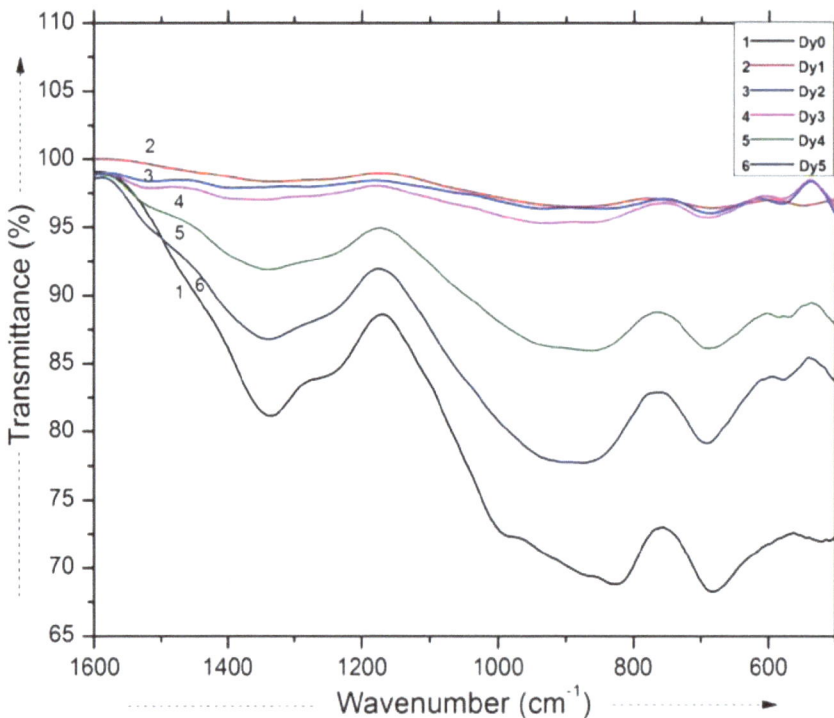

Fig. (3). FTIR spectra of the DyNaSrB glasses.

With increasing Dy_2O_3 concentration, a new band at about 930 cm^{-1} appears, which then shifts to 900 cm^{-1}. This band can be assigned to the boroxol rings, just like the holmium ions in borate glasses [21]. In all of the samples, no extra band is visible, indicating that the presence of rare-earth ions has no major effect on the structure of the borate glasses. The conversion of BO_3 to BO_4 units and the generation of non-bridging oxygen, which results in a more compact structure, can be linked to an increase in intensity.

ANALYSIS OF ErNaSrB GLASS SERIES

X-ray Diffraction

The XRD patterns recorded for ErNaSrB glass system are shown in Fig. (4). The amorphous character of the prepared materials is indicated by broad bands peaking around 27° and 45°. When there are certain fixed interatomic distances in the sample material, only then the X-ray scattering pattern shows maxima and minima [10]. The structure of borate glass is mostly made up of BO_3 and BO_4 units, with sodium and strontium filling in the interstitial gaps within the glass matrix [22]. On closer inspection, it is observed that the peaks narrow slightly, indicating that an increase in rare-earth concentration induces localized devitrification of the prepared glass samplesaround the rare-earth. Because the alterations in the XRD patterns are comparable to nanopowders, this is referred to as nano-crystallization. With increasing concentrations of erbium oxide, the peaks shift towards a lower 2θ value, indicating an increase in the lattice parameter. The peak shifts to a larger 2θ value at 1.0 mol% erbium oxide concentration, indicating a fall in the lattice parameter, which could be due to an increase in the Coulomb interaction force. The peak shift is also caused by the inhomogeneous stress distribution in the glass matrix. These findings suggest that erbium ions are substitutingone of the glass ingredients, most likely strontium ions. The charge of the rare earth ion is compensated by the anion vacancies in the nano-crystalline mass, which exhibit compelling physical properties. Because of their high melting temperatures, erbium ions, like dysprosium ions, act as nucleation centres, resulting in nano-crystallization. The average crystallite size is estimated using a modified Scherrer formula, which allows the least square technique to be applied to reduce the error [23]. The average crystallite size increases from 0.43 nm (at 0.2 mol% Er_2O_3) to 0.50 nm (at 0.8 mol% Er_2O_3) and then reduces to 0.47 nm(at 1.0 mol% Er_2O_3), as measured by XRD. The atomic radius of the erbium atom is larger than that of other atoms in the glass structure, resulting in bigger average crystallite size. Crystallite size falls at 1.0 mol% since it attempts to crystallize quickly at this concentration [24].

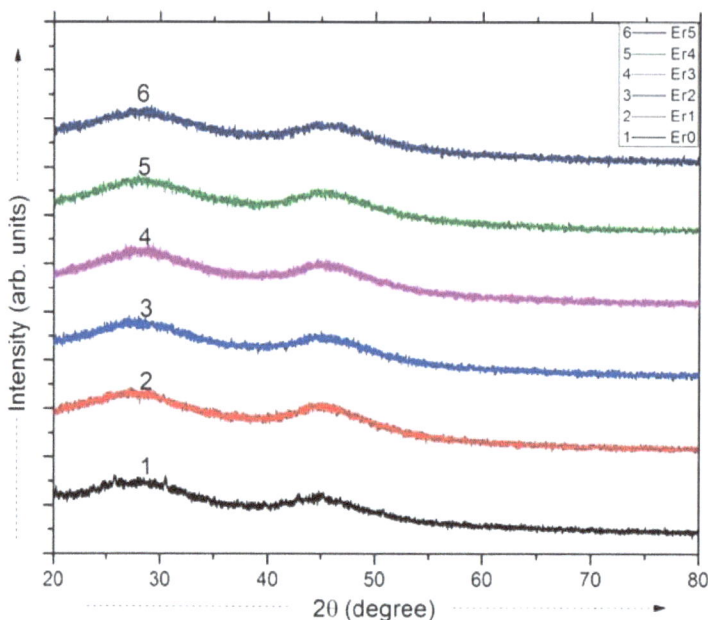

Fig. (4). XRD pattern of ErNaSrB glass system.

Physical Parameters

Table **5** lists the various calculated physical parameters of prepared ErNaSrB glasses. Density, as one of the key metrics for analyzing structural changes in the glass network in the presence of different additives, depends on thestructural compactness, geometrical configuration changes, coordination number, and other factors [25]. From Fig. (**5**), it is observed that the ρ_{cal} increases with the increasing concentration of Er_2O_3, whereas ρ_{exp} firstly decreases and then shows an increasing trend with further addition of Er_2O_3 up to 0.8 mol%, and the conclusions drawn from XRD pattern also favored the density results. On the other hand, the molar volume shows an opposite trend to ρ_{exp}. Increased concentrations of Er_2O_3 at the expense of B_2O_3 lead to increases indensity. Erbium has a higher atomic mass (167.259 g mol^{-1}) than boron (10.811 g mol^{-1}), resulting in a higher density and average molecular weight. The inclusion of the rare-earth ion also increases the oxygen packing density, which contributes to the glass structure's compactness. The larger ionic radii of the dopant Er_2O_3, as compared to the glass forming B_2O_3, also contributes to the initial increase in molar volume. The decrease in average boron-boron separation also confirms the compactness of the network with the increasing Er_2O_3 concentration [26].

Table 5. Physical parameters of ErNaSrB glass system.

Parameters	Glass codes					
	Er0	Er1	Er2	Er3	Er4	Er5
$\rho_{exp}(gcm^{-3})$	2.744	2.527	2.638	2.669	2.782	2.707
$\rho_{cal}(gcm^{-3})$	2.647	2.658	2.670	2.683	2.695	2.699
M (g)	71.490	72.116	72.742	73.367	73.993	74.619
$V_m(cm^3)$	26.053	28.538	27.567	27.485	26.594	27.968
O (g−atm/l)	92.119	84.098	87.036	87.309	90.235	85.812
Λ_{th}	0.637	0.638	0.639	0.640	0.641	0.642
N ($\times 10^{22}$ ionscm^3)	-	0.4220	0.8737	1.3145	1.8115	2.1531
Vmb	43.422	47.248	45.341	44.911	43.172	45.110
$<d_{B-B}> \times 10^{-8}$cm	4.162	4.281	4.222	4.209	4.154	4.215
$r_p(Å)$	-	2.494	1.956	1.707	1.534	1.448
$r_i(Å)$	-	6.188	4.855	4.237	3.807	3.594
$F \times 10^{16}$	-	0.482	0.783	1.028	1.273	1.429

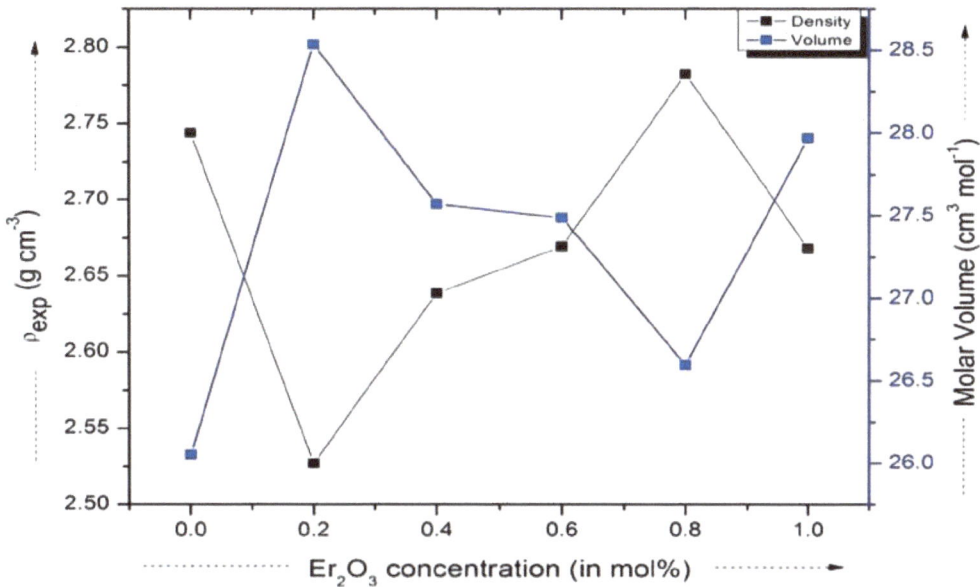

Fig. (5). Variation of ρ_{exp} and molar volume of ErNaSrB glasses.

The polaron radius, r_p, decreases with the increase in Er_2O_3 concentration, which is due to the increase in the value of ion concentration (N). The field strength around the Er^{3+} ion increases as a result of this [27, 28]. The inter-nuclear distance of prepared NaSrB glasses reduces as well, resulting in a more compact boron network. With the addition of Er_2O_3, the average boron-boron separation, $<d_{B-B}>$ similarly decreases.

Fourier Transform Infrared Spectroscopy

Since the B-O vibrations are active in the 450-1600 cm^{-1} range, the FT-IR spectra of the glasses with various concentrations are displayed in this range in Fig. (**6**). The asymmetric B-O str*etc.*hing of trigonal BO_3 units is assigned to the first band in the spectra around 1530 cm^{-1}. The presence of pyroborate and orthoborate groups comprising BO_3 can be seen in the second band at about 1350 cm^{-1} [29]. The B-O str*etc.*hing of tetrahedral BO_4 units islinked to the strong band around 1034 cm^{-1}. The bending vibrations of B-O-B linkages cause the band at 669 cm^{-1}, and the vibrations of sodium in their oxygen linkage produce the band around 592 cm^{-1} [30]. Up to 0.6 mol% Er_2O_3 concentration, the absorption band around 1530 cm^{-1} is enhanced due to asymmetric B-O str*etc.*hing of trigonal BO_3 units with one NBO. The absorption intensity of the band around 1034 cm^{-1} reduces up to 0.6 mol% due to the B-O str*etc.*hing of tetrahedral BO_4 units, and theintensity increases as Er^{3+} concentration increases. Because Er^{3+} ions have three electron vacancies, they take electrons from the oxygen atom, causing bonds to break and NBOs to form [31]. This indicates that the BO_4 units are converted into BO_3 units up to 0.6 mol%, after which the BO_3 units are converted back to BO_4, as evidenced by the density variation. According to the literature, asymmetric BO_3 and tetrahedral BO_4 units are denser than symmetric BO_3 units [32]. During the glass formation, the asymmetric BO_3 units are created in the presence of the rare-earth. As a result, the Er^{3+} ions increase the amount of asymmetric BO_3 units per NBO, resulting in a denser molecular structure. There is also a slight shift in the meta centre towards higher/lower wavenumbers [33].

ANALYSIS OF NdNaSrB GLASS SERIES

X-ray Diffraction

The synthesized NdNaSrB glass samples were transparent and bubble-free, with no evidence of phase transition. The lack of any sharp crystalline peak in the XRD spectra indicates the presence of short-range order in the lattice structure and confirms the glass system's amorphous nature. However, two wide humps around 27° and 45° obtained are due to the presence of some well-defined inter-atomic distances within the glass matrix, as shown in Fig. (**7**) [11]. The earlier studies suggest that these two broad humps are due to $NaBO_3$ and $SrBO_3$ structural groups [22]. No remarkable change has been observed in the XRD pattern with the addition of neodymium oxide except for two broad humps around 25-35° and 40-50°. The oxygen in the glass matrix forms trigonal and tetrahedral coordination around boron atoms, whereas sodium and strontium exist as Na^+ and Sr^{2+}, respectively. The total number of oxygens is high, and the incorporation of Na^+ and Sr^{2+} ions causes the various holes in the boron-oxygen network to become occupied. The structural groups, $NaBO_3$ and $SrBO_3$, are formed as a result of this [34]. The Nd^{3+} ions must be replacing the strontium ions due to their high field strength, and a periodic structure surrounding the rare-earth ion is being formed. As a result of the Nd^{3+} ions, nano-crystallization occurs, resulting in the formation of nano-scale cristobalite. The peak shift and change in peak size with increasing Nd_2O_3 concentrations were observed after carefully inspecting the X-ray diffraction pattern. The peak moves to a higher 2θ value and the peak width broadens up to 0.6 mol%concentration of neodymium oxide. This is due to a decrease in the lattice parameter, as Nd^{3+} ions act as nucleation centres owing to their high melting point, resulting in local devitrification around them. The average crystallite size was calculated using the Scherrer formula, and it was found to be reduced from 0.69 nm to 0.55 nm with the addition of Nd_2O_3.

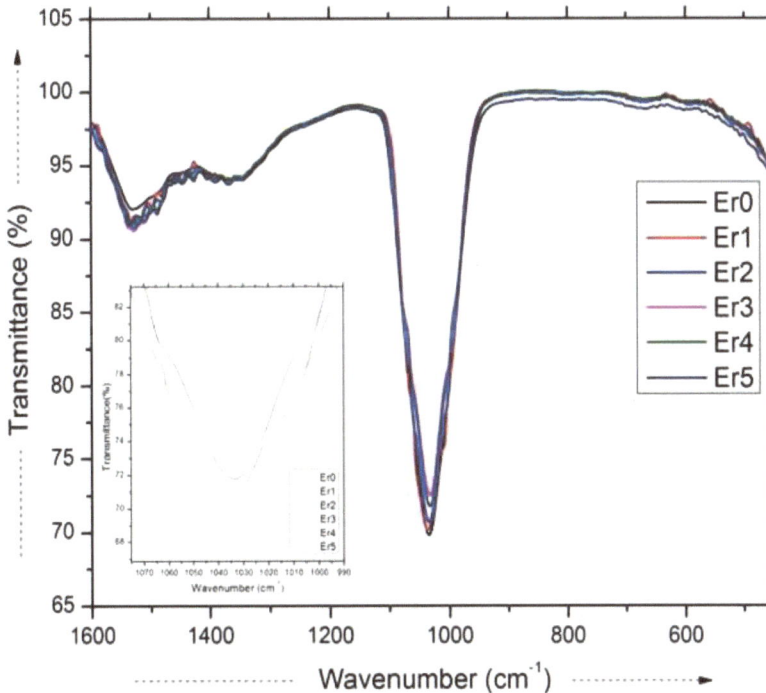

Fig. (6). FTIR spectra of ErNaSrB glasses.

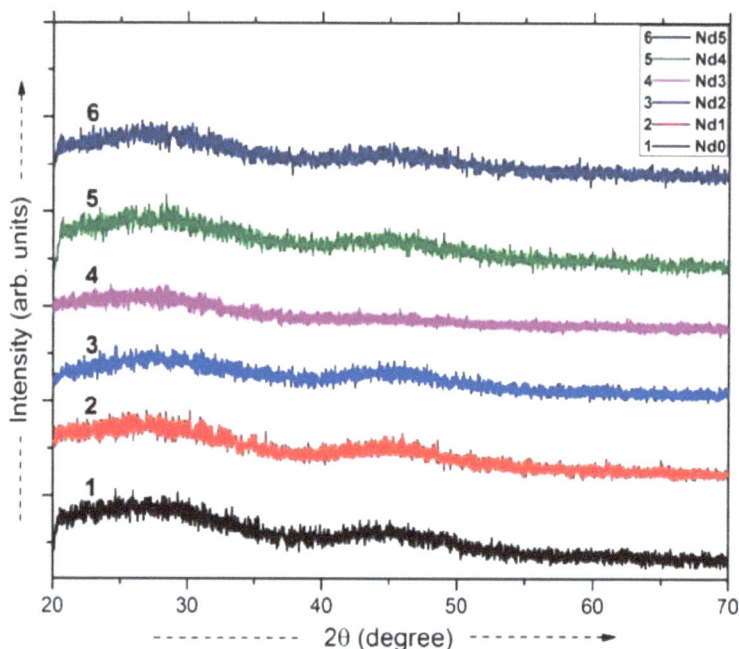

Fig. (7). XRD pattern of NdNaSrB glass system.

Physical Parameters

Table **6** lists the different parameters of the NdNaSrB glass system that were calculated to investigate the structural dependence on the chemical composition of the prepared glasses. ρ_{cal} shows a linearly increasing trend for the entire glass system, whereas ρ_{exp} first decreases up to a concentration of 0.6 mol% neodymium oxide, and then it increases for concentrations of 0.8 mol% and 1 mol% neodymium oxide. ρ_{exp} and ρ_{cal} have values in the range of 2.584-2.800 gcm^{-3} and 2.647-2.693 gcm^{-3}, respectively. The molar volume (V_m) and ρ_{exp} variation with rare-earth oxide concentration are shown in Fig. (**8**). As previously discussed, the V_m increases initially and then decreases as the concentration of rare-earth oxide increases. Density and molar volume depend upon the different structural groups present in the glass matrix, which varies the tightness of the structure [35]. Nd_2O_3 is being increased at the expense of B_2O_3, so due to the high molecular weight of Nd_2O_3 (336.47 g mol^{-1}) as compared to B_2O_3 (69.62), there should be an increase in density as in the case of ρ_{cal}. However, the drop in ρ_{exp} with increased neodymium oxide concentration is principally due to structural changes in the glass matrix. V_m increases up to 0.6 mol% concentration of Nd_2O_3 due to the conversion of tetrahedral BO_4 units to trigonal BO_3 units, according to the molar volume measurements. As a result of the less densely formed BO_3 units, the density decreases and the molar volume increases [36]. After 0.6 mol%concentration of neodymium oxide, ρ_{exp} follows the general trend of increasing with dopant concentration.

Table 6. Physical parameters of NdNaSrB glass system.

Parameters	Glass codes					
	Nd0	Nd1	Nd2	Nd3	Nd4	Nd5
ρ (gcm^{-3})	2.740	2.709	2.676	2.584	2.603	2.800
$\rho_{cal.}$(gcm^{-3})	2.647	2.655	2.665	2.674	2.684	2.693
M(g)	71.490	72.024	72.558	73.091	73.625	74.159
V_m(cm^3)	26.091	26.587	27.114	28.286	28.284	26.485
O (g-atm/l)	91.984	90.269	89.086	84.846	84.851	90.615
N ($\times 10^{22}$ionscm^3)	-	0.453	0.888	1.277	1.703	2.273
V^b_m	43.485	44.018	44.595	46.218	45.915	42.717
$<d_{B-B}>10^{-8}$cm	4.162	4.181	4.199	4.249	4.240	4.139
r_p(Å)	-	2.435	1.946	1.724	1.566	1.422
r_i(Å)	-	6.043	4.829	4.278	3.886	3.530
F$\times 10^{16}$	-	0.505	0.792	1.009	1.223	1.483
Λ_{th}	0.637	0.639	0.641	0.642	0.644	0.646

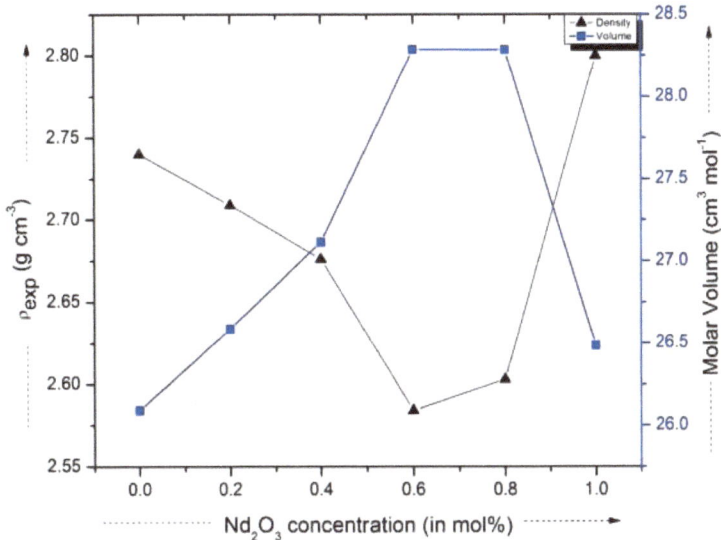

Fig. (8). ρ_{exp} and molar volume variation of NdNaSrB glasses as a function of neodymium oxide concentration.

The above results are also favored by oxygen packing density, O, and average boron-boron separation, $<d_{B-B}>$, as they give an image of the structural compactness in the glass matrix. Furthermore, because the internuclear distance, r_i, and the polaron radius, r_p, are inversely proportional to the ion concentration (N), both display a decreasing pattern as the N increases. As the number of ions increases with the dopant concentration, the decrease in r_p results in an increase in field strength, F [37].

Fourier Transform Infrared Spectroscopy

FTIR spectroscopy has been carried out to recognize the different structural groups present in the prepared glasses; the recorded spectra are shown in Fig. (9). Various bands corresponding to different wavenumbers can be identified due to the bending and str*etc.*hing vibrations of glass particles. The spectra shown in Fig. (9) are in the range of 1600-450 cm^{-1}, as all the relevant glass constituents' vibrations are active in this region. Three prominent frequency regions, 450-800 cm^{-1}, 800-1200 cm^{-1}, and 1200-1600 cm^{-1} are observed. It is also noticeable that the IR spectrum of glass samples accommodates some shoulder bands. The different sub-bands obtained in the high wavenumber region 1200-1600 cm^{-1} are due to the vibrations of various BO_3 structural units. The bands in the range 800-1200 cm^{-1} are also divided into different sub-bands, which are due to the str*etc.*hing vibrations of diborate linkages and B-O str*etc.*hing vibrations of tetrahedral BO_4 units. Due to the bending vibrations of B-O-B links, a single band exists near 690 cm^{-1}. Owing to the existence of metal cations within the glass matrix, the last obtained band occurred at a wavenumber smaller than 600 cm^{-1}. With increasing concentrations of rare-earth ions, the band associated with metallic cations shifted to a lower wavenumber. As previously stated, rare-earth ions are replacing one of the glass elements, most likely strontium, and the bond length increases due tothe larger size of the rare-earth atom. The band around 600 cm^{-1} shifts to a lower wavenumber as a result of thisincrease in bond length. The XRD results showing the existence of nanocrystallization are supported by the continual increase in the intensity of this peak up to 0.6 mol%rare-earth concentration. Furthermore, the addition of neodymium oxide caused some structural modification, and depolymerization of the glass network occurs as neodymium oxide concentration increases. The tetrahedral BO_4 units are broken and converted to BO_3 units up to a dopant content of 0.6 mol% [36]. When BO_4 is converted to BO_3, non-bridging oxygens (NBOs) are generated, and the resulting trigonal BO_3 units with NBOs are less dense than tetrahedral BO_4 units. This is the reason for the initial drop in density and increase in molar volume up to a concentration of neodymium oxide of 0.06 mol%. Furthermore, after a concentration of 0.6 mol%, the formation of dense tetrahedral BO_4 units causes an increase in density.

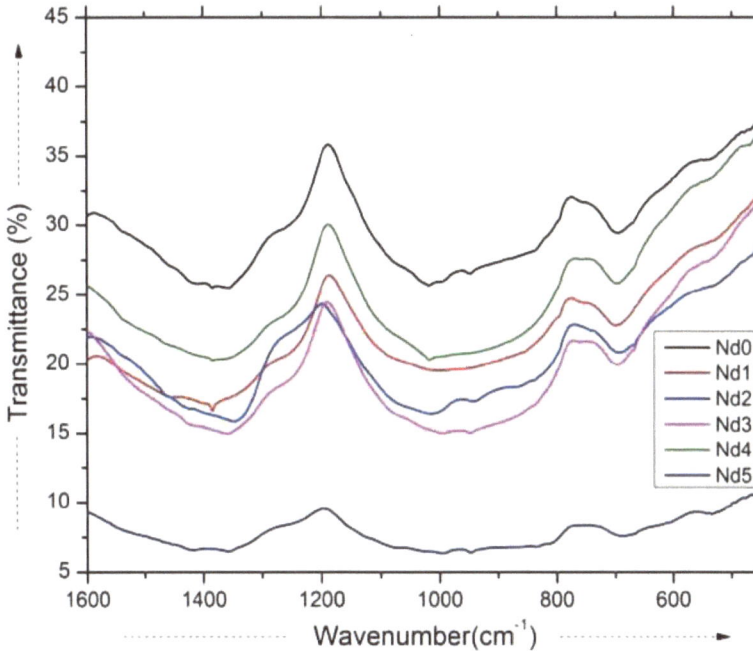

Fig. (9). FTIR spectra of the prepared NdNaSrB glasses.

CONCLUSION

The nanocrystallization induced by the addition of rare-earth ions in the standard borate glass system has been analyzed in this study. The rare-earth doped borate glasses have been synthesized by the standard melt quenching technique. The amorphous nature of all the glass samples was confirmed by XRD analysis. The XRD patterns revealed two broad peaks of $NaBO_3$ and $SrBO_3$ structures. Because of their high melting temperatures, rare earth ions act as nucleation centres, resulting in nano-crystallization. The average crystallite size is determined from the XRD patterns using a modified Scherrer formula, and the least square technique is used to minimize theuncertainty. Because the mass of REs is greater than that of B_2O_3, density and average molecular weight both increased when the concentration of rare-earth ions increased. Changes in coordination and geometrical configuration can also be attributed to an increase in thedensity values. The formation of non-bridging oxygens (NBOs) with increasing rare earth concentrations has been linked to the irregular response of the molar volume. Other physical parameters, including ion concentration, polaron radius, and inter-nuclear distance, have been computed. The vibration bands in the FTIR spectra corresponded to the str*etc.*hing of trigonal BO_3, tetragonal BO_4, and bending of various B-O-B bonds. Depolymerization of the glass network occurs as the concentration of rare earth oxides increases. The enhancement in the intensity of FTIR bands can be related to the conversion of BO_3 to BO_4 units and the production of the NBO, which results in a more compact structure. The reported results show that the borate glasses have a prominent role in glass research because when these are doped with different modifiers and RE^{3+} ions; interesting and unique results have been obtained.

REFERENCES

[1]   Rao, K.J. *Structural Chemistry of Glasses*; Elsevier Science, **2002**.

[2]   Marshall, J. *Glass source book*; Chartwell House, **1990**.

[3]   Alajerami, S.M.; Hashim, S.; Hassan, W.M.S.; Ramli, A.T.; Kasim, A. Optical properties of lithium magnesium borate glasses doped with Dy^{3+} and Sm^{3+}ions. *Physica B,* **2012**, *1026*, 159-167.

[4]   Lim, T.Y.; Wagiran, H.; Hussin, R.; Hashim, S.; Saeed, M.A. Physical and optical properties of dysprosium ion doped strontium borate glasses. *Physica B,* **2014**, *451*, 63-67.
[http://dx.doi.org/10.1016/j.physb.2014.06.028]

[5]   Bajaj, N.S.; Omanwar, S.K. **2014**.

[6]   Kumar, D.; Rao, S.M.; Singh, S.P. Effect of Er^{3+} on NaSrB glass: thermoluminescence and structural analysis. *Appl. Phys., A Mater. Sci. Process.,* **2019**, *125*(1), 38.
[http://dx.doi.org/10.1007/s00339-018-2319-5]

[7]   Desurvire, E.; Zervas, M.N. Erbium doped fiber amplifiers: principles and applications. *Phys. Today,* **1995**, *48*(2), 56-58.
[http://dx.doi.org/10.1063/1.2807915]

[8]   Jayasankar, C.K.; Babu, P. Optical properties of Sm^{3+} ions in lithium borate and lithium fluoroborate glasses. *J. Alloys Compd.,* **2000**, *307*(1-2), 82-95.
[http://dx.doi.org/10.1016/S0925-8388(00)00888-4]

[9]   Mhareb, M.H.A; Hashim, S. Hashim S., Ghoshal S.K, Alajerami Y.S.M., Saleh M.A., Maqableh M.M.A., TamchekN.,Optical and erbium ion concentration correlation in lithiummagnesium borate glass, optik 126 (2015) 3638.

[10]   Marzouk, M.A.; ElBatal, F.H.; Morsi, R.M.M. Optical and FTIR absorption spectra of Ce_2O-doped cadmium borate glasses and effect of gamma irradiation. *Silico,* **2017**, (9), 105-110.
[http://dx.doi.org/10.1007/s12633-015-9400-x]

[11]   Madhukar Reddy, C.; Deva Prasad Raju, B.; John Sushma, N.; Dhoble, N.S.; Dhoble, S.J. A review on optical and photoluminescence studies of RE^{3+} (RE=Sm, Dy, Eu, Tb and Nd) ions doped LCZSFB glasses. *Renew. Sustain. Energy Rev.,* **2015**, *51*, 566-584.
[http://dx.doi.org/10.1016/j.rser.2015.06.025]

[12]   Obayes, H.K.; Wagiran, H.; Hussin, R.; Saeed, M.A. A new strontium/copper co-doped lithium borate glass composition with improved dosimetric features. *J. Lumin.,* **2016**, *176*, 202-211.
[http://dx.doi.org/10.1016/j.jlumin.2016.03.024]

[13]   Obayes, H.K.; Wagiran, H.; Hussin, R.; Saeed, M.A. Strontium ions concentration dependent modifications on structural and optical features of $Li_4Sr(BO_3)_3$ glass. *J. Mol. Struct.,* **2016**, *1111*, 132-141.
[http://dx.doi.org/10.1016/j.molstruc.2016.01.088]

[14]   Kumar, D.; Rao, S.M.; Singh, S.P. Structural, optical and thermoluminescence study of Dy^{3+} ion doped sodium strontium borate glass. *J. Non-Cryst. Solids,* **2017**, *464*, 51-55.
[http://dx.doi.org/10.1016/j.jnoncrysol.2017.03.029]

[15]   Tang, G.; Yang, Z.; Luo, L.; Chen, W. Optical properties and local structure of Dy^{3+}-doped chalcogenide and chalcohalide glasses. *J. Rare Earths,* **2008**, *26*(6), 889-894.
[http://dx.doi.org/10.1016/S1002-0721(09)60027-2]

[16]   Sandhya Rani, P.; Singh, R. Electron spin resonance and magnetization studies of ZnO–TeO2–Fe2O3 glasses. *J. Phys. Chem. Solids,* **2013**, *74*(2), 338-343.
[http://dx.doi.org/10.1016/j.jpcs.2012.10.009]

[17]   Gunhakoon, P.; Thongklom, T.; Sopapan, P.; Laopaiboon, J.; Laopaiboon, R.; Jaiboon, O. Influence of WO_3 on elastic and structural properties of barium-borate-bagasse-cassava rhizome glass system. *Mater. Chem. Phys.,* **2020**, *243*, 122587.
[http://dx.doi.org/10.1016/j.matchemphys.2019.122587]

[18]   Mhareb, M.H.A.; Hashim, S.; Ghoshal, S.K.; Alajerami, Y.S.M.; Saleh, M.A.; Dawaud, R.S.; Razak, N.A.B.; Azizan, S.A.B. Impact of Nd^{3+} ions on physical and optical properties of Lithium Magnesium Borate glass. *Opt. Mater.,* **2014**, *37*, 391-397.
[http://dx.doi.org/10.1016/j.optmat.2014.06.033]

[19]   Gedam, R.S.; Ramteke, D.D. Electrical and optical properties of lithium borate glasses doped with Nd_2O_3. *J. Rare Earths,* **2012**, *30*(8), 785-789.
[http://dx.doi.org/10.1016/S1002-0721(12)60130-6]

[20]   Gaafar, M.S.; El-Aal, N.S.A.; Gerges, O.W.; El-Amir, G. Elastic properties and structural studies on some zinc-borate glasses derived from ultrasonic, FT-IR and X-ray techniques. *J. Alloys Compd.,* **2009**, *475*(1-2), 535-542.
[http://dx.doi.org/10.1016/j.jallcom.2008.07.114]

[21]   Kamitsos, E.I.; Karakassides, M.A.; Chryssikos, G.D.

[22]   Culea, E.; Ristoiu, T.; Bratu, I. Magnetic and structural behaviour of some borate glasses containing holmium ions. *Mater. Sci. Eng. B,* **1999**, *57*(3), 259-261.
[http://dx.doi.org/10.1016/S0921-5107(98)00422-X]

[23]   Warren, B.E. X-ray diffraction study of the structure of glass. *Chem. Rev.,* **1940**, *26*(2), 237-255.
[http://dx.doi.org/10.1021/cr60084a007]

[24]   Monshi, A.; Foroughi, M.R.; Monshi, M.R. **2012**.

[25]   Henderson, S.I.; Mortensen, T.C.; Underwood, S.M.; van Megen, W. Effect of particle size distribution on crystallisation and the glass transition of hard sphere colloids. *Physica A,* **1996**, *233*(1-2), 102-116.
[http://dx.doi.org/10.1016/S0378-4371(96)00153-7]

[26]   Rada, S.; Dehelean, A.; Culea, E. FTIR and UV–VIS spectroscopy investigations on the structure of the europium–lead–tellurate glasses. *J. Non-Cryst. Solids,* **2011**, *357*(16-17), 3070-3073.
[http://dx.doi.org/10.1016/j.jnoncrysol.2011.04.013]

[27]   Bhatia, V.; Kumar, D.; Kumar, A.; Mehta, V.; Chopra, S.; Vij, A.; Rao, S.M.D.; Singh, S.P. Mixed transition and rare earth ion doped borate glass: structural, optical and thermoluminescence study. *J. Mater. Sci. Mater. Electron.,* **2019**, *30*(1), 677-686.
[http://dx.doi.org/10.1007/s10854-018-0336-y]

[28]   Azizan, S.A.; Hashim, S.; Razak, N.A.; Mhareb, M.H.A.; Alajerami, Y.S.M.; Tamchek, N. Physical and optical properties of Dy^{3+}: $Li_2O–K_2O–B_2O_3$ glasses. *J. Mol. Struct.,* **2014**, *1076*, 20-25.
[http://dx.doi.org/10.1016/j.molstruc.2014.07.032]

[29]   Laopaiboon, R.; Thumsa-ard, T.; Bootjomchai, C. The thermoluminescence properties and determination of trapping parameters of soda lime glass doped with erbium oxide. *J. Lumin.,* **2018**, *197*, 304-309.
[http://dx.doi.org/10.1016/j.jlumin.2018.01.039]

[30]   Gautam, C.; Yadav, A.K.; Singh, A.K. *A review on infrared spectroscopy of borate glasses with*

effects of different Additives,International Scholarly Research Network; ISRN Ceramics, **2012**, p. 428497.

[31] Kumar, D.; Bhatia, V.; Rao, S.M.; Chen, C-L.; Kaur, N.; Singh, S.P. Synthesis of N a S r B : N d $^{3+}$ glass system for the analysis of structural, optical and thermoluminescence properties. *Mater. Chem. Phys.,* **2020**, *243*, 122546.
[http://dx.doi.org/10.1016/j.matchemphys.2019.122546]

[32] Noorazlan, A.; Kamari, H.M.; Zulkefly, S.S.; Mohamad, D.W. Effect of erbium nanoparticles on optical properties of zinc borotellurite glass system. *J. Nanomater.,* **2013**, 940-917.

[33] Saddeek, Y.B.; Aly, K.A.; Bashier, S.A. Optical study of lead borosilicate glasses. *Physica B,* **2010**, *405*(10), 2407-2412.
[http://dx.doi.org/10.1016/j.physb.2010.02.055]

[34] Genge, M.J.; Jones, A.P.; Price, G.D. An infrared and Raman study of carbonate glasses: implications for the structure of carbonatite magmas. *Geochim. Cosmochim. Acta,* **1995**, *59*(5), 927-937.
[http://dx.doi.org/10.1016/0016-7037(95)00010-0]

[35] Johnson, P.D.; Williams, F.E. Specific magnetic susceptibilities and related properties of manganese - activated Zinc Fluoride. *J. Chem. Phys.,* **1950**, *18*(3), 323-326.
[http://dx.doi.org/10.1063/1.1747625]

[36] Marzouk, M.A. Optical characterization of some rare earth ions doped bismuth borate glasses and effect of gamma irradiation. *J. Mol. Struct.,* **2012**, *1019*, 80-90.
[http://dx.doi.org/10.1016/j.molstruc.2012.03.041]

[37] Gedam, R.S.; Ramteke, D.D. Electrical and optical properties of lithium borate glasses doped with terbium and dysprosium. *J. Non. Cryst.,* **2013**, *367*, 58-69.
[http://dx.doi.org/10.1016/j.jnoncrysol.2013.02.018]

CHAPTER 13

Functionalization of Carbon Nanotubes for Sensing Applications

Anshul Kumar Sharma[2], Manreet Kaur Sohal[1] and **Aman Mahajan[1,*]**

[1] *Department of Physics, Materials Science Laboratory, Guru Nanak Dev University, Amritsar-143005, Punjab, India*

[2] *Centre for Sustainable Habitat, Guru Nanak Dev University, Amritsar 143005, Punjab, India*

Abstract: Carbon nanotubes (CNTs) composed of sp^2 carbon units oriented as one rolled-up graphene have provided exceptional advances in the design of chemical sensors for environmental and health monitoring. The remarkable properties of CNTs, such as high active surface area, chemical inertness, high strength, high electrical conductivity, excellent thermal stability, and low charge-transfer resistance, have made them a potential candidate for the detection of various explosive, combustible, and toxic gases, such as hydrogen sulfide (H_2S), nitrogen oxides (NO_x), ozone (O_3), and halogens (Br_2, Cl_2, and I_2). However, CNT-based sensor shows issues like low sensitivity and slow response/recovery time due to minimum charge transfer between the pristine CNTs and target analytes. The functionalization of CNTs with metal oxides, noble metal nanoparticles, and organic semiconductors not only improves the gas sensing parameters but also enhances their selectivity toward a particular type of target analyte due to the better charge transfer between the composite and gas analytes. This book chapter focuses on the ways to create CNT-based sensors exhibiting selective responses to different target analytes, future developments in the field of chemical sensors, and the viewpoint of their commercialization.

Keywords: Carbon nanotubes, Chemiresistive sensor, Metal phthalocyanine, Sensing mechanism, X-ray photoelectron spectroscopy.

INTRODUCTION

Since their discovery, carbon nanotubes (CNTs) have achieved tremendous consideration due to their extraordinary physical, chemical, structural, mechanical, and electronic properties [1]. CNTs are cylinder-shaped macromolecules, and the walls of these tubes are made up of a hexagonal lattice of carbon atoms analogous to the atomic planes of graphite [2]. These CNTs can

* **Corresponding author Aman Mahajan:** Department of Physics, Materials Science Laboratory, Guru Nanak Dev University, Amritsar-143005, Punjab, India; Email: aman.phy@gndu.ac.in

Karamjit Singh Dhaliwal (Ed.)

be either single-walled or multi-walled, where single-walled carbon nanotubes (SWCNTs) possess a single carbon layer with diameters ranging from 0.4 to 3 nm, while multi-walled carbon nanotubes (MWCNTs) consists of various layers of carbon (Fig. **1**) that can reach diameters of up to 100 nm [3, 4]. The CNTs can be synthesized using chemical vapor deposition, arc discharge, or laser ablation methods [3].

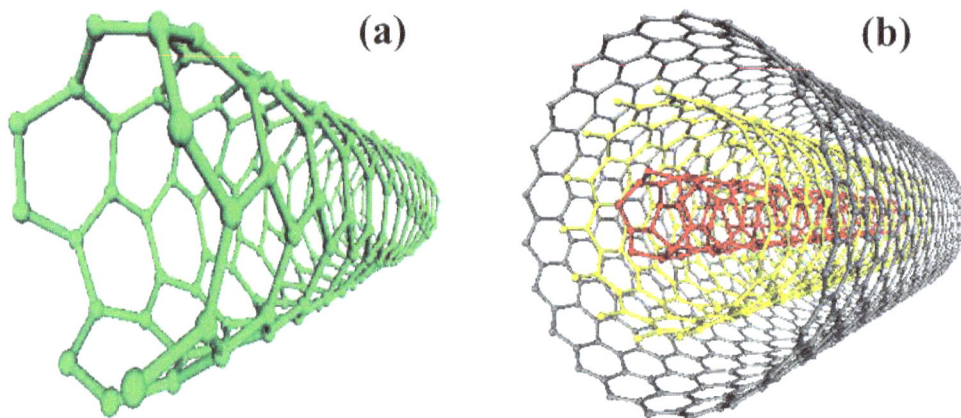

Fig. (1). Basic structures of **(a)** single-walled carbon nanotubes and **(b)** multi-walled carbon nanotubes.

Moreover, depending on the diameter, direction, and functionalization of graphene sheets, CNTs can exhibit metallic, semi-metallic, or semiconducting properties [5, 6]. However, in MWCNTs, a weaker coupling bond between the individual cylinders occurs in comparison to SWCNTs. In addition, due to the 1D structure of CNTs, ballistic transportation of electrons takes place in them. The thermal conductivity of the CNTs at room temperature is found to be <3000 W/(m.K), which is higher than that of diamond and the basal plane of graphite. Owing to these unique properties, a variety of potential applications of CNTs, such as energy storage [7], chemical sensors [8], field emitters in display technology [9], electrodes for rechargeable batteries [10], resistors, and interconnects [11], have been explored in recent years. Among these potential applications, chemical sensors based on CNTs have recently attracted a great deal of attention due to their small size, good response, and low operating temperature. In general, for good and competent gas sensing systems, there are many basic criteria, such as (i) high selectivity, (ii) small power consumption, (iii) low analyst consumption, (iv) fast action in terms of response and recovery time, and (v) stability in performances. It is worth mentioning that commonly used materials in gas sensing applications, including semiconductor metal oxides, vapor-sensitive polymers, and some other structured materials, such as porous silicon [12], often work at high temperatures and suffer from cross-sensitivity issues. Recently, progress and improvement in nanotechnology have demonstrated a high potential to develop cheap, sensitive, and portable sensors with very small power utilization. In view of this, sensors based on nanomaterials, such as CNTs, nanowires, nanofibers, and nanoparticles, have been broadly investigated due to their large surface to volume ratio and hollow-type structure, which facilitates gas adsorption [13].

Based on the operating principle or output signal, the gas sensors are classified into various categories. Table **1** lists the different types of gas sensors along with their operating principles. This book chapter is limited to chemiresistive gas sensors, the detection mode of which relies on changes in the device's electrical current, resistance, or conductance. The fact that all the carbon atoms in a CNT are surface atoms, and this unique property makes them optimally suited for components of chemical sensors. Hence, it is not surprising that gas sensors made from individual nanotubes show good sensitivities even at room temperature in comparison to commercially available traditional semiconductor sensors, which generally operate above 200°C [2].

Table 1. A list of gas sensors with their operational principles.

Sensor Type	Sensor Configuration	Detection Principle
Solid-State Sensors	Chemiresistive	Resistance of sensor changes on exposure to analyte gas.
	Chemical field-effect transistors	Current-Voltage (I–V) curves of a field-effect transistor are sensitive to a gas when it interacts with a gate.
	Colorimetric	Change in temperature due to the oxidation process on a catalytic element.
	Potentiometric	Change in potential difference (voltage) between the working electrode and the reference electrode.
	Amperometric	Diffusion limited current of an ionic conductor is proportional to the gas concentration.
Mass Sensitive Sensors	Acoustic	Variation in frequency of surface-acoustic waves excited on quartz or piezoelectric substrate upon absorption of gas in a suitable sorption layer (*e.g.,* metals, polymers).
	Micro-electromechanical systems-based sensors	On adsorption of gas, mechanical bending of micro or nano cantilevers occurs.
	Surface Plasmon Resonance (SPR)	The amount of bound gas molecules is observed by a change in SPR signals, which is proportional to the refractive index.
Optical Sensors	Optodes	Based on the measurement of change in optical properties, such as absorbance, reflectance, luminescence, Raman, and others.

Furthermore, semiconductor-based chemiresistive gas sensors, due to their simple structure, low cost, lightweight, and ease of handedness, have attracted tremendous attention in the sensor industry. The operating principle of these sensors is based on the fact that their resistance increases or decreases upon exposure to analyte gas or vapors. An increase or decrease in resistance depends upon the majority of charge carriers of sensor material (n- or p-type) and the nature of the test gas (oxidizing or reducing). A schematic representation of the gas sensor and typical response curve is shown in Figs. (**2**) and (**3**), respectively.

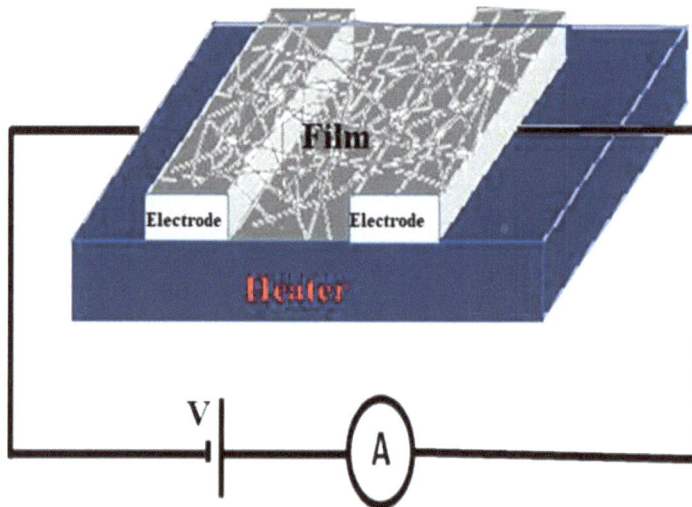

Fig. (2). Schematic representation of a chemiresistive gas sensor.

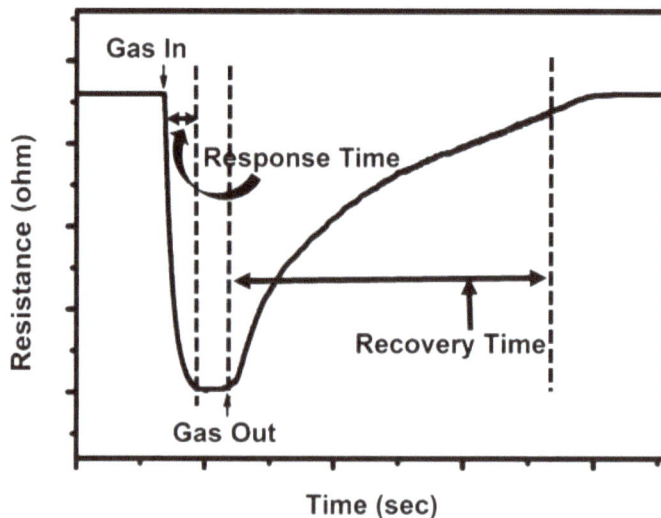

Fig. (3). Schematic representation response curve of a chemiresistive gas sensor.

BASIC FEATURES OF GAS SENSOR

To characterize the performance of gas sensors, the set of technical parameters required are sensor response (S), sensitivity (S'), repose time, recovery time and selectively. Details of these parameters are listed below:

Sensor Response (S)

It is defined as the relative change in resistance when exposed to gas and has been calculated using the eq. (**1**) [14].

$$S(\%) = \frac{R_g - R_a}{R_a} \times 100 \tag{1}$$

Here, R_g and R_a exhibit the sensor resistance in gas and air environment, respectively.

Sensitivity (S')

It is defined as the rate of change of sensor response per analyte concentration eq. (**2**) [15].

$$S'(\%) = \frac{R_g - R_a}{R_a} \times C \times 100 \tag{2}$$

Here, R_g and R_a are the sensor resistance in gas and air environment, respectively, and C is concentration.

Response Time

It is the time required for the sensor's resistance to attain 90% of the saturation value after the gas is introduced into the test chamber [16]. The minimum value of response time is suggestive of a good sensor.

Recovery Time

It is the time required for the sensor to recover 10% of the original resistance after the removal of gas from the test chamber [16]. A small value of recovery time is an indicator of a good sensor.

Selectivity

It determines the selective response of the gas sensor toward a specific target gas among a group of target gases and is the capacity of a sensor to discriminate between gases in a mixture.

This chapter presents an overview of the characterization and application of carbon-based materials for the fabrication of cost-effective and ppb-level gas sensors. The sensing characteristics of both the pristine and functionalized CNTs have been discussed herein. The chapter has been systematically divided into different sections covering CNTs and their need for pristine and functionalized CNTs-based gas sensors.

NEED FOR CNTs-BASED GAS SENSORS

In sensors and sensor arrays, the goal is to develop sensors that can detect the presence of substances at appropriate concentration levels with very precise monitoring. Such research would ultimately have a strong impact on different segments, such as food processing, environmental remediation, agriculture, *etc.* Recently, sensor research has become more materials-oriented, and stress has been given to advanced functional nanomaterials that serve as specific sensing layers. Nanomaterials are chosen over their bulk counterparts due to their high surface-to-volume ratio and the unique properties associated with their nanoscale dimensions. Nanomaterials include nanoparticles, nanospheres, nanowires, nanotubes, nanoflowers, nanobelts, *etc.* Among nanomaterials, carbon-based nanomaterials like fullerenes, graphene, and CNTs have shown tremendous potential in bringing sensor technology to the next level. Furthermore, among carbonaceous materials, CNTs stand out as the most promising material for deployment in electronic sensing platforms due to their superior chemical and electronic properties [17]. Furthermore, CNTs possess great potential for being employed both as a part of the transducer element as well as a functional receptor element in an electronic device. It is well known that the electrical properties of CNTs are very sensitive to the amount of charge transfer between the sensing layer and target gas and also to chemical doping by a variety of molecules. On exposing p-type semiconducting CNTs to an oxidizing or reducing analyte, the conductance of CNTs varies due to a change in the concentration of the main charge carriers, *i.e.,* holes, in nanotubes. Nevertheless, slow charge transfers between the pristine CNTs and gas analyte results in numerous issues, like low sensitivity and slow response/recovery of sensors. To overcome these limitations, the functionalization of CNTs with materials, such as metal oxides, conductive polymers, noble metal nanoparticles, and organic semiconductors, is performed.

Functionalization not only improves the gas sensing parameters but also enhances their selectivity toward specific test gas [18, 19]. It is to be noted that the adsorption of analyte molecules on the sensor surface leads to a change in the surface energy due to the occurrence of an intermolecular interaction force between them [21]. CNTs are found to form bundles by adhering to each other due to strong van der Waals interaction. The distinct sites present on these bundles act as adsorption sites for analyte molecules, and the adsorption properties of these sites depend on the diameter of neighboring tubes, size of gas molecules, binding energy, and availability of the sites [22]. Nevertheless, depending on the type of interaction with test gas molecules, the gas adsorption processes can be physisorption, chemisorption, or a combination of both [21, 23, 24]. Both the reversible and irreversible reactions occur simultaneously on the sensing surface in the chemisorption process, whereas the change in the energy occurs due to the electron transfer between the nanocomposite-based sensing area and the measured gas, as shown in the band diagram (Fig. **4**) in physisorption process [21]. Furthermore, in the physisorption process, the rate of gas adsorption is faster when the gas is purged but varies slowly with further exposure to gas. Additionally, the activation energy in the adsorption sites is not constant due to the presence of both the metallic and semiconducting nature of CNTs [21, 23]. Furthermore, electron transfer between the sensing area and analyte gas has been studied using X-ray photoelectron spectroscopy (XPS). Recently, we have revealed that on analyte exposure, a peak shift in the spectrum of core level C-1s, O-1s, F-1s, N-1s, and Co-2p of the hybrid sample occurs [20] (Fig. **5**). The prominent shift in the Co-2p core level of the peak toward the higher binding energy side confirms that charge transfer interaction occurs upon the adsorption of analyte molecules onto CNTs [20].

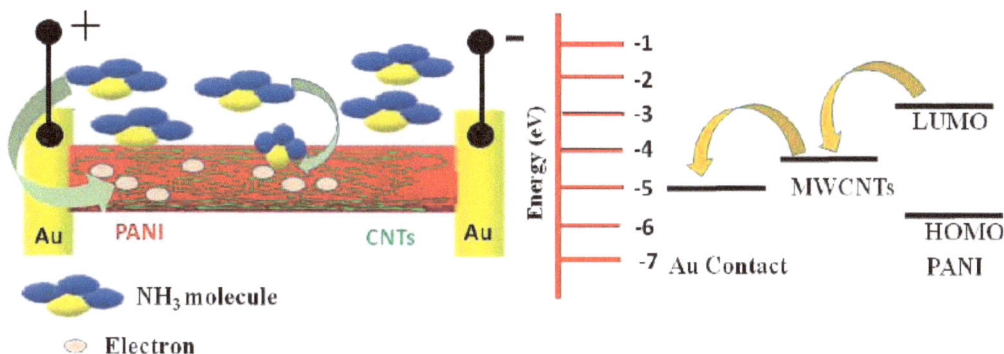

Fig. (4). Schematic representation of sensing mechanism of ammonia adsorption.

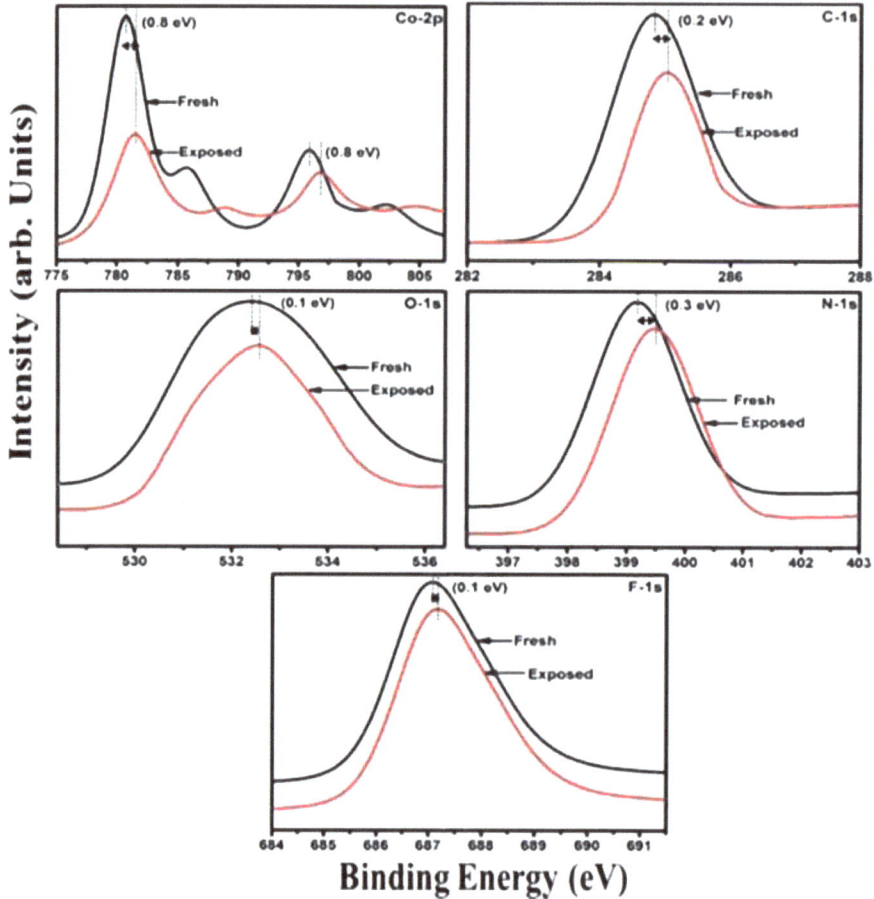

Fig. (5). XPS spectra of sensor performed with and without gas analyte [20]. Published by The Royal Society of Chemistry.

SENSING CHARACTERISTICS OF PRISTINE CNTs-BASED SENSORS

The utilization of single semiconducting SWCNTs as a fast and sensitive chem-FETs sensor at ambient temperatures was first reported in 2000 by Kong and co-workers [25]. First, individual semiconducting SWCNTs were deposited *via* chemical vapor deposition (CVD) on SiO_2/Si substrate and subsequently, Ti/Au-based electrical contacts were photolithographically patterned over SWCNTs. The conductance of the semiconducting SWCNTs was found to vary dramatically upon exposure to NO_2 and NH_3. The conductance increased by three orders of magnitude on exposure to 200 ppm of NO_2 (V_g = +4 V), whereas it decreased by two orders of magnitude within two minutes on exposure to 10,000 ppm of NH_3 (V_g = 0 V).

Kordás *et al.* [26] employed a simple inkjet printing method for developing electrically conductive CNT patterns on paper and plastic surfaces, and the fabricated sensor was found to be sensitive to various substances, such as water, ammonia, and methanol. Valentini *et al.* [27] reported chemiresistive NO_2 sensors based on CNT mats with excellent reversibility and reproducibility. These plasma-enhanced CVD synthesized CNT mats showed a detection limit of up to 10 ppb of NO_2. In comparison, Li *et al.* [8] demonstrated that a simple and low-cost drop-casting technique used to fabricate SWCNTs-based sensors can also produce excellent sensing properties. These sensors showed high selectivity for NO_2 with a detection limit of up to 44 ppb and a fast response period. However, the recovery to the baseline was much slower, and it can be improved by using UV irradiation. Han *et al.* [28] developed a flexible NH_3 sensor based on a CNT-cellulose composite using the drop-coating technique. It was found that both the physical morphology and chemical properties of cellulose affect the sensitivity of the fabricated sensor. A faster and higher response/recovery of the layer-by-layer structure is found in comparison to composite structure due to the presence of a larger active surface area of the layer-by-layer structure in comparison to CNT–cellulose composite.

FUNCTIONALIZED CNTs-BASED GAS SENSORS

Nevertheless, the drawback of using pristine CNTs as a sensing element is the slow charge transfer between the pristine CNTs and gas analyte, which results in low sensitivity and lack of selectivity to different gaseous entities. These limitations can be circumvented by functionalizing CNTs with analyte-specific entities [29]. Depending upon the type of interaction between the analyte and CNTs, the surface functionalization of CNTs can be simply classified as covalent and non-covalent functionalization. Furthermore, covalently functionalized CNTs are based on esterification or amidation of carboxylic acid groups that are introduced on the defect sites of the CNTs. Thus, functionalization opens new avenues for the application of CNTs in numerous areas. Until now, CNTs have been functionalized with different polymers, metal nanoparticles, and phthalocyanines to improve the sensing characteristics of sensors.

Organic Polymer Functionalized CNTs-Based Gas Sensors

Organic polymers have been used as key materials in gas sensing systems. Among them, conducting polymers are regarded as the most promising materials for gas sensing because of the presence of delocalized bonds that results in their semiconducting or even highly conductive behavior. These delocalized bonds result from the presence of a conjugated backbone of alternating single and double bonds. The conducting polymers, such as polyaniline, polypyrrole, and polythiophene, are considered excellent sensing materials for the detection of different toxic gases and volatile organic compounds. These conducting polymers are of great scientific and technological importance because of their unique electrical, electronic, magnetic, and optical properties [30]. Nevertheless, non-conductive polymers have also found their applicability as active materials for gas sensors. It is worth mentioning that the bulk dissolution of gas into the film causes changes in their physical properties and is not readily detectable. To get an easier detection process, the integration of polymers with functional sensing elements has been performed, and such doped conducting polymers are synthesized by adding the analyte binding species. Nonetheless, the introduction of the polymer molecules on CNTs improves their dispersion. Modification of CNTs with polymers can be done through covalent attachment and non-covalent attachment. However, conducting polymers/CNTs composites have fascinated considerable interest in different device applications not only because the CNTs can improve the electrical and mechanical properties of the polymer but also due to the possession of properties of individual components with a synergistic effect in fabricated composites [30]. Zhang *et al.* [31] investigated poly-(m-aminobenzene sulfonic acid) functionalized SWCNTs and showed enhanced sensing characteristics toward NH_3 and NO_2 in comparison to SWCNTs-COOH. They synthesized SWCNT-PABS by treating carboxylic acid groups with oxalyl, and an increase in sensing response toward NH_3 with good recovery time was found due to protonation–deprotonation of the PABS functional group, which was responsible for the change in the density of charge carriers in the SWCNTs. Alizadeh *et al.* demonstrated a composite prepared using a mixing of CNTs with a nano-sized molecularly imprinted polymer that acted as a recognition element and polymethyl methacrylate, which acted as an adhesive substance for the selective and sensitive determination of ethanol vapor [32]. Some of the polymer/CNTs composite sensors reported in literature are listed in Table **2**.

Table 2. Sensing performance of certain polymer/CNT-based sensors.

Polymer type	CNTs type	Analytes	Detection limit (ppm)	Response time (s)	References
PEI	SWCNTs	NO	0.005	70	[33]
PMMA	MWCNTs	Dichloromethane, chloroform, acetone	N/S	2–5	[34]
Polyaniline	SWCNTs	NH_3	0.050	600	[35]
Polyaniline	MWCNTs	Triethylamine	0.500	200–400	[36]
PEDOT: PSS	MWCNTs	Ethanol	200	13	[29]
PPY	MWCNTs	CO_2	250	30	[37]
PMMA	MWCNT	Ethanol	0.5	60	[32]

Metal Nanoparticle/Nanocluster Functionalized CNTs-Based Gas Sensors

Recently, many studies have been carried out for the functionalization of CNTs' surface [REMOVED HYPERLINK FIELD]with metal nanoclusters. Penza *et al.* [38] demonstrated CNTs surface-functionalized with metal nanoclusters (Pt, Ru, and Ag) prepared using magnetron sputtering. Metal functionalization remarkably enhanced the NO_2 gas sensitivity with the detection of 1 ppm NO_2 at room temperature and 0.1 ppm NO_2 at 150°C by metal-functionalized CNT-sensors. The maximum NO_2 gas sensitivity was found for the Ag-CNT device. Zilli *et al.* [39] have reported room temperature hydrogen gas sensors based on Pd-decorated MWCNTs nanocomposite thin films. Pd-decorated CNTs nanocomposite films and pristine CNTs presented a similar sensing behavior, while the response of the oxidized Pd-CNTs-based films was double. Furthermore, the sensing behavior of the Pd-decorated oxidized CNTs nanocomposites was modified by varying the H_2 concentration. The sensor magnitude of response at room temperature reached a saturation-like state with a maximum associated value of 2.15% for an average H_2 concentration above 350 ppm. Table **3** lists some of the metal nanoparticles decorated CNTs-based gas sensors reported in literature.

Table 3. Sensing performance of selected metal/CNT sensors.

Metal	CNTs type	Targeted analytes	Detection limit (ppm)	Response time (s)	References
Au, Pt	MWCNTs	NO_2	0.100	<600	[40]
Pd	SWCNTs	H_2	100	600	[41]
Ag	MWCNTs	CO	0.047	N/S	[42]
Au	SWCNTs	NO_2	0.004	N/S	[43]
Pd	SWCNTs	CH_4	6	120–240	[44]

Metallophthalocyanine Functionalized CNTs-Based Gas Sensors

Metallophthalocyanines (MPcs) have always been attractive choices for the functionalization of CNTs due to their synergic interaction with CNTs through π-π interactions [46, 47]. The MPcs have emerged as excellent sensing materials for gas sensors due to their conjugated macrocyclic units [48]. Liang *et al.* [14] developed substituted MPc/MWCNT hybrid (TFPMPc/MWCNT, M = Co, Zn, Cu, Pb, Pd, and Ni) sensors where the central metal atoms played an important role in the high sensitivity and selectivity of the sensor toward NH_3. They reported that binding energies of the MPc-NH_3 system are responsible for the response in the order Co>Zn>Cu>Pb>Pd ⫯Ni for fabricated hybrid sensors toward NH_3. We have previously reported the nanostructured growth of substituted MPcs toward

Cl_2 with detection limits of up to 5 ppb [49]. Kaya *et al.* [50] fabricated pyrene-substituted Pcs with SWCNTs for sensing applications. The central metal in MPc molecules played an important role in gas detection, and the response of different SWCNT/MPc-py hybrid films toward NH_3 was very good.

Recently, we developed hybrids of functionalized CNTs with hexadecafluorinated metal phthalocyanines ($F_{16}MPc$, M = Cu, Zn, Co) for the detection of Cl_2 (Figs. **6** and **7**) [20, 45, 51]. The response of the sensor was found to be 63% for the $F_{16}CoPc$/MWCNTs-COOH hybrid sensor and 21.28% for the $F_{16}ZnPc$/MWCNTs-COOH hybrid sensor toward 2 ppm of Cl_2, while the theoretical detection limit of the sensor was found to be 0.05 and 0.06 ppb, respectively. The response of $F_{16}MPc$/SWCNTs-COOH hybrids toward Cl_2 was found to be greater than that of $F_{16}MPc$/MWCNTs-COOH hybrids due to some superior features of SWCNTs, such as extremely small diameter, stronger inter-tube attraction, and larger specific surface area compared to the MWCNTs [20, 52]. The response of SWCNTs hybrids decreased in the order $F_{16}CoPc>F_{16}ZnPc>F_{16}CuPc$, which may be due to the larger ionic radius of the central ion and resulting interactions between Cl_2 and different central ions (Fig. **8**).

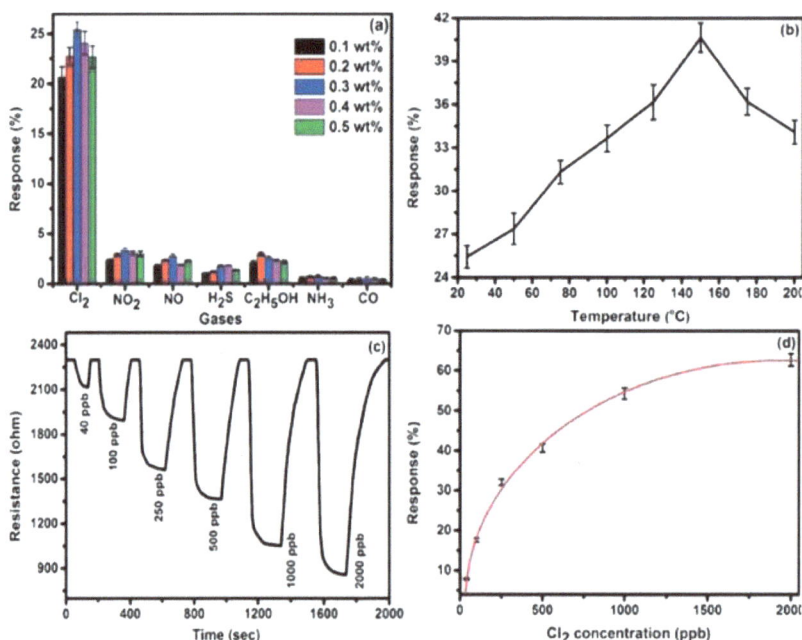

Fig. (6). (a) Selectivity histogram of $F_{16}CoPc$/MWCNTs-COOH sensors for different test gases; **(b)** Response as a function of temperature for 500 ppb of analyte concentration; **(c)** Response curves of the sensor for different concentrations of analyte gas; and **(d)** Calibration curve of the sensor [20]. Published by The Royal Society of Chemistry.

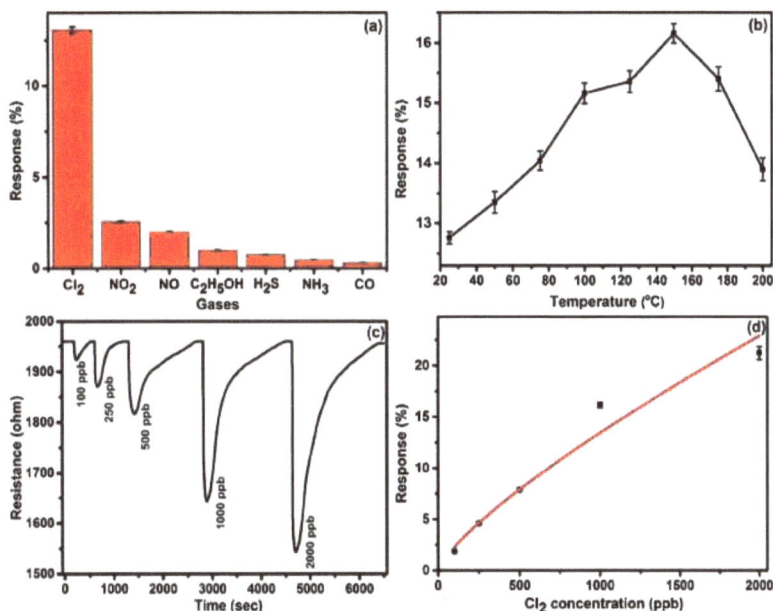

Fig. (7). (a) Selectivity histogram of $F_{16}ZnPc$/MWCNT-COOH sensor for 1 ppm of various test gases; **(b)** Response as a function of temperature for 1 ppm of analyte concentration; **(c)** Response curves of the sensor for different concentrations of analyte at optimum temperature; and **(d)** Calibration curve of the sensor [45]. Published by The Royal Society of Chemistry.

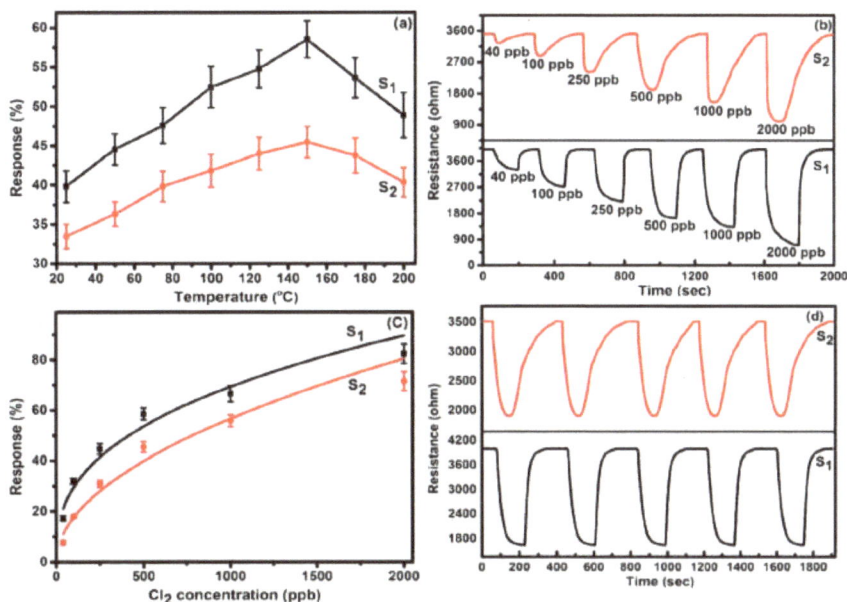

Fig. (8). (a) The response of sensors as a function of temperature to 500 ppb of gas analyte; **(b)** Response curves of sensors for different doses of analyte at 150°C; **(c)** Variation in the response of sensors with analyte concentration; and **(d)** Reproducibility of the response curve of the sensor to 500 ppb of analyte at 150°C [52]. Published by The Royal Society of Chemistry.

Verma *et al.* [53] demonstrated MPc functionalized CNTs for H_2O_2 vapor detection. The composites of MPcs and f-MWCNTs were synthesized by first suspending f-MWCNTs in 30 ml toluene and then sonicated to partially disperse bundled MWCNTs and subsequently adding weighed quantities of copper-phthalocyanine (CuPc), vanadium-phthalocyanine (VPc), and cobalt-phthalocyanine (CoPc) to the sonicated suspension of f-MWCNTs.

The sensors fabricated from composites comprising of f-MWCNTs and CuPc were found to exhibit a ~25% decrease, while the VPc-based sensor showed only a ~5% decrease in resistance on exposure to H_2O_2 vapors. Whereas the sensor fabricated from composites of f-MWCNTs and CoPc exhibited ~4% increase in resistance on exposure to H_2O_2.

CONCLUSION

This book chapter summarizes the configuration and potential application of CNTs in gas sensors. Here, the charge transfer between adsorbed gas molecules and CNTs leads to variations in the resistance of CNTs that contribute to their gas-sensing mechanism. However, CNTs-based chemiresistors showed several issues like low sensitivity, slow response/recovery time due to less charge transfer between the pristine CNTs and gas molecules, *etc*. It is concluded that the functionalized CNTs-based gas sensing materials have gained more attention in terms of their sensing performances, such as stability, repeatability, and recovery speed, in comparison to pristine CNTs. Moreover, based on the preparation technique, the property and behavior of the sensors varied considerably, which is crucial from the point of view of the stability of the CNTs-based devices. It is, therefore, necessary to synthesize identical and reproducible CNTs for their application in all areas. Nonetheless, it is very difficult to synthesize pure and ideal CNTs with minimum defects. Moreover, precise control over the growth or dispersion of CNTs on surfaces is another problem.

ACKNOWLEDGMENTS

The authors are thankful to the Council of Scientific and Industrial Research (CSIR), New Delhi, and UGC, New Delhi, India, for providing financial funding through the RUSA 2.0 scheme in support of the present research work. One of the authors, A.K. Sharma, is grateful to "The Centre for Sustainable Habitat," Guru Nanak Dev University, Amritsar, Punjab, India for awarding a post-doctoral fellowship.

REFERENCES

[1]   Obirai, J.C.; Hunter, G.; Dutta, P.K. Multi-walled carbon nanotubes as high temperature carbon

monoxide sensors. *Sens. Actuators B Chem.,* **2008**, *134*(2), 640-646.
[http://dx.doi.org/10.1016/j.snb.2008.06.005]

[2]   Balasubramanian, K.; Burghard, M. Chemically functionalized carbon nanotubes. *Small,* **2005**, *1*(2), 180-192.
[http://dx.doi.org/10.1002/smll.200400118] [PMID: 17193428]

[3]   Ebbesen, T.W.; Ajayan, P.M. Large-scale synthesis of carbon nanotubes. *Nature,* **1992**, *358*(6383), 220-222.
[http://dx.doi.org/10.1038/358220a0]

[4]   Arepalli, S. Laser ablation process for single-walled carbon nanotube production. *J. Nanosci. Nanotechnol.,* **2004**, *4*(4), 317-325.
[http://dx.doi.org/10.1166/jnn.2004.072] [PMID: 15296222]

[5]   Saito, R.; Fujita, M.; Dresselhaus, G.; Dresselhaus, M.S. Electronic structure of chiral graphene tubules. *Appl. Phys. Lett.,* **1992**, *60*(18), 2204-2206.
[http://dx.doi.org/10.1063/1.107080]

[6]   McEuen, P.L.; Fuhrer, M.S.; Hongkun Park, Single-walled carbon nanotube electronics. *IEEE Trans. Nanotechnol.,* **2002**, *1*(1), 78-85.
[http://dx.doi.org/10.1109/TNANO.2002.1005429]

[7]   Mallakpour, S.; Soltanian, S. Surface functionalization of carbon nanotubes: fabrication and applications. *RSC Advances,* **2016**, *6*(111), 109916-109935.
[http://dx.doi.org/10.1039/C6RA24522F]

[8]   Li, J.; Lu, Y.; Ye, Q.; Cinke, M.; Han, J.; Meyyappan, M. Carbon nanotube sensors for gas and organic vapor detection. *Nano Lett.,* **2003**, *3*(7), 929-933.
[http://dx.doi.org/10.1021/nl034220x]

[9]   Hernández-Fernández, P.; Montiel, M.; Ocón, P.; de la Fuente, J.L.G.; García-Rodríguez, S.; Rojas, S.; Fierro, J.L.G. Functionalization of multi-walled carbon nanotubes and application as supports for electrocatalysts in proton-exchange membrane fuel cell. *Appl. Catal. B,* **2010**, *99*(1-2), 343-352.
[http://dx.doi.org/10.1016/j.apcatb.2010.07.005]

[10]   Landi, B.J.; Ganter, M.J.; Cress, C.D.; DiLeo, R.A.; Raffaelle, R.P. Carbon nanotubes for lithium ion batteries. *Energy Environ. Sci.,* **2009**, *2*(6), 638-654.
[http://dx.doi.org/10.1039/b904116h]

[11]   Mittal, J.; Lin, K.L. Carbon nanotube-based interconnections. *J. Mater. Sci.,* **2017**, *52*(2), 643-662.
[http://dx.doi.org/10.1007/s10853-016-0416-4]

[12]   Fenner, R.; Zdankiewicz, E. Micromachined water vapor sensors: a review of sensing technologies. *IEEE Sens. J.,* **2001**, *1*(4), 309-317.
[http://dx.doi.org/10.1109/7361.983470]

[13]   Wang, Y.; Yeow, J.T.W. A review of carbon nanotubes-based gas sensors. *J. Sens.,* **2009**, *2009*, 1-24.
[http://dx.doi.org/10.1155/2009/493904]

[14]   Liang, X.; Chen, Z.; Wu, H.; Guo, L.; He, C.; Wang, B.; Wu, Y. Enhanced NH_3-sensing behavior of 2,9,16,23-tetrakis(2,2,3,3-tetrafluoropropoxy) metal(II) phthalocyanine/multi-walled carbon nanotube hybrids: An investigation of the effects of central metals. *Carbon,* **2014**, *80*, 268-278.
[http://dx.doi.org/10.1016/j.carbon.2014.08.065]

[15]   Tsiulyanu, D.; Tsiulyanu, A.; Liess, H.D.; Eisele, I. Characterization of tellurium-based films for NO_2 detection. *Thin Solid Films,* **2005**, *485*(1-2), 252-256.
[http://dx.doi.org/10.1016/j.tsf.2005.03.045]

[16]   Wang, B.; Wu, Y.; Wang, X.; Chen, Z.; He, C. Copper phthalocyanine noncovalent functionalized single-walled carbon nanotube with enhanced NH_3 sensing performance. *Sens. Actuators B Chem.,* **2014**, *190*, 157-164.
[http://dx.doi.org/10.1016/j.snb.2013.08.066]

[17]   Wang, Y.; Hu, N.; Zhou, Z.; Xu, D.; Wang, Z.; Yang, Z.; Wei, H.; Kong, E.S.W.; Zhang, Y. Single-walled carbon nanotube/cobalt phthalocyanine derivative hybrid material: preparation, characterization and its gas sensing properties. *J. Mater. Chem.*, **2011**, *21*(11), 3779-3787.
       [http://dx.doi.org/10.1039/c0jm03567j]

[18]   Leghrib, R.; Felten, A.; Pireaux, J.J.; Llobet, E. Gas sensors based on doped-CNT/SnO_2 composites for NO_2 detection at room temperature. *Thin Solid Films*, **2011**, *520*(3), 966-970.
       [http://dx.doi.org/10.1016/j.tsf.2011.04.186]

[19]   Sharma, A.K.; Kumar, P.; Saini, R.; Bedi, R.K.; Mahajan, A. Kinetic response study in chemiresistive gas sensor based on carbon nanotube surface functionalized with substituted phthalocyanines. *AIP Conf. Proc.*, **2016**, *1728*(1), 020493.
       [http://dx.doi.org/10.1063/1.4946544]

[20]   Sharma, A.K.; Mahajan, A.; Bedi, R.K.; Kumar, S.; Debnath, A.K.; Aswal, D.K. CNTs based improved chlorine sensor from non-covalently anchored multi-walled carbon nanotubes with hexa-decafluorinated cobalt phthalocyanines. *RSC Advances*, **2017**, *7*(78), 49675-49683.
       [http://dx.doi.org/10.1039/C7RA08987B]

[21]   Han, T.; Nag, A.; Chandra Mukhopadhyay, S.; Xu, Y. Carbon nanotubes and its gas-sensing applications: A review. *Sens. Actuators A Phys.*, **2019**, *291*, 107-143.
       [http://dx.doi.org/10.1016/j.sna.2019.03.053]

[22]   Agnihotri, S.; Mota, J.P.B.; Rostam-Abadi, M.; Rood, M.J. Adsorption site analysis of impurity embedded single-walled carbon nanotube bundles. *Carbon*, **2006**, *44*(12), 2376-2383.
       [http://dx.doi.org/10.1016/j.carbon.2006.05.038]

[23]   Kumar, D.; Kumar, I.; Chaturvedi, P.; Chouksey, A.; Tandon, R.P.; Chaudhury, P.K. Study of simultaneous reversible and irreversible adsorption on single-walled carbon nanotube gas sensor. *Mater. Chem. Phys.*, **2016**, *177*, 276-282.
       [http://dx.doi.org/10.1016/j.matchemphys.2016.04.028]

[24]   Abdulla, S.; Mathew, T.L.; Pullithadathil, B. Highly sensitive, room temperature gas sensor based on polyaniline-multiwalled carbon nanotubes (PANI/MWCNTs) nanocomposite for trace-level ammonia detection. *Sens. Actuators B Chem.*, **2015**, *221*, 1523-1534.
       [http://dx.doi.org/10.1016/j.snb.2015.08.002]

[25]   Kong, J.; Franklin, N.R.; Zhou, C.; Chapline, M.G.; Peng, S.; Cho, K.; Dai, H. Nanotube molecular wires as chemical sensors. *Science*, **2000**, *287*(5453), 622-625.
       [http://dx.doi.org/10.1126/science.287.5453.622] [PMID: 10649989]

[26]   Kordás, K.; Mustonen, T.; Tóth, G.; Jantunen, H.; Lajunen, M.; Soldano, C.; Talapatra, S.; Kar, S.; Vajtai, R.; Ajayan, P.M. Inkjet printing of electrically conductive patterns of carbon nanotubes. *Small*, **2006**, *2*(8-9), 1021-1025.
       [http://dx.doi.org/10.1002/smll.200600061] [PMID: 17193162]

[27]   Valentini, L.; Cantalini, C.; Armentano, I.; Kenny, J.M.; Lozzi, L.; Santucci, S. Investigation of the NO sensitivity properties of multiwalled carbon nanotubes prepared by plasma enhanced chemical vapor deposition. *J. Vac. Sci. Technol. B*, **2003**, *21*(5), 1996-2000.
       [http://dx.doi.org/10.1116/1.1599858]

[28]   Han, J.W.; Kim, B.; Li, J.; Meyyappan, M. A carbon nanotube based ammonia sensor on cellulose paper. *RSC Advances*, **2014**, *4*(2), 549-553.
       [http://dx.doi.org/10.1039/C3RA46347H]

[29]   Alshammari, A.S.; Alenezi, M.R.; Lai, K.T.; Silva, S.R.P. Inkjet printing of polymer functionalized CNT gas sensor with enhanced sensing properties. *Mater. Lett.*, **2017**, *189*, 299-302.
       [http://dx.doi.org/10.1016/j.matlet.2016.11.033]

[30]   Rajesh, ; Ahuja, T.; Kumar, D. Recent progress in the development of nano-structured conducting polymers/nanocomposites for sensor applications. *Sens. Actuators B Chem.*, **2009**, *136*(1), 275-286.

[http://dx.doi.org/10.1016/j.snb.2008.09.014]

[31]   Zhang, T.; Mubeen, S.; Bekyarova, E.; Yoo, B.Y.; Haddon, R.C.; Myung, N.V.; Deshusses, M.A. Poly(m-aminobenzene sulfonic acid) functionalized single-walled carbon nanotubes based gas sensor. *Nanotechnology,* **2007**, *18*(16), 165504.
[http://dx.doi.org/10.1088/0957-4484/18/16/165504]

[32]   Alizadeh, T.; Rezaloo, F. A new chemiresistor sensor based on a blend of carbon nanotube, nano-sized molecularly imprinted polymer and poly methyl methacrylate for the selective and sensitive determination of ethanol vapor. *Sens. Actuators B Chem.,* **2013**, *176*, 28-37.
[http://dx.doi.org/10.1016/j.snb.2012.08.049]

[33]   Kuzmych, O.; Allen, B.L.; Star, A. Carbon nanotube sensors for exhaled breath components. *Nanotechnology,* **2007**, *18*(37), 375502.
[http://dx.doi.org/10.1088/0957-4484/18/37/375502]

[34]   Abraham, J.K.; Philip, B.; Witchurch, A.; Varadan, V.K.; Reddy, C.C. A compact wireless gas sensor using a carbon nanotube/PMMA thin film chemiresistor. *Smart Mater. Struct.,* **2004**, *13*(5), 1045-1049.
[http://dx.doi.org/10.1088/0964-1726/13/5/010]

[35]   Zhang, T.; Nix, M.B.; Yoo, B.Y.; Deshusses, M.A.; Myung, N.V. Electrochemically functionalized single-walled carbon nanotube gas sensor. *Electroanalysis,* **2006**, *18*(12), 1153-1158.
[http://dx.doi.org/10.1002/elan.200603527]

[36]   Zhang, T.; Mubeen, S.; Myung, N.V.; Deshusses, M.A. Recent progress in carbon nanotube-based gas sensors. *Nanotechnology,* **2008**, *19*(33), 332001.
[http://dx.doi.org/10.1088/0957-4484/19/33/332001] [PMID: 21730614]

[37]   Kumar, U.; Yadav, B.C.; Haldar, T.; Dixit, C.K.; Yadawa, P.K. Synthesis of MWCNT/PPY nanocomposite using oxidation polymerization method and its employment in sensing such as CO2 and humidity. *J. Taiwan Inst. Chem. Eng.,* **2020**, *113*, 419-427.
[http://dx.doi.org/10.1016/j.jtice.2020.08.026]

[38]   Penza, M.; Rossi, R.; Alvisi, M.; Signore, M.A.; Cassano, G.; Dimaio, D.; Pentassuglia, R.; Piscopiello, E.; Serra, E.; Falconieri, M. Characterization of metal-modified and vertically-aligned carbon nanotube films for functionally enhanced gas sensor applications. *Thin Solid Films,* **2009**, *517*(22), 6211-6216.
[http://dx.doi.org/10.1016/j.tsf.2009.04.009]

[39]   Zilli, D.; Bonelli, P.R.; Cukierman, A.L. Room temperature hydrogen gas sensor nanocomposite based on Pd-decorated multi-walled carbon nanotubes thin films. *Sens. Actuators B Chem.,* **2011**, *157*(1), 169-176.
[http://dx.doi.org/10.1016/j.snb.2011.03.045] [PMID: 21666780]

[40]   Penza, M.; Cassano, G.; Rossi, R.; Alvisi, M.; Rizzo, A.; Signore, M.A.; Dikonimos, T.; Serra, E.; Giorgi, R. Enhancement of sensitivity in gas chemiresistors based on carbon nanotube surface functionalized with noble metal (Au, Pt) nanoclusters. *Appl. Phys. Lett.,* **2007**, *90*(17), 173123.
[http://dx.doi.org/10.1063/1.2722207]

[41]   Mubeen, S.; Zhang, T.; Yoo, B.; Deshusses, M.A.; Myung, N.V. Palladium nanoparticles decorated single-walled carbon nanotube hydrogen sensor. *J. Phys. Chem. C,* **2007**, *111*(17), 6321-6327.
[http://dx.doi.org/10.1021/jp067716m]

[42]   Sharafeldin, I.; Garcia-Rios, S.; Ahmed, N.; Alvarado, M.; Vilanova, X.; Allam, N.K. Metal-decorated carbon nanotubes-based sensor array for simultaneous detection of toxic gases. *J. Environ. Chem. Eng.,* **2020**, 104534.

[43]   Young, P.; Lu, Y.; Terrill, R.; Li, J. High-sensitivity NO$_2$ detection with carbon nanotube-gold nanoparticle composite films. *J. Nanosci. Nanotechnol.,* **2005**, *5*(9), 1509-1513.
[http://dx.doi.org/10.1166/jnn.2005.323] [PMID: 16193966]

[44]   Lu, Y.; Li, J.; Han, J.; Ng, H.T.; Binder, C.; Partridge, C.; Meyyappan, M. Room temperature methane detection using palladium loaded single-walled carbon nanotube sensors. *Chem. Phys. Lett.,* **2004,** *391*(4-6), 344-348.
[http://dx.doi.org/10.1016/j.cplett.2004.05.029]

[45]   Sharma, A.K.; Mahajan, A.; Bedi, R.K.; Kumar, S.; Debnath, A.K.; Aswal, D.K. Non-covalently anchored multi-walled carbon nanotubes with hexa-decafluorinated zinc phthalocyanine as ppb level chemiresistive chlorine sensor. *Appl. Surf. Sci.,* **2018,** *427,* 202-209.
[http://dx.doi.org/10.1016/j.apsusc.2017.08.040]

[46]   Song, W.; He, C.; Dong, Y.; Zhang, W.; Gao, Y.; Wu, Y.; Chen, Z. The effects of central metals on the photophysical and nonlinear optical properties of reduced graphene oxide–metal(II) phthalocyanine hybrids. *Phys. Chem. Chem. Phys.,* **2015,** *17*(11), 7149-7157.
[http://dx.doi.org/10.1039/C4CP05963H] [PMID: 25691138]

[47]   Pişkin, M.; Can, N.; Odabaş, Z.; Altındal, A. Toluene vapor sensing characteristics of novel copper(ii), indium(iii), mono-lutetium(iii) and tin(iv) phthalocyanines substituted with 2,6-dimethoxyphenoxy bioactive moieties. Journal of Porphyrins and Phthalocyanines, **2018,** *22*(01n03), 189-197.

[48]   Yazıcı, A.; Ünüş, N.; Altındal, A.; Salih, B.; Bekaroğlu, Ö. Phthalocyanine with a giant dielectric constant. *Dalton Trans.,* **2012,** *41*(13), 3773-3779.
[http://dx.doi.org/10.1039/c2dt11902a] [PMID: 22310939]

[49]   Saini, R.; Mahajan, A.; Bedi, R.K.; Aswal, D.K.; Debnath, A.K. Room temperature ppb level Cl$_2$ detection and sensing mechanism of highly selective and sensitive phthalocyanine nanowires. *Sens. Actuators B Chem.,* **2014,** *203,* 17-24.
[http://dx.doi.org/10.1016/j.snb.2014.06.081]

[50]   Kaya, E.N.; Tuncel, S.; Basova, T.V.; Banimuslem, H.; Hassan, A.; Gürek, A.G.; Ahsen, V.; Durmuş, M. Effect of pyrene substitution on the formation and sensor properties of phthalocyanine-single walled carbon nanotube hybrids. *Sens. Actuators B Chem.,* **2014,** *199,* 277-283.
[http://dx.doi.org/10.1016/j.snb.2014.03.101]

[51]   Sharma, A.K.; Mahajan, A.; Saini, R.; Bedi, R.K.; Kumar, S.; Debnath, A.K.; Aswal, D.K. Reversible and fast responding ppb level Cl$_2$ sensor based on noncovalent modified carbon nanotubes with Hexadecafluorinated copper phthalocyanine. *Sens. Actuators B Chem.,* **2018,** *255,* 87-99.
[http://dx.doi.org/10.1016/j.snb.2017.08.013]

[52]   Sharma, A.K.; Mahajan, A.; Kumar, S.; Debnath, A.K.; Aswal, D.K. Tailoring of the chlorine sensing properties of substituted metal phthalocyanines non-covalently anchored on single-walled carbon nanotubes. *RSC Advances,* **2018,** *8*(57), 32719-32730.
[http://dx.doi.org/10.1039/C8RA05529G] [PMID: 35547684]

[53]   Verma, A.L.; Saxena, S.; Saini, G.S.S.; Gaur, V.; Jain, V.K. Hydrogen peroxide vapor sensor using metal-phthalocyanine functionalized carbon nanotubes. *Thin Solid Films,* **2011,** *519*(22), 8144-8148.
[http://dx.doi.org/10.1016/j.tsf.2011.06.034]

CHAPTER 14

Graphene Nanoribbons and Doped Graphene

Nancy[1] and **Babita Rani**[1,*]

[1] *Department of Physics, Punjabi University, Patiala-147002, Punjab, India*

Abstract: Graphene has been an interesting material for scientists and engineers by virtue of its remarkable properties. It has unique electronic properties with zero bandgap at the Dirac point. The absence of bandgap in graphene limits its application in electronics. The formation of graphene nanoribbons and substitutional doping of graphene are the methods to manipulate the geometric and hence electronic structure of graphene. Starting from the geometric and electronic properties of graphene, this chapter involves a discussion on the geometric and electronic structure of graphene nanoribbons and substitutionally doped graphene systems based on first principles studies.

Keywords: Doping, Electronic properties, Graphene nanoribbons, Graphene.

INTRODUCTION

Graphene [1, 2] is a one-atom-thick monolayer of carbon atoms arranged in a two-dimensional honeycomb lattice, as shown in Fig. (**1**). Carbon ($2s^2 2p^2$) has four valence electrons in s and p orbitals. The sp^2 hybridization between one s and two p orbitals leads to a trigonal planar structure with the formation of σ bonds between carbon atoms. The resulting C-C bond length and C-C-C bond angle are 1.42 Å and 120°, respectively. The unaffected p orbital remains perpendicular to the planar structure. It can bind covalently with neighbouring carbon atoms, leading to π-bonding.

Fig. (1). Pristine graphene.

* **Corresponding author Babita Rani:** Department of Physics, Punjabi University, Patiala-147002, Punjab, India; Email: dr.babita@pbi.ac.in

Karamjit Singh Dhaliwal (Ed.)

The unit cell of graphene contains two carbon atoms denoted by A and B, as shown in Fig. (**2**).

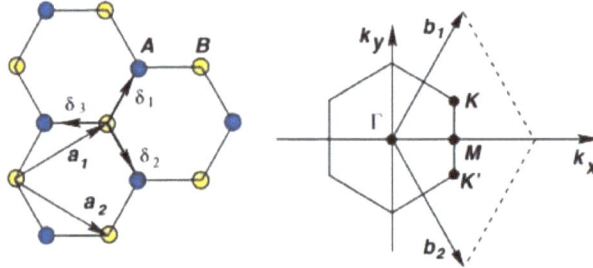

Fig. (2). (a) Lattice structure and **(b)** first Brillouin zone of graphene (Adopted from reference 3).

The primitive translation vectors are:

$$\mathbf{a}_1 = \frac{a}{2}\left(\sqrt{3}, 1\right)\mathbf{a}_2 = \frac{a}{2}\left(\sqrt{3}, -1\right) \tag{1}$$

where a= 2.46 Å is the lattice constant.

The vectors drawn from one carbon atom to its three nearest neighbours in real space are:

$$\boldsymbol{\delta}_1 = \frac{a}{2\sqrt{3}}\left(1, \frac{1}{\sqrt{3}}\right)\boldsymbol{\delta}_2 = \frac{a}{2\sqrt{3}}\left(1, -\sqrt{3}\right)\boldsymbol{\delta}_3 = -\frac{a}{\sqrt{3}}(1,0) \tag{2}$$

The reciprocal lattice vectors are:

$$\mathbf{b}_1 = \frac{2\pi}{\sqrt{3}a}\left(1, \sqrt{3}\right)\mathbf{b}_2 = \frac{2\pi}{\sqrt{3}a}\left(1, -\sqrt{3}\right) \ (3) \tag{3}$$

The first Brillouin zone of graphene is a hexagon. Due to symmetry, six points at the edges of the hexagon reduce to three special points denoted by Γ, K and M. These points are also known as high symmetry points. The band structure of graphene is plotted along the lines joining these points. Two points K and K' are non-equivalent points known as Dirac points [3].

The position of these points in momentum space is:

$$\Gamma = (0, 0) \tag{4}$$

$$\mathbf{K} = \frac{2\pi}{\sqrt{3}a}\left(1, \frac{1}{\sqrt{3}}\right)\mathbf{K}' = \frac{2\pi}{\sqrt{3}a}\left(1, -\frac{1}{\sqrt{3}}\right) \tag{5}$$

$$\mathbf{M} = (\frac{2\pi}{\sqrt{3}a}, 0) \tag{6}$$

Since its experimental isolation in 2004 [4], graphene has attracted the great attention of various researchers because of its extraordinary properties [5, 6]. It has a high electron mobility of about 15,000 cm²/Vs at a charge density of 10^{12} cm⁻² [7] and high electrical conductivity [8] at room temperature, owing to its perfect crystal quality. It exhibits anomalous quantum Hall effect [9] and ambipolar charge transport at room temperature [10]. Electrons in graphene behave as massless relativistic fermions, which satisfy the Dirac equation [11]. The electron wave function has non-zero Berry's phase of π as a consequence of unique electronic properties of graphene.

The low energy band structure of graphene is depicted by conical forms of conduction and valence bands located near K and K' points of the Brillouin zone, as shown in Fig. (3). The π band is completely filled and π^* is completely empty and meets at these k-points, forming the singular point at the Fermi level [3, 6]. The two-dimensional energy dispersion relation is linear in crystal momentum near the Fermi level and given as:

$$E = {}^{+}_{-}\hbar\upsilon_f k = \pm\upsilon_f p \tag{7}$$

Where $\hbar$ is reduced Planck's constant, V_f is Fermi velocity and given by $V_f \approx c / 300 \approx 10^6$ m/s (c is the speed of light), k is the wave vector measured from one of the Dirac points, and p is crystal momentum.

Due to the linear dispersion relation, one can expect the quasiparticles in graphene to behave differently as compared to those in semiconductors and metals, where energy dispersion is approximated by a parabolic curve.

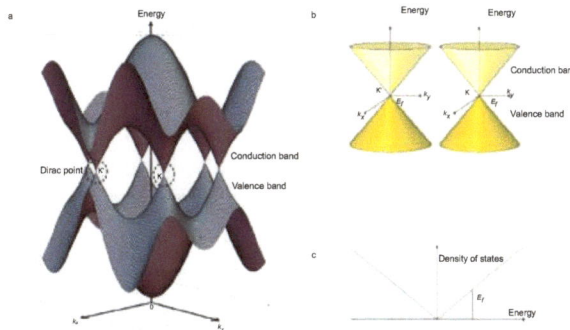

Fig. (3). (a) Energy bands near the Fermi level in the graphene **(b)** conic energy bands near the vicinity of the K and K' points **(c)** density of states near the Fermi level E_f (Adopted from reference 6).

According to the special theory of relativity, the energy of a relativistic particle is given as:

$$E = \sqrt{m^2 c^4 + c^2 p^4} \tag{8}$$

If the rest mass, m, is set as zero, equation 8 gives the dispersion relation as E=c|p|, which is equivalent to that given by equation 7 when υ_f is replaced with c.

From equation 7, k can be obtained as:

$$k = \frac{E}{\hbar \upsilon_f} \tag{9}$$

The number of states per unit area with a k value specific to the Dirac point is n(k) = $k^2/2\pi$.

Density of states is given as:

$$D(E) = \frac{dn(E)}{dE} = \frac{1}{\hbar \upsilon_f} \frac{dn(k)}{dk} \tag{10}$$

$$D(E) = \frac{1}{\hbar \upsilon_f} \cdot \frac{k}{\pi} = \frac{E}{\pi (\hbar \upsilon_f)^2} \tag{11}$$

Density of states vanishes at Dirac point. The absence of density of state at Fermi energy makes graphene a semimetal.

Conduction and valence bands in graphene meet each other at non-equivalent points, which makes it a zero-bandgap semiconductor. The absence of bandgap in graphene limits its application in electronics. In the literature, various methods are found to tune the bandgap of graphene from zero to some finite value so that the devices employing it may have improved semiconducting properties. These methods include the formation of graphene nanoribbons (GNRs) with edge effects [12], modified graphene with heteroatom substitutional doping [13], substrate-induced methods [14], functionalization via covalent and non-covalent approaches [15], application of electric field [4], defect engineering [16], *etc.* In this chapter, the discussion has been confined to GNRs and heteroatom substitutional doped graphene as methods for manipulating structural and hence electronic properties of graphene.

GRAPHENE NANORIBBONS (GNRs)

GNRs [17, 18] are small strips of graphene with a finite width (less than 50 nm). These are quasi-one-dimensional structures obtained by the termination of graphene sheets with smooth edges. Depending upon the edge structures, GNRs are classified as an armchair (AGNRs) and zigzag (ZGNRs) nanoribbons. Fig. (4) shows two highly symmetric zigzag and armchair edges (directions) in graphene, which have an orientation difference of $\frac{\pi}{6}$.

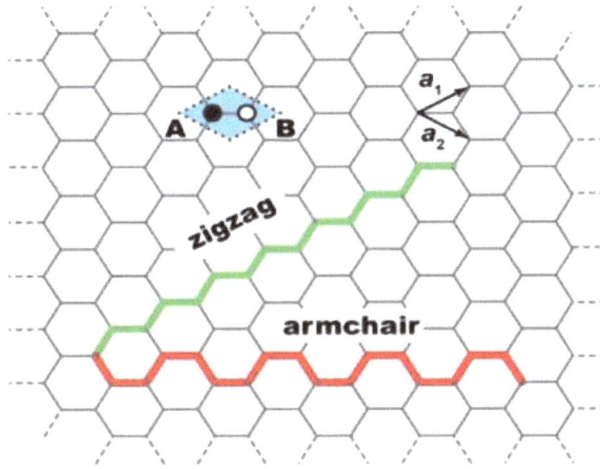

Fig. (4). Zigzag and armchair edges in the honeycomb lattice of graphene (Adopted from reference 19).

Armchair Graphene Nanoribbons (AGNRs)

The unit cell of AGNRs, along with their atomic structure, is shown in Fig. (5a). AGNRs possess armchair edge structure along the periodicity (longitudinal) direction. These have the same corresponding edge structure on both sides across the width, N_a, which is determined by a number of dimer lines and is referred to as N_a-AGNRs. In other words, AGNRs have longitudinal axes in armchair direction and zigzag termini. The smallest width for AGNRs is N_a=3 containing 6 carbon atoms in the unit cell. The structure repeats itself with a period of $\sqrt{3}a$ along the armchair longitudinal axis and the length of 1D Brillouin zone (Fig. **5b**) is given as:

$$d_a = \frac{2\pi}{\sqrt{3}a} \qquad\qquad (12)$$

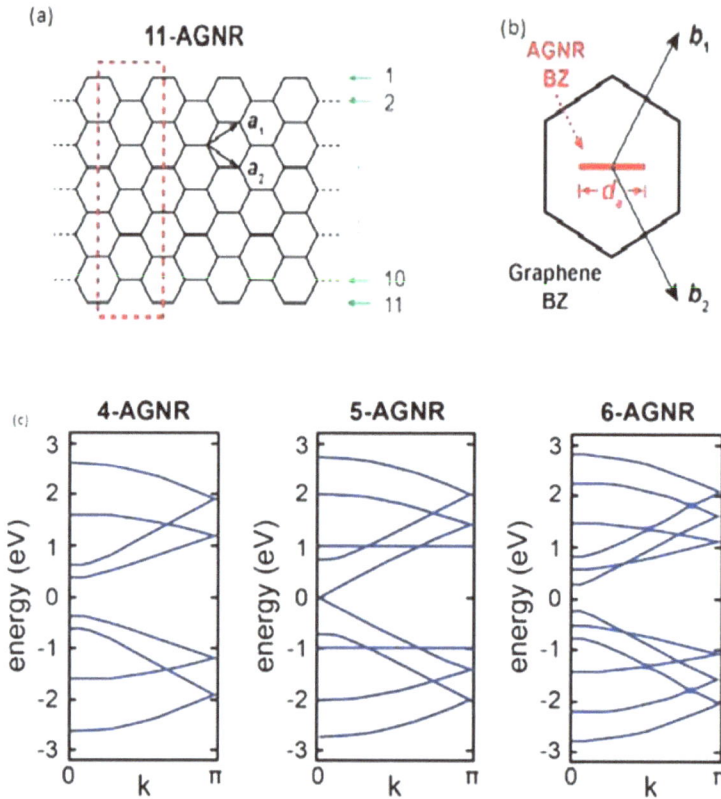

Fig. (5). (a) Atomic structure of 11-AGNRs; the red rectangle shows its 1-D unit cell, and numbers (marked at right) count the dimer lines **(b)** 1-D Brillouin zone of AGNRs with length d_a and **(c)** Band structure of 4, 5, 6-AGNRs (Adopted from reference 12).

The band structure of AGNRs can be obtained by projecting a 2D graphene band structure onto AGNR's 1D Brillouin zone. AGNRs have three typical families corresponding to $N_a = 3p$, $3p+1$ and $3p+2$ (p is any positive integer). They can exhibit both metallic as well as semiconducting states, depending on their width. According to tight-binding calculations, AGNRs with $N_a = 3p+2$ have no bandgap between the lowest conduction band and the highest valence band at k=0 and are metallic. On the other hand, a non-zero bandgap exists at k=0 for AGNRs with $N_a=3p$ or $N_a=3p+1$, indicating their semiconducting nature [20]. Contrary to the above, density functional theory calculations predict non-zero bandgap for all the families [21]. Generally, the bandgap of AGNRs varies inversely as their width is within the same family. For the same p, energy bandgap (Δ) varies as $\Delta p+1>\Delta p>\Delta p+2$. Fig. **(5c)** represents the band structure of AGNRs with $N_a= 4$, 5 and 6 calculated with tight-binding theory [12].

Zigzag Graphene Nanoribbons (ZGNRs)

The unit cell of ZGNRs, along with their atomic structure, is shown in Fig. (**6a**). ZGNRs have longitudinal axis in zigzag direction and armchair termini. The width N_z of GNRs is determined by the number of zigzag lines from one edge to another. The numbers N= 1, 2, 3.....6 marked on the top of Fig. (**6a**) count the zigzag lines across the width of the ribbon. The red rectangle in Fig. (**6a**) indicates the unit cell of ZGNRs. The smallest width of ZGNRs is N =2, and the corresponding 1-D unit cell contains 4 carbon atoms. The structure of ZGNRs repeats itself with a period of 'a' along the zigzag axis. The 1-D Brillouin zone of ZGNRs (Fig. **6b**) is rotated through $\pi/2$ with respect to that of AGNRs, as their longitudinal axes have the same relative orientation. The length of the Brillouin zone is given as:

$$d_z = 2\pi/a \tag{13}$$

The band structure of ZGNRs can be predicted by projecting the graphene band structure on 1-D ZGNR's Brillouin zone. Fig. (**6c**) shows the band structure of 4-, 5- and 6-ZGNRs based on tight-binding calculations [12]. In all these ZGNRs, the lowest conduction band and the highest valence band remain always degenerated at $k = \pi$, giving rise to its non-bonding character and sharp peaks in the density of states. Both bands get flattened near the zone boundary$(\frac{2\pi}{3} < k < \pi)$. ZGNRs possess an edge localised state at the zone boundary, which is completely absent in AGNRs. The two special bands get flattened with an increase in the ribbon width. The highest valence band below the centre band shows a rise and comes close to each other as N_z increases [22]. ZGNRs are metallic irrespective of the width of the ribbon.

Chiral Graphene Nanoribbons (CGNRs)

CGNRs [23] are those nanoribbons whose edges are neither zigzag nor armchair, as shown in Fig. (**7**). Their geometries are complex, with a mixture of an armchair and zigzag-shaped edges. The red box shows the unit cell of CGNRs.

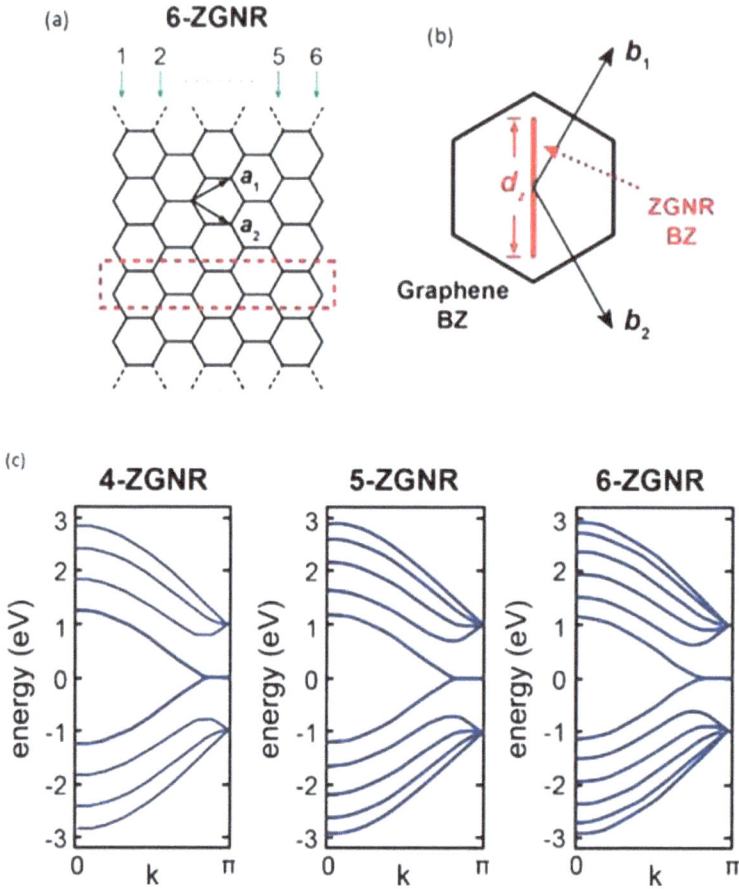

Fig. (6). (a) Atomic structure 6-ZGNRs; the red rectangle shows its 1-D unit cell and numbers (marked at top) count the zigzag lines **(b)** 1D Brillouin zone of ZGNRs with length d_z and **(c)** Band structure of 4-, 5-, 6-ZGNRs (Adopted from reference 12).

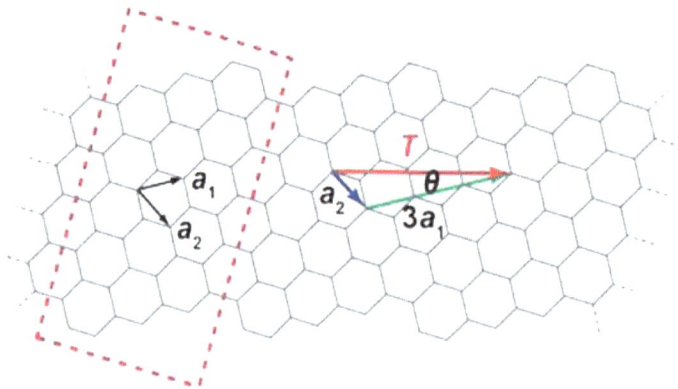

Fig. (7). CGNRs with translation vector **T** (Adopted from reference 19).

The chiral translation vector joins one lattice point to another and is given as:

$$\mathbf{T} = n\mathbf{a}_1 + m\mathbf{a}_2 \tag{14}$$

Here, two positive integers (n, m) as n>m define the chirality.

In Fig. (7), the translation vector is $\mathbf{T}=3\mathbf{a}_1+\mathbf{a}_2$. The chiral angle θ is defined as the angle between $\mathbf{T}$ and the closest zigzag orientation (the direction of a1). CGNRs correspond to the possible combination of n and m with a chiral angle varied as $0 \le \theta \le \frac{\pi}{6}$. The relationship between (n, m) and θ is given as:

$$\tan\theta = \frac{\sqrt{3}m}{2n+m} \tag{15}$$

Bandgap in CGNRs depends on both chiral angle θ and ribbon width [24]. CGNRs exhibit metallic behavior for ribbon width $N_c = 3p+1$ (p is a positive integer). On the other hand, 3p and 3p+2 families show semiconducting characteristics [25].

HETEROATOM SUBSTITUTIONAL DOPED GRAPHENE

Heteroatom substitutional doping occurs when some carbon atom(s) of a graphene lattice are substituted by other atom(s) having different valence electrons [13]. Doping helps in modulating the structural and electronic properties of graphene. Furthermore, bandgap opening in graphene depends upon the nature of the dopant, the site of substitution and the concentration of dopant atoms. In this section, the discussion has been confined to first principles studies based on structural and electronic properties of graphene doped with alkaline earth metals (Be, Mg, Ca, Sr and Ba), B, N, Al, Si, P, S, Ge, Ga, As and Se. The site and concentration-specific effects on the bandgap opening of graphene are also discussed.

Alkaline Earth Metals (Be, Mg, Ca, Sr and Ba)

Geometric structures of beryllium ($2s^2$), magnesium ($3s^2$), calcium ($4s^2$), strontium ($5s^2$) and barium ($6s^2$) doped single layer of graphene are shown in Fig. (8). Substitutional doping of an alkaline earth metal atom in graphene causes electron transfer from the dopant to the substrate [26]. The respective formation energies of doped systems are -5.868eV, -4.448eV, -6.045eV, -8.18eV and -10.61eV for dopant concentration (4.35 at%) of Be, Mg, Ca, Sr and Ba [27]. Be, Mg, and Ca-doped graphene are thermodynamically more stable than Sr and Ba-doped systems. It is because of the fact that atomic size increases as one moves down the group. Atoms with larger radii cause more structural deformations, leading to the instability of graphene. The bond distance between carbon and alkaline earth metal in doped systems is in the range of 1.54-1.94 Å, depending upon the size of the dopant atom. Be and Mg atoms cause the Fermi level to move to the valence band and are regarded as p-type dopants. These dopants induce a bandgap of 0.3eV and 0.6eV, respectively. Ca dopant shows metallic behaviour while Sr and Ba shift the Fermi level to the conduction band and are regarded as n-type dopants with an induced bandgap of 0.2eV and 0.8eV, respectively [27]. Fig. (**9**) shows the band structure of alkaline earth metal doped single layer of graphene.

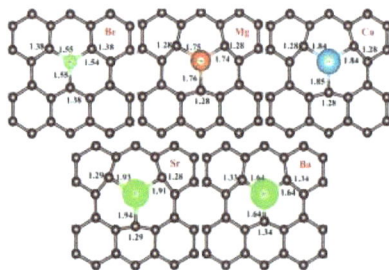

Fig. (8). Geometric structures of alkaline earth metal doped single layer graphene (Adopted from reference 27).

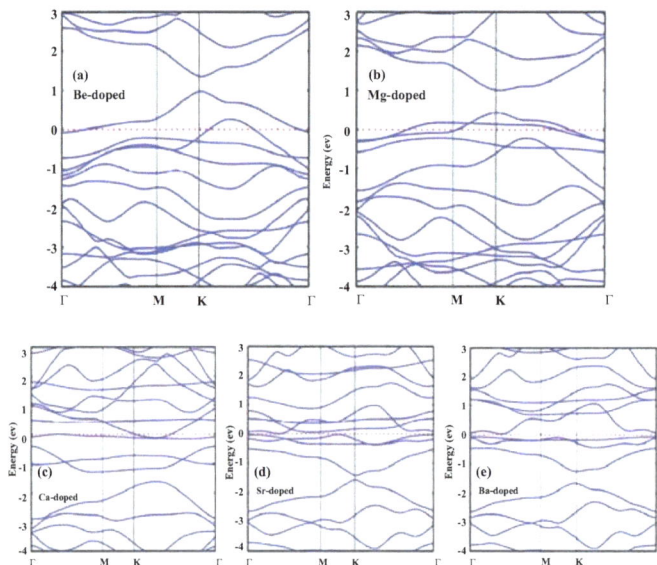

Fig. (9). Band structure of alkaline earth metal doped single layer of graphene (Adopted from reference 27).

Other Dopants (B, N, O, Al, Si, S, P, Ge, Ga, As and Se)

Boron $(2s^2 2p^1)$ is one of the most commonly used substitutional dopants for designing the bandgap of graphene. Its atomic size is almost similar to carbon, having one electron less than that of carbon. Boron forms a strong bond with the nearest C atoms with a B-C bond length of 1.477 Å. The formation energy of B-doped graphene with a dopant concentration of ~2.08 at % is 5.55eV [28]. Boron doping preserves the planar honeycomb structure of graphene, as a result of which linear energy dispersion is not destroyed near the Dirac point. The Fermi level shifts to the valence band, resulting in p-doped graphene [29]. Single-atom doping of boron (2 at% dopant concentration) causes a bandgap of 0.14eV in graphene [30]. Furthermore, the bandgap value is proportional to the dopant concentration. Bandgap increases from 0.14 eV to 0.23 eV and 0.54 eV to 1.88 eV as the concentration of dopant is changed from 2% to 12% [30] and 6.25% to 25% [31], respectively. Moreover, not only the concentration but site-specific substitution also plays an important role in the bandgap opening of graphene [29]. Single boron atom substitution does not cause any structural deformation, but with an increase in the concentration of boron atoms, structural distortion of graphene is observed [30]. Fig. (**10**) shows the geometric structure of boron atom doped graphene and its band structure.

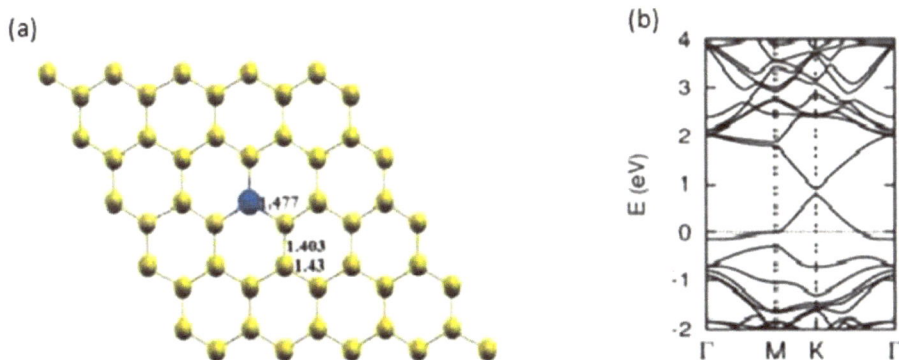

Fig. (10). (a) Geometric structure of boron-doped graphene and **(b)** its band structure (Adopted from reference 30).

Nitrogen $(2s^2 2p^3)$, the next neighbour to carbon, is also used as a common dopant of graphene. Its substitutional doping preserves the planar structure of graphene [32]. Nitrogen shows mainly three types of configurations in doped graphene, including graphitic-N, pyridinic-N, and pyrrolic-N. Graphitic-N is a simple substitutional configuration in which nitrogen is bonded to three carbon atoms. In pyridinic-N, a monovacancy is accompanied by three nitrogen atoms in which each nitrogen atom is bonded to two carbon atoms as part of a six-membered ring, while in pyrrolic-N, nitrogen bonds to two carbon atoms as a part of a five-membered ring. Fig. (**11**) shows the various types of nitrogen bonding in graphene. The graphitic-N leads to n-type bonding, while pyridinic-N and pyrrolic-N lead to p-type doping in graphene [28, 33]. The formation energy of N-doped graphene with a dopant concentration of ~2.08 at% is 4.44eV [28]. Fig. (**12**) shows the geometric structure of nitrogen atom doped graphene and its band structure. The N-C bond length in N-doped graphene is 1.398 Å [28]. Single-atom substitution of nitrogen induces a bandgap of 0.14 eV at a dopant concentration of 2% [30]. Bandgap varies from 0.14eV to 0.22 eV and 0.21eV to 0.60eV by changing the concentration from 2% to 12% [30] and 3.13% to 12.50% [34], respectively. Fig. (**13**) shows a variation of bandgap with an increase in the concentration of B and N dopants.

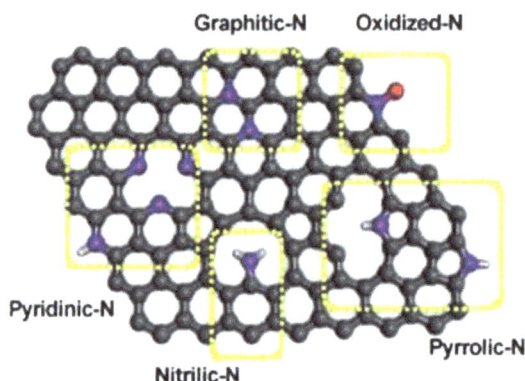

Fig. (11). Different configurations of nitrogen dopants in graphene (Adopted from reference 33).

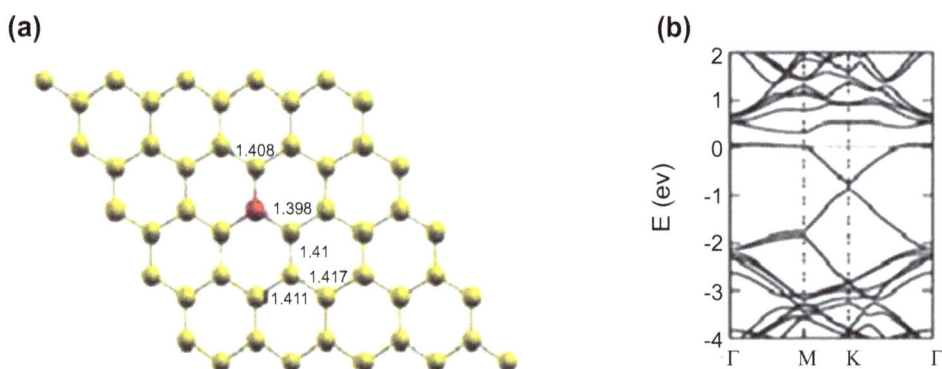

Fig. (12). (a) Geometric structure of nitrogen-doped graphene and **(b)** its band structure (Adopted from reference 30).

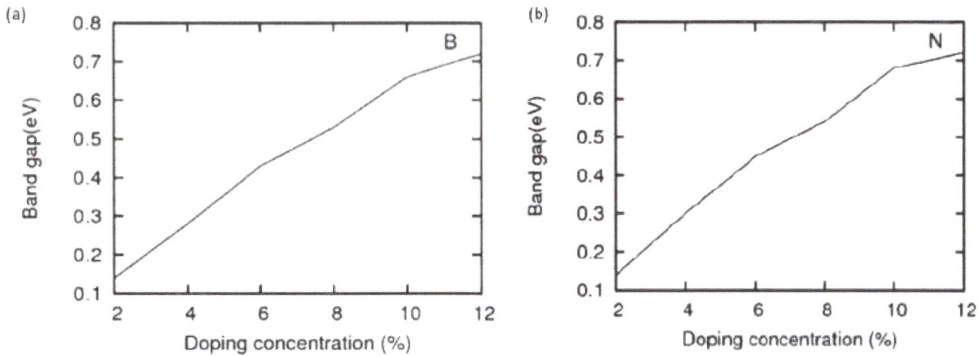

Fig. (13). Variation of bandgap with change in concentration of **(a)** boron and **(b)** nitrogen dopants (Adopted from reference 30).

Oxygen $(2s^2 2p^4)$ atom has two electrons more than carbon.. Two unpaired electrons of oxygen dopant are shared by three neighbouring carbon atoms of graphene, forming weak sigma bonds with a C-O bond length of 1.389 Å [35] Fig. (**14a**). Single oxygen dopant causes negligible structural distortion of the substrate. Oxygen dopant with concentrations of 2 and 3.125 at% induces a bandgap of about 0.50 eV [29, 35]. The site and concentration-specific dopant effects on bandgap opening are also observed [36]. On substituting the dopant atom at rectangular and hexagonal sites of graphene Fig. (**15**), different bandgap values are observed. Rectangular doping configuration shows a linear increase in bandgaps, while hexagonal doping sites lead to an exponential increase in bandgaps by increasing the concentration of dopants Fig. (**15**). Site-specific non-zero bandgaps (maximum value= 1.68 eV) can be obtained by increasing the concentration of dopants from 3.125% to 12.5% [36]. However, an increase in the concentration of dopants causes a decrease in cohesive energy and hence the structural stability of doped graphene.

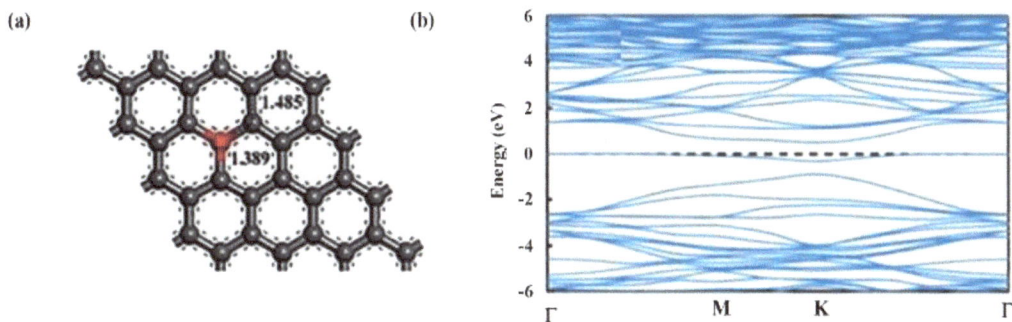

Fig. (14). (a) Geometric structure of oxygen doped graphene and **(b)** its band structure (Adopted from reference 35).

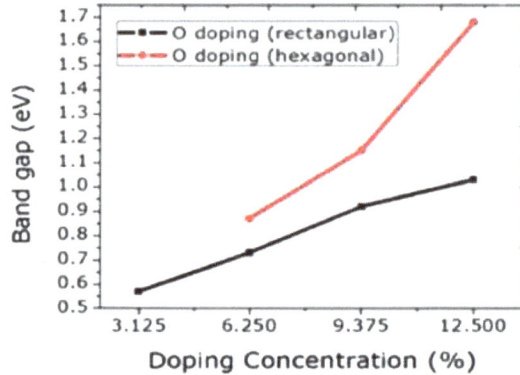

Fig. (15). Induced bandgap with the variation of oxygen dopant concentration at rectangular (R) and hexagonal (H) sites (Adopted from reference 36).

Doping of aluminium ($3s^2 3p^1$) results in structural distortion of graphene. Al loses its electrons to carbon atoms of graphene and forms a weak bond with a C-Al bond length of 1.74 Å [37]. However, the C-C bond length shrinks to 1.39 Å because of the larger covalent radius of Al. The doped system has large formation energy of approximately 10 eV at a dopant concentration of 3.125% [38]. The resulting p-type doped system induces a bandgap of 0.40 eV [37]. Bandgap varies from 0.40 eV to 1.480 eV, with an increase in dopant concentration from 3.125% to 12.5%, respectively [39]. Fig. (**16**) shows the geometric structure of aluminium-doped graphene and its band structure.

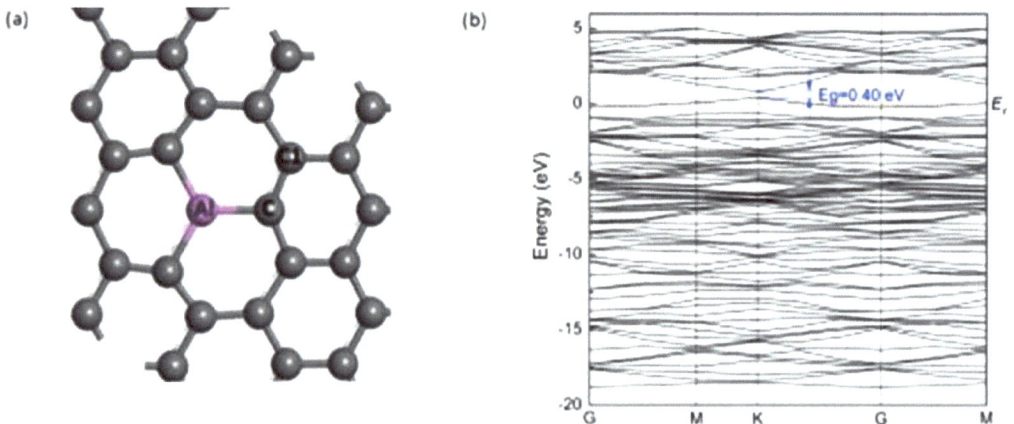

Fig. (16). (a) Geometric structure of aluminium doped graphene and **(b)** its band structure (Adopted from reference 37).

Silicon ($3s^2 3p^2$) is a semiconductor with a bandgap of 1.14 eV [40]. It has the same number of valence electrons in the outer shell as those of C. It binds strongly with the nearest three C atoms through sp^2 hybridization after doping. Structural distortion of Si-doped graphene is because of a larger atomic radius of the dopant. The doped system has a Si-C bond length of 1.65 Å, and the change in C-C bond length is 0.096 Å. Fig. (**17**) shows the geometric structure of Si-doped in graphene and its band structure. Single-atom substitution of graphene at a doping concentration of 3.125% opens the bandgap of 0.783 eV [41]. In a study [41], a variation of 0.3 eV in bandgap is observed with an increase in dopant concentration from 3.125% to 9.37%.

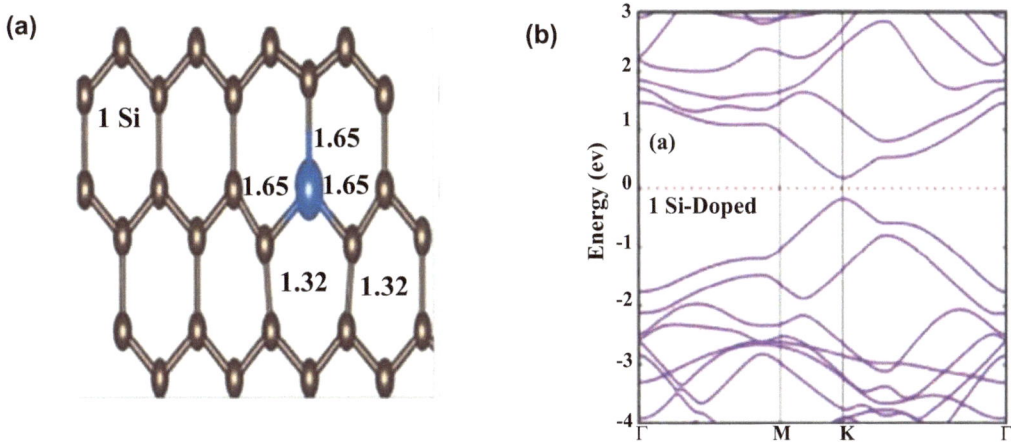

Fig. (**17**). (**a**) Geometric structure of silicon doped graphene and (**b**) its band structure (Adopted from reference 41).

Sulphur (3s^{2}3p^4) doped graphene has large formation energy of 0.90 eV at a dopant concentration of 2%, which indicates the difficulty of preparing this system [42]. Theoretically, sulphur causes large structural distortions of doped graphene. The corresponding C-S bond length is 1.78 Å. S atom acts as a donor impurity, and the Fermi level of the doped system moves to the conduction band [43]. A bandgap of 0.13 eV is induced in the doped system at a dopant concentration of 2% [42]. Bandgap varies from 0.22 eV to 0.57 eV with variation in dopant concentration from 0.78% to 3.125% [38]. Doping at non-equivalent sites induces more bandgap opening than the other sites of graphene [44]. Fig. (**18**) shows the geometric structure of S-doped graphene and its band structure.

(a)   **(b)**

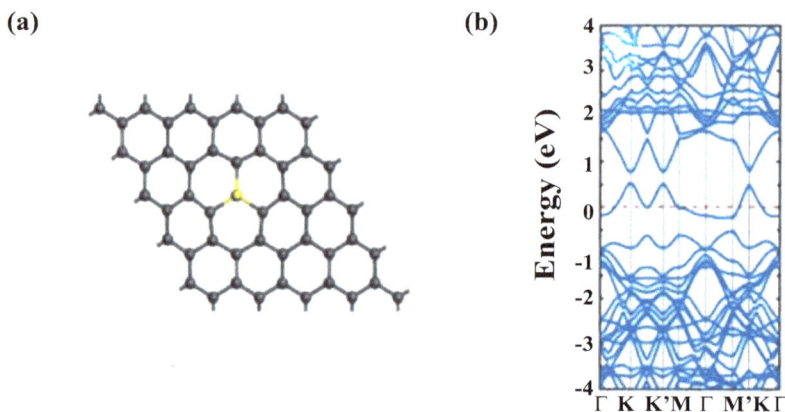

Fig. (18). (a) Geometric structure of sulphur doped graphene and (b) its band structure (Adopted from reference 42).

N-doped monovacancy defects of graphene act as p-doped systems. On substituting nitrogen atoms with sulphur atom(s), n-doped systems are obtained [45]. Table **1** shows the bandgap values for these systems.

Table 1. Bandgaps for N and S doped graphene (Adopted from reference 47).

	3N-gra	2N1S-gra	1N2S-gra	3S-gra
Bandgap (eV)	0.473	0.350	0.275	0.255

Phosphorous ($3s^23p^3$) atom does not fit into the graphene lattice due to its large size. Instead, it protrudes out of the crystal plane of graphene. The corresponding C-P bond length is 1.76 Å, which is greater than the C-C bond length (1.42 Å). The formation energy of P-doped system (dopant concentration of ~1.39 at%) is 0.89 eV, which indicates that P atom can be easily doped in graphene structure [46]. Single-atom doping of P (dopant concentration of 1.38%) induces a bandgap of 0.067eV. The bandgap value increases to 0.0728 eV and 0.0874 eV by changing the site of substitution at the same dopant concentration [47]. The bandgap variations from 0.166 eV to 1.262 eV and 0.34 eV to 0.67 eV are reported by changing the dopant concentration from 3.125% to 15.625% [48] and 0.78% to 3.13% [38], respectively. Fig. (**19**) shows the geometric structure of P-doped graphene and its band structure.

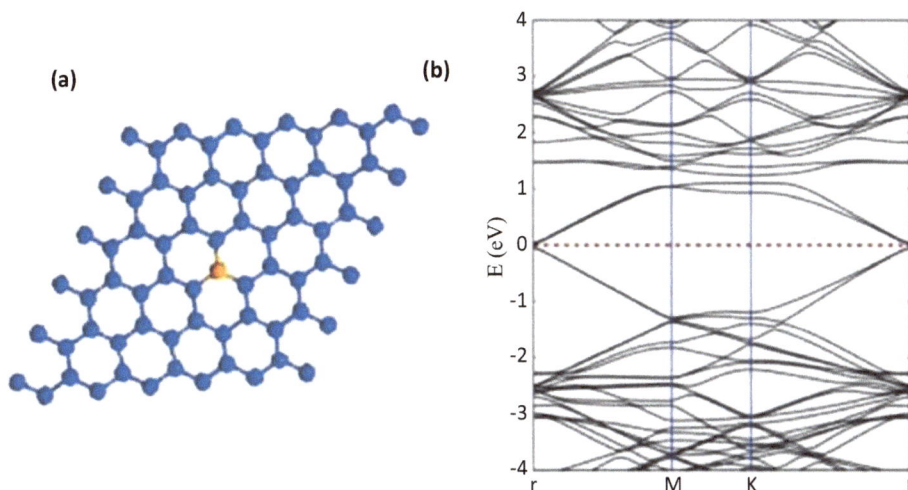

Fig. (19). (a) Geometric structure of P-doped graphene and **(b)** its structure (Adopted from reference 47).

The large covalent radius of the germanium $(4s^2 4p^2)$ atom results in the structural deformation of graphene. The formation energy of Ge-doped graphene with a dopant concentration of ~2.5% is -0.96 eV. Thus, the resulting system is thermodynamically favourable [49]. The bond length of Ge–C is 1.67 Å. Ge doping (5.5 at%) opens a bandgap of 0.71 eV in graphene [50]. An increase in dopant concentration increases the energy difference between the valence and conduction band and hence the bandgap value. The bandgap varies from 0.71 eV to 0.20 eV and from 0.8 eV to 1.1 eV, with an increase in concentration from 5.5% to 12.5% [50] and 2.5% to 7.5% [49], respectively. Fig. (**20**) shows the geometric structure of Ge-doped graphene and its band structure.

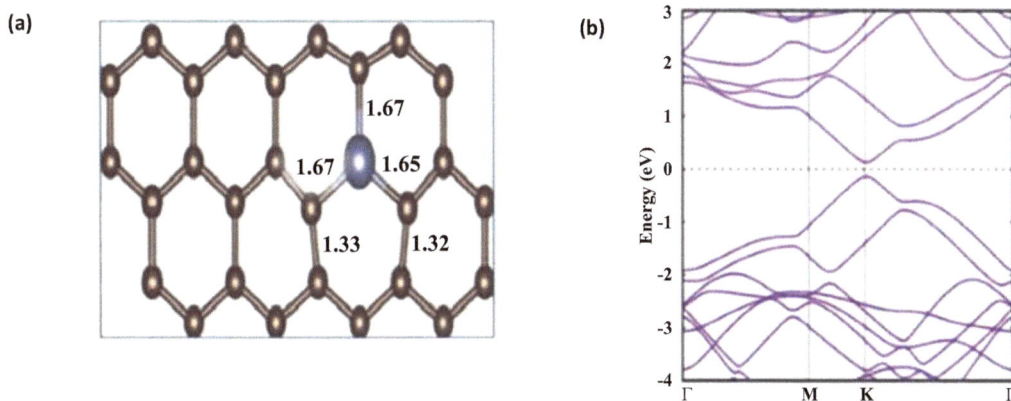

Fig. (20). (a) Geometric structure of Ge doped graphene and **(b)** its band structure (Adopted from reference 49).

Although gallium ($4s^2 4p^1$) dopant causes some structural distortion, the geometry of doped graphene remains essentially planar (Fig. **21a**) [51]. Ga-doped graphene (dopant concentration of 3.125 at%) has large formation energy of 2.81eV [52]. The bond length of Ga-C is 1.71 Å, and the change in C-C bond length is 0.04 Å. On substitutional doping of Ga, the Dirac point of graphene shifts up, and the Fermi level shifts to the valence band. This indicates p-type doping similar to boron doping. Ga dopant with 0.625 at % concentration induces a bandgap of 0.49 eV in graphene (Fig. **21b**). Energy bandgap changes from 0.49 eV to 0.17 eV and 0.1 eV to 0.2 eV by a change in concentration of dopants from 0.625% to 1.875% [51] and 0.781% to 3.125% [52], respectively. Also, Ga dopant shows metallic behaviour in a doped system for a particular concentration [52].

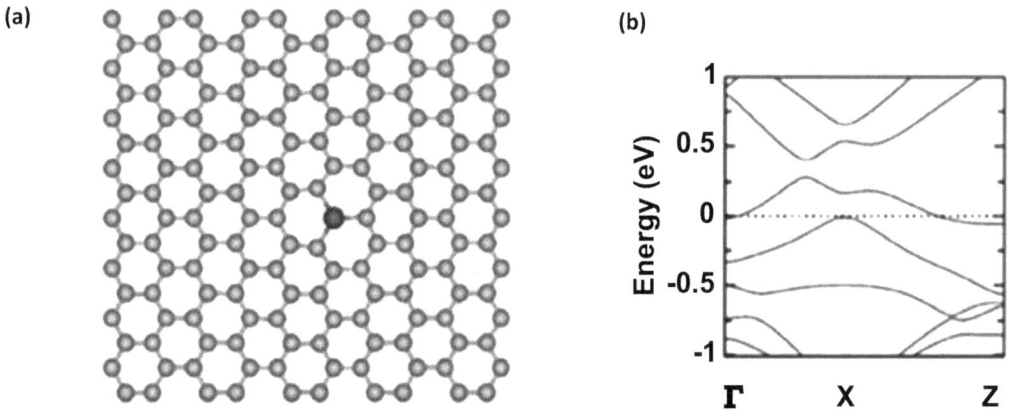

Fig. (21). (a) Geometric structure of Ga-doped graphene and (b) its band structure (Adopted from reference 51).

Arsenic ($4s^2 4p^3$) does not fit into the sp^2 hybridized frame of carbon atoms of graphene due to its large radius but protrudes out of the graphene plane [39], as shown in Fig. (**22a**). The corresponding As-C bond length is 1.875 Å. A bandgap of 0.682 eV is observed (Fig. **22c**) for a doped system with a dopant concentration of 12.5% [39]. Varying the concentration of dopant affects the bandgap values. Bandgap values vary from 0.3 eV to 0.8 eV by changing the dopant concentration from 0.78% to 3.125% [52]. The formation energy of As-doped graphene (at 3.125% dopant concentration) is 0.47 eV [52].

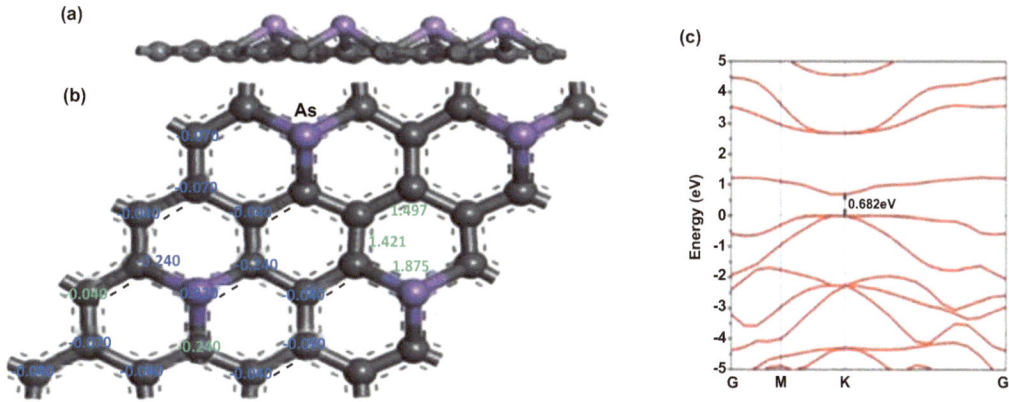

Fig. (22). (a) Side and (b) top view of geometric structure and (c) band structure of As-doped graphene (Adopted from reference 39).

Selenium ($4s^2 4p^4$) atom does not fit into the sp^2 hybridized structure of graphene, as shown in Fig. (23). The bond length Se-C is 1.896 Å. The formation energy of Se-doped graphene (3.125 at% dopant concentration) is 1.59 eV [53]. The system has a bandgap of 0.6 eV at the same dopant concentration. The bandgap varies from 0.6 eV to 0.2 eV with a change in concentration of dopant from 3.125% to 0.78% [52].

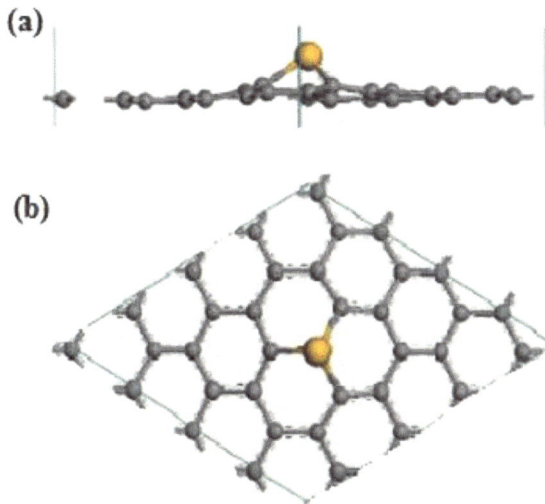

Fig. (23). (a) Side and (b) top view of the geometric structure of Se-doped graphene (Adopted from reference 53).

CONCLUSION

In this chapter, the geometric and electronic properties of graphene nanoribbons and substitutional doped graphene have been discussed along with graphene. Graphene has extraordinary electronic properties and is a zero bandgap semiconductor. The absence of bandgap in graphene limits its application in electronics. Formation of graphene nanoribbons and heteroatom substitutional doping of graphene are the methods to open up an energy bandgap of graphene and hence widen its application in nanoelectronics.

REFERENCES

[1]   Geim, A.K.; Novoselov, K.S. The rise of graphene. In Nanoscience and Technology: a Collection of Reviews from Nature Journals, **2010**; pp. 11-19.

[2]   Geim, A.K. Graphene: Status and Prospects. *Science,* **2009**, *324*(5934), 1530-1534.
       [http://dx.doi.org/10.1126/science.1158877] [PMID: 19541989]

[3]   Castro Neto, A.H.; Guinea, F.; Peres, N.M.R.; Novoselov, K.S.; Geim, A.K. The electronic properties of graphene. *Rvmp,* **2009**, *81*(1), 109-162.

[4]   Novoselov, K.S.; Geim, A.K.; Morozov, S.V.; Jiang, D.; Zhang, Y.; Dubonos, S.V.; Grigorieva, I.V.; Firsov, A.A. Electric field effect in atomically thin carbon films. *Science,* **2004**, *306*(5696), 666-669.
       [http://dx.doi.org/10.1126/science.1102896] [PMID: 15499015]

[5]   Novoselov, K.S.; Fal'ko, V.I.; Colombo, L.; Gellert, P.R.; Schwab, M.G.; Kim, K. A roadmap for graphene. *Nature,* **2012**, *490*(7419), 192-200.
       [http://dx.doi.org/10.1038/nature11458] [PMID: 23060189]

[6]   Ando, T. The electronic properties of graphene and carbon nanotubes. *NPG Asia Mater.,* **2009**, *1*(1), 17-21.
       [http://dx.doi.org/10.1038/asiamat.2009.1]

[7]   Bolotin, K.I.; Sikes, K.J.; Jiang, Z.; Klima, M.; Fudenberg, G.; Hone, J.; Kim, P.; Stormer, H.L. Ultrahigh electron mobility in suspended graphene. *Solid State Commun.,* **2008**, *146*(9-10), 351-355.
       [http://dx.doi.org/10.1016/j.ssc.2008.02.024]

[8]   Chen, H.; Müller, M.B.; Gilmore, K.J.; Wallace, G.G.; Li, D. Mechanically strong, electrically conductive, and biocompatible graphene paper. *Adv. Mater.,* **2008**, *20*(18), 3557-3561.
       [http://dx.doi.org/10.1002/adma.200800757]

[9]   Ostrovsky, P.M.; Gornyi, I.V.; Mirlin, A.D. Theory of anomalous quantum Hall effects in graphene. *Phys. Rev. B Condens. Matter Mater. Phys.,* **2008**, *77*(19), 195430.
       [http://dx.doi.org/10.1103/PhysRevB.77.195430]

[10]   Heersche, H.B.; Jarillo-Herrero, P.; Oostinga, J.B.; Vandersypen, L.M.K.; Morpurgo, A.F. Bipolar supercurrent in graphene. *Nature,* **2007**, *446*(7131), 56-59.
       [http://dx.doi.org/10.1038/nature05555] [PMID: 17330038]

[11]   Novoselov, K.S.; Geim, A.K.; Morozov, S.V.; Jiang, D.; Katsnelson, M.I.; Grigorieva, I.V.; Dubonos, S.V.; Firsov, A.A. Two-dimensional gas of massless Dirac fermions in graphene. *Nature,* **2005**, *438*(7065), 197-200.
       [http://dx.doi.org/10.1038/nature04233] [PMID: 16281030]

[12]   Nakada, K.; Fujita, M.; Dresselhaus, G.; Dresselhaus, M.S. Edge state in graphene ribbons: Nanometer size effect and edge shape dependence. *Phys. Rev. B Condens. Matter,* **1996**, *54*(24), 17954-17961.
       [http://dx.doi.org/10.1103/PhysRevB.54.17954] [PMID: 9985930]

[13]   Liu, H.; Liu, Y.; Zhu, D. Chemical doping of graphene. *J. Mater. Chem.,* **2011**, *21*(10), 3335-3345.

[http://dx.doi.org/10.1039/C0JM02922J]

[14]   Zhou, S.Y.; Gweon, G.H.; Fedorov, A.V.; First, P.N.; de Heer, W.A.; Lee, D.H.; Guinea, F.; Castro Neto, A.H.; Lanzara, A. Substrate-induced bandgap opening in epitaxial graphene. *Nat. Mater.,* **2007**, *6*(10), 770-775.
       [http://dx.doi.org/10.1038/nmat2003] [PMID: 17828279]

[15]   Boukhvalov, D.W.; Katsnelson, M.I. Chemical functionalization of graphene. *J. Phys. Condens. Matter,* **2009**, *21*(34), 344205.
       [http://dx.doi.org/10.1088/0953-8984/21/34/344205] [PMID: 21715780]

[16]   Jafri, S.H.M.; Carva, K.; Widenkvist, E.; Blom, T.; Sanyal, B.; Fransson, J.; Eriksson, O.; Jansson, U.; Grennberg, H.; Karis, O.; Quinlan, R.A.; Holloway, B.C.; Leifer, K. Conductivity engineering of graphene by defect formation. *J. Phys. D Appl. Phys.,* **2010**, *43*(4), 045404.
       [http://dx.doi.org/10.1088/0022-3727/43/4/045404]

[17]   Wang, Z.F.; Li, Q.; Zheng, H.; Ren, H.; Su, H.; Shi, Q.W.; Chen, J. Tuning the electronic structure of graphene nanoribbons through chemical edge modification: A theoretical study. *Phys. Rev. B Condens. Matter Mater. Phys.,* **2007**, *75*(11), 113406.
       [http://dx.doi.org/10.1103/PhysRevB.75.113406]

[18]   Wurm, J.; Wimmer, M.; Adagideli, İ.; Richter, K.; Baranger, H.U. Interfaces within graphene nanoribbons. *New J. Phys.,* **2009**, *11*(9), 095022.
       [http://dx.doi.org/10.1088/1367-2630/11/9/095022]

[19]   Chen, Y.C. *Exploring graphene nanoribbons using scanning Probe Microscopy and Spectroscopy,* **2014**.

[20]   Son, Y.W.; Cohen, M.L.; Louie, S.G. Energy gaps in graphene nanoribbons. *Phys. Rev. Lett.,* **2006**, *97*(21), 216803.
       [http://dx.doi.org/10.1103/PhysRevLett.97.216803] [PMID: 17155765]

[21]   Zheng, H.; Wang, Z.F.; Luo, T.; Shi, Q.W.; Chen, J. Analytical study of electronic structure in armchair graphene nanoribbons. *Phys. Rev. B Condens. Matter Mater. Phys.,* **2007**, *75*(16), 165414.
       [http://dx.doi.org/10.1103/PhysRevB.75.165414]

[22]   Kan, E.; Li, Z.; Yang, J.; Hou, J.G. Half-metallicity in edge-modified zigzag graphene nanoribbons. *J. Am. Chem. Soc.,* **2008**, *130*(13), 4224-4225.
       [http://dx.doi.org/10.1021/ja710407t] [PMID: 18331034]

[23]   Yazyev, O.V.; Capaz, R.B.; Louie, S.G. Theory of magnetic edge states in chiral graphene nanoribbons. *Phys. Rev. B Condens. Matter Mater. Phys.,* **2011**, *84*(11), 115406.
       [http://dx.doi.org/10.1103/PhysRevB.84.115406]

[24]   Tao, C.; Jiao, L.; Yazyev, O.V.; Chen, Y.C.; Feng, J.; Zhang, X.; Capaz, R.B.; Tour, J.M.; Zettl, A.; Louie, S.G.; Dai, H.; Crommie, M.F. Spatially resolving edge states of chiral graphene nanoribbons. *Nat. Phys.,* **2011**, *7*(8), 616-620.
       [http://dx.doi.org/10.1038/nphys1991]

[25]   Jiang, Z.; Song, Y. Band gap oscillation and novel transport property in ultrathin chiral graphene nanoribbons. *Physica B,* **2015**, *464*, 61-67.
       [http://dx.doi.org/10.1016/j.physb.2015.02.003]

[26]   Serraon, A.C.F.; Del Rosario, J.A.D.; Abel Chuang, P.Y.; Chong, M.N.; Morikawa, Y.; Padama, A.A.B.; Ocon, J.D. Alkaline earth atom doping-induced changes in the electronic and magnetic properties of graphene: a density functional theory study. *RSC Advances,* **2021**, *11*(11), 6268-6283.
       [http://dx.doi.org/10.1039/D0RA08115A] [PMID: 35423162]

[27]   Rafique, M.; Shuai, Y.; Tan, H.P.; Hassan, M. Manipulating intrinsic behaviors of graphene by substituting alkaline earth metal atoms in its structure. *RSC Advances,* **2017**, *7*(27), 16360-16370.
       [http://dx.doi.org/10.1039/C7RA01406F]

[28]   Rani, B.; Jindal, V.K.; Dharamvir, K. Energetics of a Li Atom adsorbed on B/N doped graphene with

monovacancy. *J. Solid State Chem.,* **2016**, *240*, 67-75.
[http://dx.doi.org/10.1016/j.jssc.2016.05.014]

[29]   Wu, M.; Cao, C.; Jiang, J.Z. Light non-metallic atom (B, N, O and F)-doped graphene: a first-principles study. *Nanotechnology,* **2010**, *21*(50), 505202.
[http://dx.doi.org/10.1088/0957-4484/21/50/505202] [PMID: 21098927]

[30]   Rani, P.; Jindal, V.K. Designing band gap of graphene by B and N dopant atoms. *RSC Advances,* **2013**, *3*(3), 802-812.
[http://dx.doi.org/10.1039/C2RA22664B]

[31]   Ullah, S.; Hussain, A.; Syed, W.; Saqlain, M.A.; Ahmad, I.; Leenaerts, O.; Karim, A. Band-gap tuning of graphene by Be doping and Be, B co-doping: a DFT study. *RSC Advances,* **2015**, *5*(69), 55762-55773.
[http://dx.doi.org/10.1039/C5RA08061D]

[32]   Kaykılarlı, C.; Uzunsoy, D.; Parmak, E.D.Ş.; Fellah, M.F.; Çakır, Ö.Ç. Boron and nitrogen doping in graphene: an experimental and density functional theory (DFT) study. *Nano Express,* **2020**, *1*(1), 010027.
[http://dx.doi.org/10.1088/2632-959X/ab89e9]

[33]   Granzier-Nakajima, T.; Fujisawa, K.; Anil, V.; Terrones, M.; Yeh, Y.T. Controlling nitrogen doping in graphene with atomic precision: Synthesis and characterization. *Nanomaterials (Basel),* **2019**, *9*(3), 425.
[http://dx.doi.org/10.3390/nano9030425] [PMID: 30871112]

[34]   Olaniyan, O.; Maphasha, R.E.; Madito, M.J.; Khaleed, A.A.; Igumbor, E.; Manyala, N. A systematic study of the stability, electronic and optical properties of beryllium and nitrogen co-doped graphene. *Carbon,* **2018**, *129*, 207-227.
[http://dx.doi.org/10.1016/j.carbon.2017.12.014]

[35]   Goudarzi, M.; Parhizgar, S.S.; Beheshtian, J. Electronic and optical properties of vacancy and B, N, O and F doped graphene: DFT study. *Opto-Electron. Rev.,* **2019**, *27*(2), 130-136.
[http://dx.doi.org/10.1016/j.opelre.2019.05.002]

[36]   Ullah, S.; Hussain, A.; Sato, F. Rectangular and hexagonal doping of graphene with B, N, and O: a DFT study. *RSC Advances,* **2017**, *7*(26), 16064-16068.
[http://dx.doi.org/10.1039/C6RA28837E]

[37]   Zhou, X.; Zhao, C.; Wu, G.; Chen, J.; Li, Y. DFT study on the electronic structure and optical properties of N, Al, and N-Al doped graphene. *Appl. Surf. Sci.,* **2018**, *459*, 354-362.
[http://dx.doi.org/10.1016/j.apsusc.2018.08.015]

[38]   Denis, P.A. Band gap opening of monolayer and bilayer graphene doped with aluminium, silicon, phosphorus, and sulfur. *Chem. Phys. Lett.,* **2010**, *492*(4-6), 251-257.
[http://dx.doi.org/10.1016/j.cplett.2010.04.038]

[39]   Wang, Y.; Wang, W.; Zhu, S.; Yang, G.; Zhang, Z.; Li, P. Theoretical studies on the structures and properties of doped graphenes with and without an external electrical field. *RSC Advances,* **2019**, *9*(21), 11939-11950.
[http://dx.doi.org/10.1039/C9RA00326F] [PMID: 35517038]

[40]   Kaminski, N. State of the art and the future of wide band-gap devices. **2009**.

[41]   Rafique, M.; Shuai, Y.; Hussain, N. First-principles study on silicon atom doped monolayer graphene. *Physica E,* **2018**, *95*, 94-101.
[http://dx.doi.org/10.1016/j.physe.2017.09.012]

[42]   Qu, Y.; Ding, J.; Fu, H.; Chen, H.; Peng, J. Investigation on tunable electronic properties of semiconducting graphene induced by boron and sulfur doping. *Appl. Surf. Sci.,* **2021**, *542*, 148763.
[http://dx.doi.org/10.1016/j.apsusc.2020.148763]

[43]   Garcia, A.G.; Baltazar, S.E.; Castro, A.H.R.; Robles, J.F.P.; Rubio, A. Influence of S and P doping in

a graphene sheet. *J. Comput. Theor. Nanosci.,* **2008**, *5*(11), 2221-2229.
[http://dx.doi.org/10.1166/jctn.2008.1123]

[44]   Olaniyan, O.; Mapasha, R.E.; Momodu, D.Y.; Madito, M.J.; Kahleed, A.A.; Ugbo, F.U.; Bello, A.; Barzegar, F.; Oyedotun, K.; Manyala, N. Exploring the stability and electronic structure of beryllium and sulphur co-doped graphene: a first principles study. *RSC Advances,* **2016**, *6*(91), 88392-88402.
[http://dx.doi.org/10.1039/C6RA17640B]

[45]   Lee, J.; Kwon, S.; Kwon, S.; Cho, M.; Kim, K.; Han, T.; Lee, S. Tunable electronic properties of nitrogen and sulfur doped graphene: Density functional theory approach. *Nanomaterials (Basel),* **2019**, *9*(2), 268.
[http://dx.doi.org/10.3390/nano9020268] [PMID: 30781379]

[46]   Rani, B.; Bubanja, V.; Jindal, V.K. Atomistic insights into lithium adsorption and migration on phosphorus☐doped graphene. *Int. J. Quantum Chem.,* **2021**, *121*(14), e26659.
[http://dx.doi.org/10.1002/qua.26659]

[47]   Mohammed, M.H. Designing and engineering electronic band gap of graphene nanosheet by P dopants. *Solid State Commun.,* **2017**, *258*, 11-16.
[http://dx.doi.org/10.1016/j.ssc.2017.04.011]

[48]   Dai, X.S.; Shen, T.; Liu, H.C. DFT study on electronic and optical properties of graphene modified by phosphorus. *Mater. Res. Express,* **2019**, *6*(8), 085635.
[http://dx.doi.org/10.1088/2053-1591/ab29bc]

[49]   Rafique, M.; Yong, S.; Ali, I.; Ahmed, I. Germanium Atom Substitution in Monolayer Graphene: A First-principles Study *IOP Conference Series: Materials,* **2018**.

[50]   Ould Ne, M.L.; Abbassi, A.; El hachimi, A.G.; Benyoussef, A.; Ez-Zahraouy, H.; El Kenz, A. Electronic optical, properties and widening band gap of graphene with Ge doping. *Opt. Quantum Electron.,* **2017**, *49*(6), 218.
[http://dx.doi.org/10.1007/s11082-017-1024-5]

[51]   Brito, E.; Leite, L.; Azevedo, S.; Martins, J.R.; Kaschny, J.R. Simulated annealing and first-principles study of substitutional Ga-doped graphene. *Appl. Phys., A Mater. Sci. Process.,* **2019**, *125*(1), 35.
[http://dx.doi.org/10.1007/s00339-018-2330-x]

[52]   Denis, P.A. Chemical reactivity and band-gap opening of graphene doped with gallium, germanium, arsenic, and selenium atoms. *ChemPhysChem,* **2014**, *15*(18), 3994-4000.
[http://dx.doi.org/10.1002/cphc.201402608] [PMID: 25349028]

[53]   Gholizadeh, R.; Yu, Y.X. Work functions of pristine and heteroatom-doped graphenes under different external electric fields: an ab initio DFT study. *J. Phys. Chem. C,* **2014**, *118*(48), 28274-28282.
[http://dx.doi.org/10.1021/jp5095195]

CHAPTER 15

Strain-Induced 2D Materials

Isha Mudahar[1,*] and **Sandeep Kaur**[1]

[1] *Department of Basic and Applied Sciences, Punjabi University, Patiala-147 002, Punjab, India*

Abstract: In this chapter, different structural, electronic and magnetic properties of strained graphene nanoribbons are examined. All the calculations are performed by using density functional theory. Compressive stress along a nanoribbon's longer axis and tensile stress at the midpoint and perpendicular to the nanoribbon's plane are studied. There are remarkable changes in the structures, including the formation of nanoripples in the ribbons. The shape and size of the ribbons lead to variation in their electronic and magnetic properties. Strained nanoribbons show tunable magnetic properties that can be used for developing magnetic nano-switches.

Keywords: Density functional theory, Graphene nanoribbons, Magnetism, Nanoripples, Strain.

INTRODUCTION

Semiconductors form the basic components of many important devices used in the fields like electronics, sensors, computers, spintronics, *etc.* The parameter which decides the semiconducting, electronic and optical properties of different materials is bandgap. Bandgap engineering or modulation of materials has shown completely different properties from their respective constituents.

At the nanoscale, the bandgap plays a very significant role as it modifies the optoelectronic properties, leading to other exciting applications. One more factor that can modulate the bandgap is the effect of the strain. Different 2D materials show different behaviour under the influence of strain. For example, uniaxial strain is applied by either stretching or bending the substrate beneath the 2D material or by the introduction of ripples in the material. Strain engineering has recently been utilised to construct 2D semiconductor-based straintronic devices. This chapter focuses on the effect of axial strain on different properties of graphene nanoribbons.

---

[*] **Corresponding author Isha Mudahar:** Department of Basic and Applied Sciences, Punjabi University, Patiala-147 002, Punjab, India; Email: dr.ishamudahar@gmail.com

Graphene nanoribbons (GNRs) are quasi-one-dimensional nanostructures with distinct electronic, optical, and transport properties [1 - 3]. As a result, several investigations have been conducted on their prospective uses as FETs, spintronics and optoelectronic devices [4 - 6]. This area of GNRs has attracted a lot of interest as their characteristics can be controlled by applying external electric and magnetic fields, doping, defects and edge-modification [7 - 9]. GNRs have also been used in hybrid materials, such as vanadium dioxide-graphene structures [10], and have demonstrated exceptional electrochemical performance, indicating that they might be used in high-power lithium batteries as an electrode material. The finding of graphene [11] has paved the way for the discovery of other two-dimensional materials [12], which show promise in sectors like photonics [13] and nanomedicine [14]. The counterparts of graphene nanoribbons, i.e., silicene [15, 16], stanene [17], phosphorene [18], germanene [19, 20] and new monolayer materials [21], have been explored for their electronic and magnetic properties by applying density functional theory and other first principle methods.

GNRs have been successfully synthesized using a variety of top-down techniques, such as graphene etching [22], chemical vapour deposition [23], scanning probe lithography [24] and unzipping of single- [25] and multi-walled carbon nanotubes [26]. Still, their specific electrical structure cannot be completely predicted by the above-mentioned approaches; therefore, a bottom-up approach can be applied. A significant work [27] is based on some suitably designed precursor molecules which react in a selective way to produce GNRs [28, 29] and has paved the way for a number of techniques to form GNRs. Furthermore, the method has been used in solution [30] and supported on solid surfaces [31].

It has been observed that graphene has the greatest stiffness and strength among known materials and hence shows exceptional stress-strain behaviour [32, 34]. It can be bent readily to produce intricate folded shapes [35], and it can endure elastic deformations of up to 25% [32], which is far bigger than any other known material. Graphene is an excellent contender for nanomechanical systems [36] and flexible electronic devices [37] due to its superior mechanical characteristics. Measurements of the elastic response of graphene indicated that it is extremely nonlinear for stresses exceeding 10% [32]. Interpretation of the experiments in terms of a generalised nonlinear stress-strain relationship includes cubic components in strain, and nonlinear elastic coefficients are calculated using atomistic simulations [33].

Owing to the lattice mismatch [38] or the substrate's surface corrugation [39], stresses appear impulsively when graphene is put on a substrate. Moreover, intrinsic edge stress exists along the edges of graphene, which causes warping instability [40]. Strain can be purposely caused and regulated in addition to naturally occurring strain. Bending the flexible substrate can cause uniaxial strain [41]. Biaxial strain can be induced by exploiting the temperature mismatch between graphene and the substrate [43], shrinking or elongating the piezoelectric substrate with a bias voltage [42], or pushing graphene clamped on top of a hole in the substrate with an atomic force microscope tip [32].

The quantized pseudomagnetic field [44], zero-field quantum Hall effect [45], increased electron-phonon coupling [46], and shifting of Dirac cones [47] are only a few of the remarkable physical phenomena of graphene generated by the strain. While graphene's bandgap remains near zero even under enormous strain, GNRs' bandgap is extremely sensitive to both uniaxial and shear strain. Uniaxial weak strain alters the band linearly for an armchair GNR, while significant strain causes periodic oscillation of the bandgap. Shear strain, on the other hand, tends to narrow the bandgap. Strain changes the spin polarisation at the edges of ribbons, modifying their bandgap in a zigzag GNR [48].

Due to an applied uniaxial strain, the planar structure of GNRs is distorted by the production of ripples [49 - 51]. Furthermore, molecular dynamics investigations have demonstrated that applied strain may readily regulate the amplitude and orientation of ripples [52, 53]. Additionally, spin-polarized first principle calculations have revealed that zigzag GNRs are magnetic, but armchair GNRs are not [54]. The magnetic moment is generated along zigzag edges with sinusoidal deformations when strain is applied to magnetic GNRs [55].

Under the influence of strain, the structural, electrical, and magnetic characteristics of the armchair and zigzag graphene nanoribbons change. Nanoribbons with square, triangular, rectangular, or circular geometries with their core atom exposed to focused load have varied characteristics.

The geometric and electrical features of graphene and small fullerenes were investigated, and it was found that they were consistent with prior research [60 - 64]. The same parameters to get the stable structures of GNRs were used.

GNR ground-state structures were optimized for axial loads applied at the two parallel edges of rectangular GNRs and a load applied at the centre and perpendicular to a plane of square, triangular, rectangular, and circular nanoribbons. In the case of axial load, the three types of boundary conditions were considered: three, one, or two atoms of the hexagons at the edges were relocated and stayed fixed (Fig. **1**). The atoms were pushed inwards in steps of 0.2 Å along the longer edge of a 3 nm by 1 nm nanoribbon, resulting in compressive strains ranging from 1.3% to 6.7%. The free boundary conditions were applied at the two unloaded edges in all three cases of axial loads. In the case of the point load, the effects of pulling the central C atom out of the nanoribbon plane by up to 5 Å were considered.

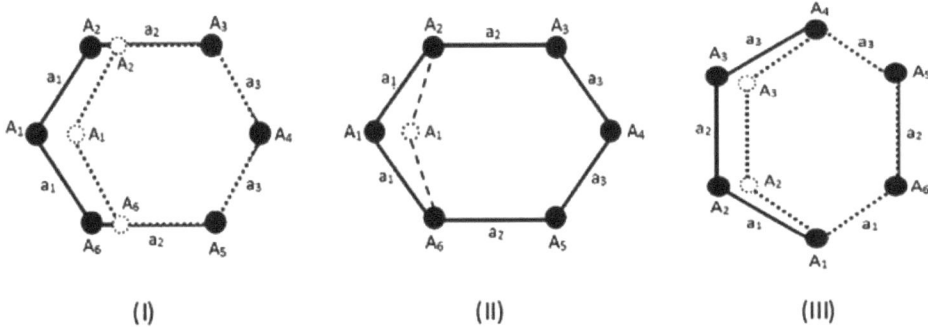

Fig. (1). Types of applied axial stress.

The ground state configurations and binding energies per atom (B.E./atom) of GNRs are listed below for each arrangement. The density of states (DOS), HOMO-LUMO gaps, total magnetic moments (TMM), and localised magnetic moments (LMM) of GNRs were calculated using spin-polarized calculations.

GRAPHENE NANORIBBONS UNDER AXIAL COMPRESSION

The following six cases are investigated in order to evaluate the edge effects on the properties of nanoribbons under axial compression: $(14,5)^{(I)}$, $(14,5)^{(II)}$, $(6,13)^{(III)}$, $(\mathbf{16},4)^{(I)}$, $(\mathbf{16},4)^{(II)}$ and $(6,13)^{(III)}$, where the notation $(N_a, N_z)^{(i)}$ designates the nanoribbon with N_a and N_z carbon atoms along the armchair and zigzag borders, respectively, under type-i boundary conditions (Fig. **2** and Fig. **3**); the edges passivated by the hydrogen atoms are marked in bold font (Fig. **4**).

Structural Properties

Table **1** shows that configurations $(14,5)^{(I)}$, $(14,5)^{(II)}$, and $(6,13)^{(III)}$ are more stable than the other three GNRs in terms of binding energies per atom [65].

Table 1. Binding energies per atom in eV for graphene nanoribbons under axial compression.

Strain (%)	$(14,5)^{(I)}$	$(14,5)^{(II)}$	$(6,13)^{(III)}$	$(\mathbf{16},4)^{(I)}$	$(\mathbf{16},4)^{(II)}$	$(6,13)^{(III)}$
0.0	8.30	8.30	7.87	7.63	7.63	8.23
1.3	8.29	8.30	7.87	7.61	7.62	8.22
2.7	8.27	8.28	7.86	7.60	7.61	8.21

(Table 1) cont.....

4.0	8.28	8.27	7.84	7.58	7.60	8.19
5.3	8.22	8.24	7.81	7.56	7.57	8.16
6.7	8.27	8.28	7.76	7.61	7.62	8.11

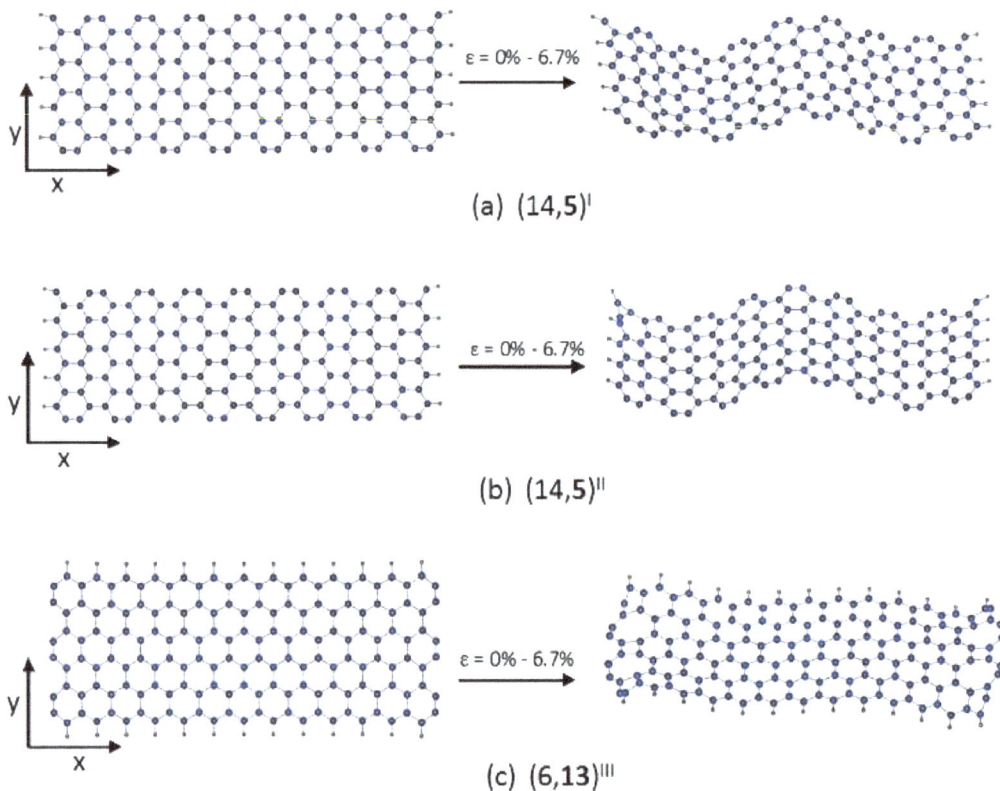

Fig. (2). Structure of nanoribbons; **(a)** $(14,5)^{I}$, **(b)** $(14, 5)^{II}$, **(c)** $(6, 13)^{III}$.

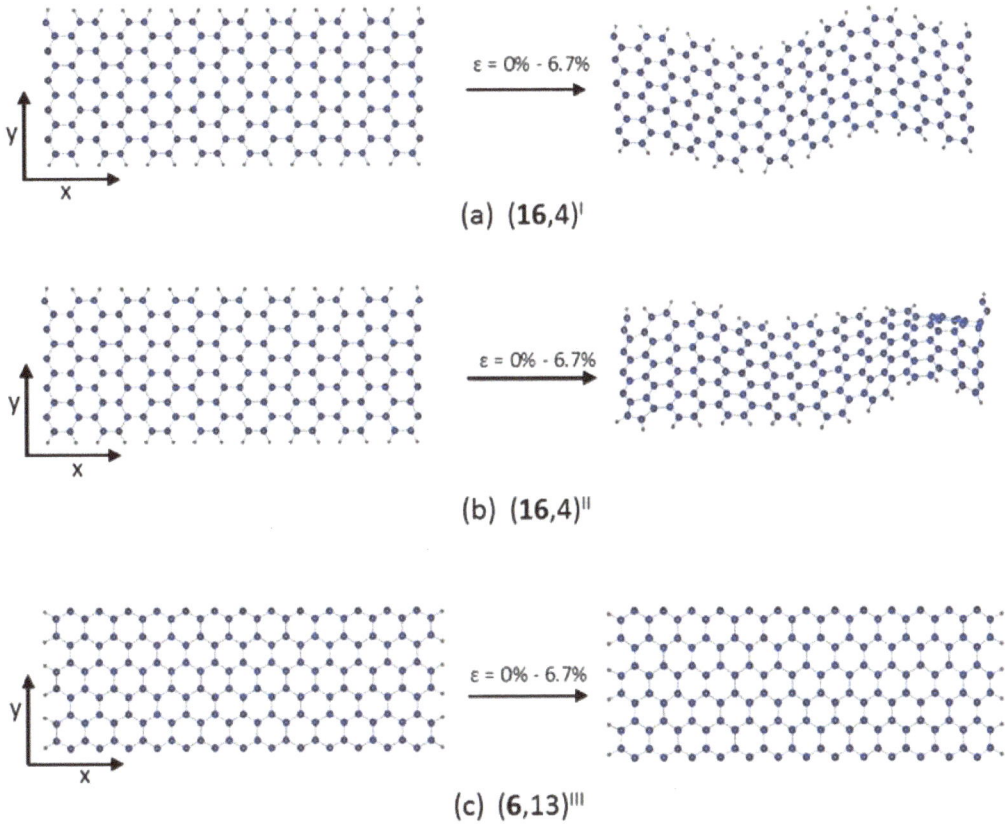

(a) (16,4)^{I}

(b) (16,4)^{II}

(c) (6,13)^{III}

Fig. (3). Structure of nanoribbons; **(a)** (16, 4)^{I}, **(b)** (16, 4)^{II}, **(c)** (6, 13)^{III}.

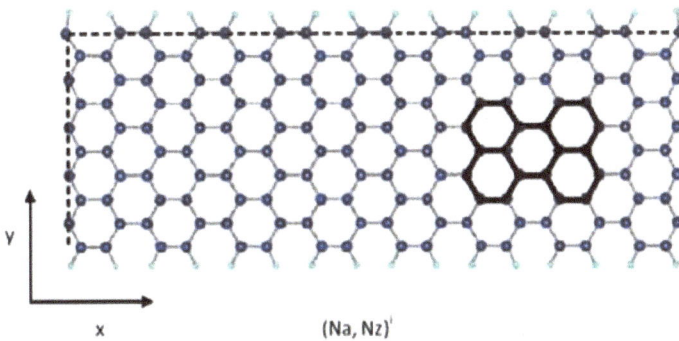

(Na, Nz)^{i}

Fig. (4). Structure of GNR, where Na and Nz are counted along dashed lines, and i is the type of applied stress. The highlighted area is shown in Fig. (**5**).

The overall stability of GNRs diminishes as strain increases. The bond lengths and bond angles, as shown in Fig. (**5**), are used to investigate the geometry of the ground state structures. The change in bond lengths and bond angles for a single vertex of a hexagon of a GNR are shown in Figs. (**6** and **7**), respectively. Bond length changes are on the order of 3%, while bond angle changes are on the order of 7%. For strains of up to 5.3 percent, bonds along the axis of the applied tension are shortened, while those perpendicular to the applied stress are stretched. With the exception of (**6**,13)$^{(III)}$, which remains planar, atom displacement in the vertical direction occurs at 5.3 percent strain, causing ripples to form. GNRs begin to take on a sinusoidal shape, as predicted by continuum mechanics theory. Vertical displacement ranges from 1.3 for (**6**,13)^{III} to 2.4 for (**16**,4)^{II} for a 6.7 percent strain. The carbon atoms distant from the boundaries tend to keep the hexagonal form of the lattice due to their freedom of mobility in the perpendicular direction to the GNR plane. Some of the bonds are broken towards the edges, forming octagonal forms [65].

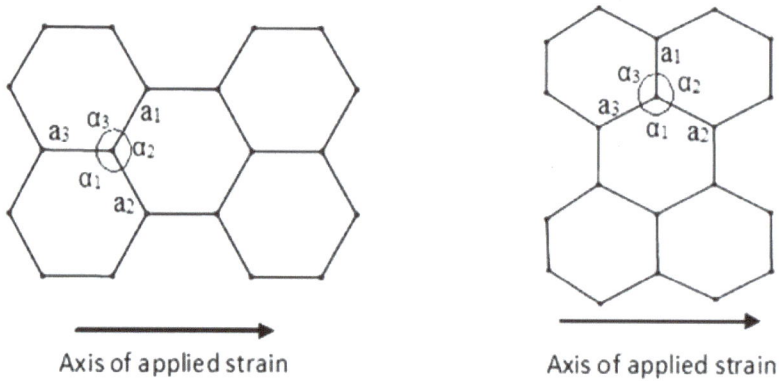

Fig. (5). Atomic structure of ribbons along armchair and zigzag edges, where a_1, a_2 and a_3 are bond lengths and α_1, α_2, and α_3 are bond angles, respectively.

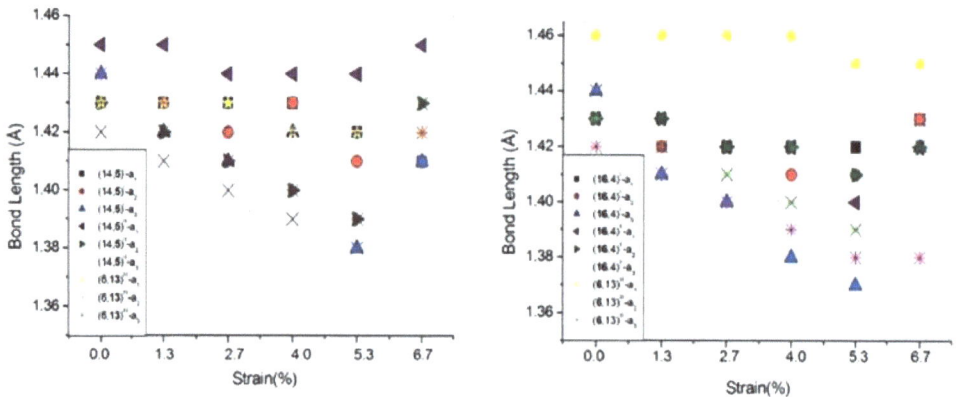

Fig. (6). Variation in bond lengths of nanoribbons with applied stress.

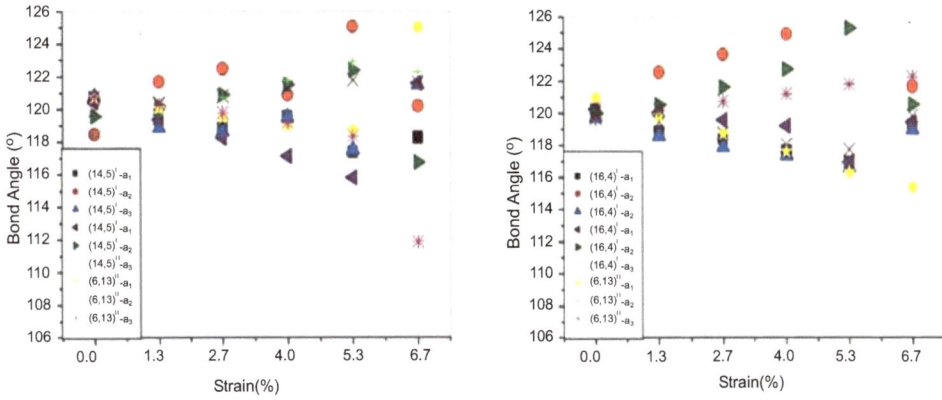

Fig. (7). Variation in bond angles of nanoribbons with applied stress.

Electronic and Magnetic Properties

Fig. (**8**) depicts the total magnetic moments as a function of strain for all GNRs considered. With the exception of (**16**,4)$^{(I)}$ and (**6**,13)$^{(III)}$, TMMs vary dramatically from the 5.3 percent strain. The states along the zigzag edges are responsible for the magnetism in GNRs [65]. When axial stress is applied, the structural distortions suppressing these states cause TMMs to drop in (**14**,5)$^{(II)}$ and (**16**,4)$^{(II)}$, whereas edge states appear in (**14**,5)$^{(I)}$ and (6, **13**)$^{(III)}$, causing TMMs to increase. Figs. (**9** and **10**) show these observations clearly.

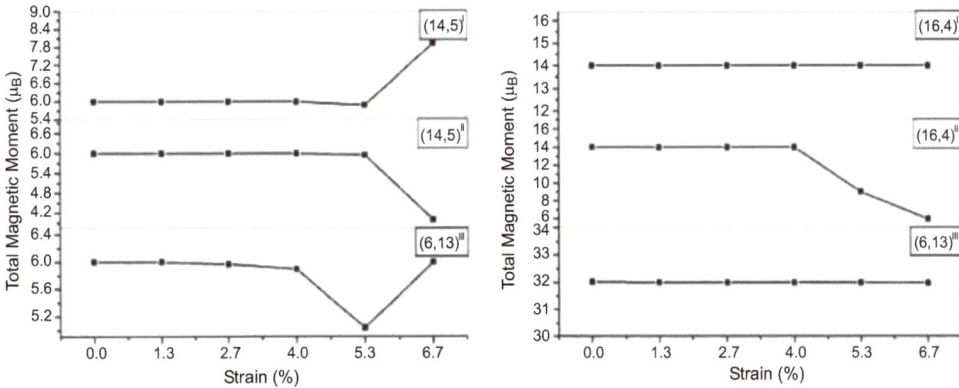

Fig. (8). Total magnetic moments for nanoribbons with applied axial stress.

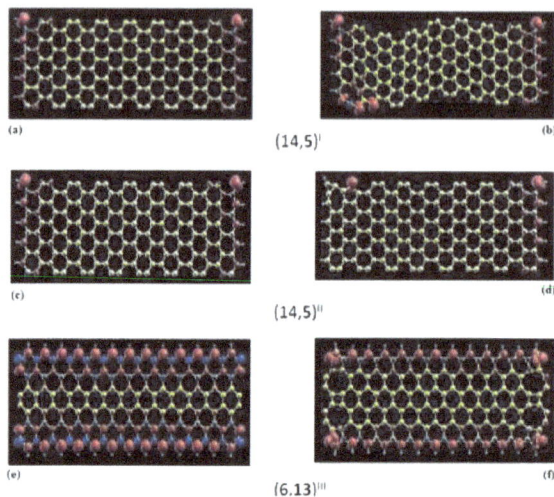

Fig. (9). Spin density maps for nanoribbons; pristine **(a)** $(14,\mathbf{5})^{\mathrm{I}}$, **(c)** $(14,\ \mathbf{5})^{\mathrm{II}}$, **(e)** $(6,\ \mathbf{13})^{\mathrm{III}}$ and strained **(b)** $(14,\mathbf{5})^{\mathrm{I}}$, **(d)** $(14,\mathbf{5})^{\mathrm{II}}$, **(f)** $(6,\ 13)^{\mathrm{III}}$.

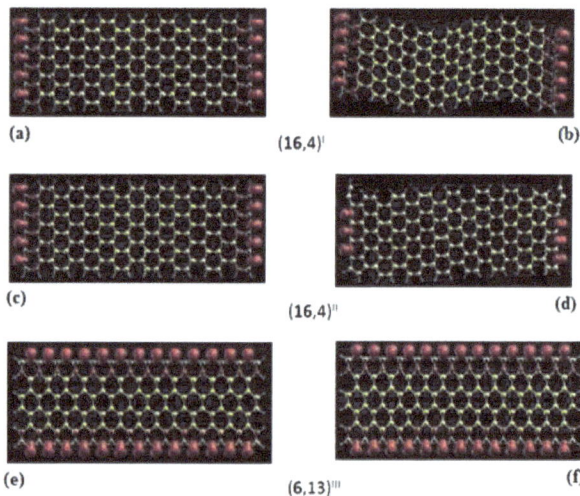

Fig. (10). Spin density maps for nanoribbons; pristine **(a)** $(\mathbf{16},\ 4)\mathrm{I}$, **(c)** $(\mathbf{16},\ 4)\mathrm{II}$, **(e)** $(6,\ 13)\mathrm{III}$ and strained **(b)** $(\mathbf{16},\ 4)\mathrm{I}$, **(d)** $(\mathbf{16},\ 4)\mathrm{II}$, **(f)** $(6,\ 13)\mathrm{III}$.

Fig. (**11**) depicts the HOMO-LUMO energy gaps for spin-up and spin-down electronic states. The HOMO-LUMO gaps in both spin up and spin down states increase linearly with strain for $(6,13)^{\mathrm{III}}$ GNR, while it varies non-uniformly in other cases. The magnetic behaviour of the GNRs under consideration is confirmed by the finite energy difference between spin-up and spin-down HOMO-LUMO gaps. The DOS plots (Figs. **12** and **13**) revealed an unbalanced distribution of spin-up and spin-down states, which leads to all GNRs' magnetic behaviour. The energy levels become available around the Fermi level when the applied strain rises to 5.3%, resulting in an increase in conductivity. GNR conductivity decreases at 5.3% strain [65].

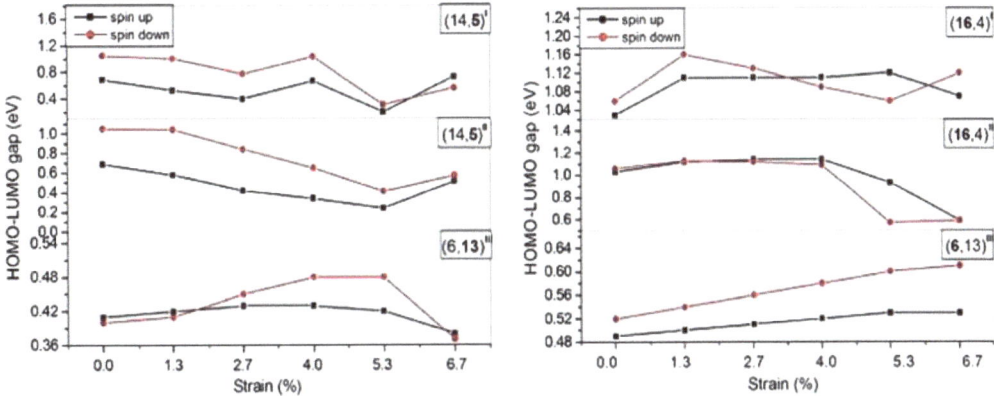

Fig. (11). HOMO-LUMO gaps in spin up and down states for nanoribbons with applied stress.

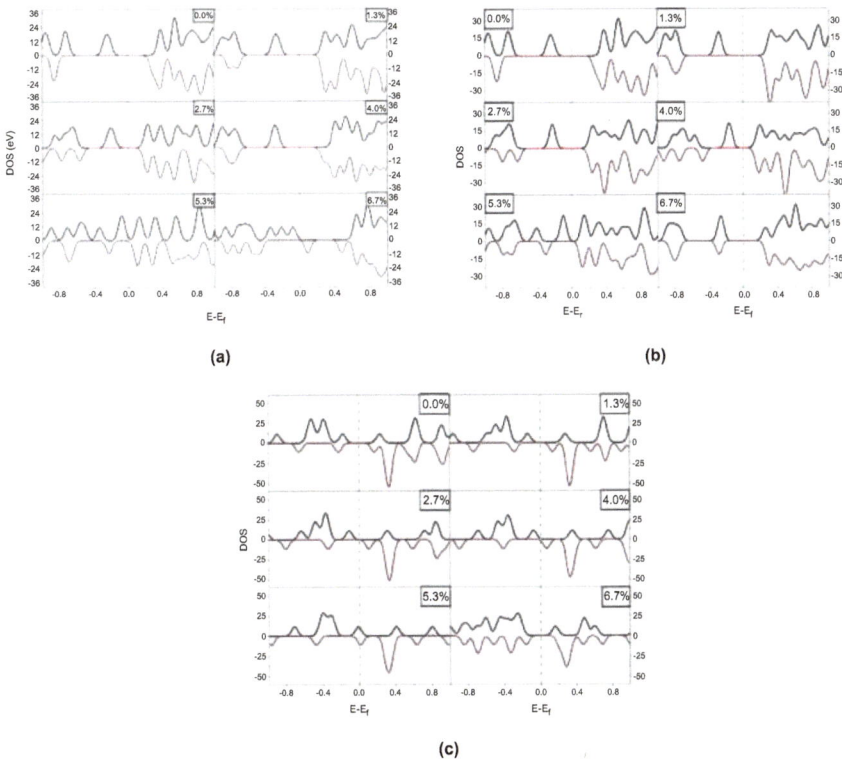

Fig. (12). Density of states for nanoribbons; **(a)** (14,5)[I], **(b)** (14, 5)[II], **(c)** (6, 13)[III]. The black and red lines correspond to DOS for the spin-up and spin-down electronic states.

GRAPHENE NANORIBBONS UNDER CONCENTRATED LOAD

Structural Properties

Point stress is given to GNRs of square, triangular, rectangular, and circular forms by pulling the centre carbon atom out of the nanoribbon plane by 5 Å. Fig. (**14**) depicts the pristine and relaxed structures after applying stress.

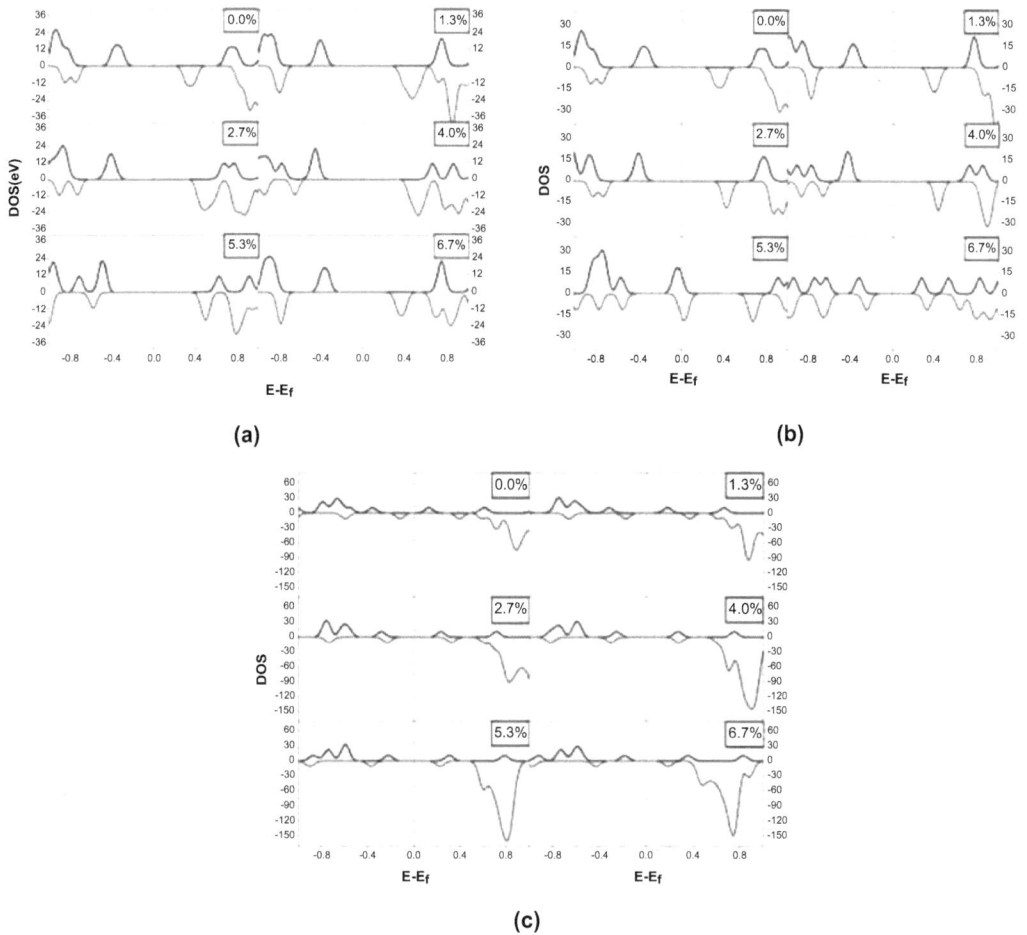

Fig. (13). Density of states for nanoribbons; (**a**) (**16**, 4)[I], (**b**) (**16**, 4)[II], (**c**) (**6**, 13)[III]. The black and red lines correspond to DOS for the spin-up and spin-down electronic states.

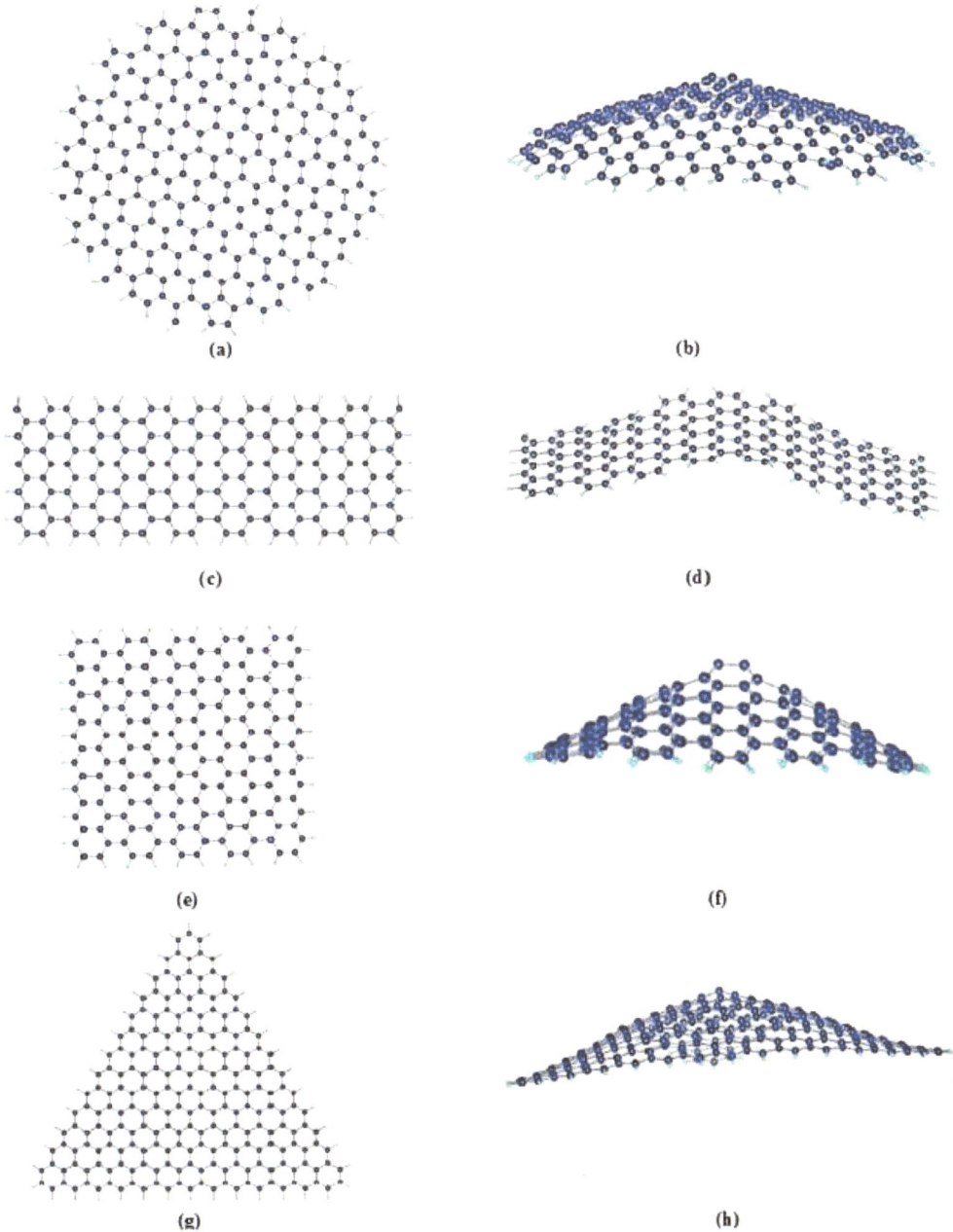

Fig. (14). Graphene naoribbons before [(a),(c),(e),g] and after [(b),(d),(f),h] applied point stress.

Table **2** shows the related structural parameters. For all forms except for square, the predicted binding energies per atom drop when the tension is applied, where the binding energy remains constant. For both pristine and stressed GNR, the C-C bond lengths of rectangular ribbon are in the range of 1.37-1.46 Å [65]. However, For the other three GNRs, C-C bond lengths rise dramatically as a function of applied strain (Table **2**). The bond angles for the rectangular ribbon change slightly as strain increases, but the bond angles for the other three GNRs change dramatically.

Table 2. Structural parameters for graphene nanoribbons under point stress.

Ribbon Shape	Pristine			Strained		
	B.E./atom (eV)	Bond Lengths (Å)	Bond Angles (°)	B.E./atom (eV)	Bond Lengths (Å)	Bond Angles (°)
Rectangle	7.42	1.37-1.46	117.89-123.12	7.42	1.38-1.46	117.83-123.02
Square	7.68	1.38-1.45	118.08-121.83	7.61	1.35-1.65	102.67-124.36
Triangle	7.81	1.40-1.44	118.73-122.38	7.79	1.39-1.54	110.97-122.92
Circle	7.92	1.35-1.48	111.73-125.35	7.88	1.34-1.56	108.63-126.28

Electronic and Magnetic Properties

All four distinct shaped GNRs are subjected to spin-polarized calculations. Fig. (**15**) depicts the TMMs and HOMO-LUMO gaps for the spin-up and spin-down states. The TMM for the circular-shaped GNR increases from 6.13 μ_B to 6.78 μ_B when strain is applied, while it remains the same in the other three cases [65].

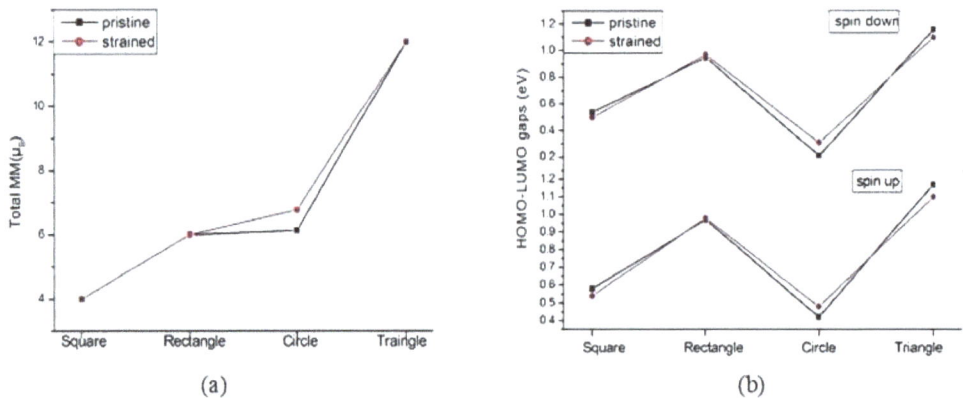

Fig. (15). Total Magnetic Moments and HOMO-LUMO gaps for nanoribbons with applied point stress.

Fig. (**16**) shows that for the square, rectangular, and circular ribbons, no local MM is generated at the central C site under strain. However, for the triangular-shaped ribbon, localised MM is induced at the centre atom of the ribbon.

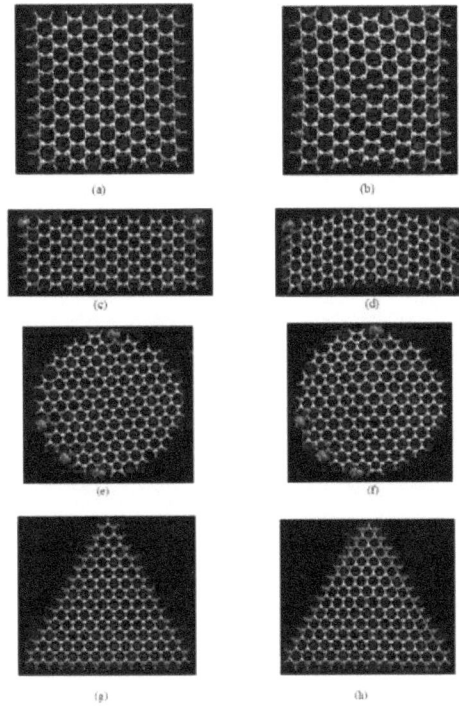

Fig. (16). Spin density maps for nanoribbons before [(a),(c),(e),(g)] and after [(b),(d),(f),(h)] applied point axial stress.

The edge atoms provide a significant contribution to TMM. In both spin-up and spin-down electronic states, the HOMO-LUMO gaps for both pristine and strained nanoribbons display a similar pattern. For circular and rectangular nanoribbons, applied strain causes an increase in HOMO-LUMO gaps, but for square and triangular ribbons, strain causes a decrease in HOMO-LUMO gaps.

Fig. (**17**) depicts the density of states for the four graphene nanoribbon configurations investigated. Plots show that uneven spin density states cause the magnetic behaviour of all four types of GNRs. DOS plots also demonstrate that as the point stress is applied, there is redistribution and spin polarization of electrons in the spin-up and down states [65]. The structural, magnetic, and electronic characteristics of graphene nanoribbons are shape-dependent and may be modified by the applied point strain, indicating that they could be helpful in spintronic devices and stress sensors.

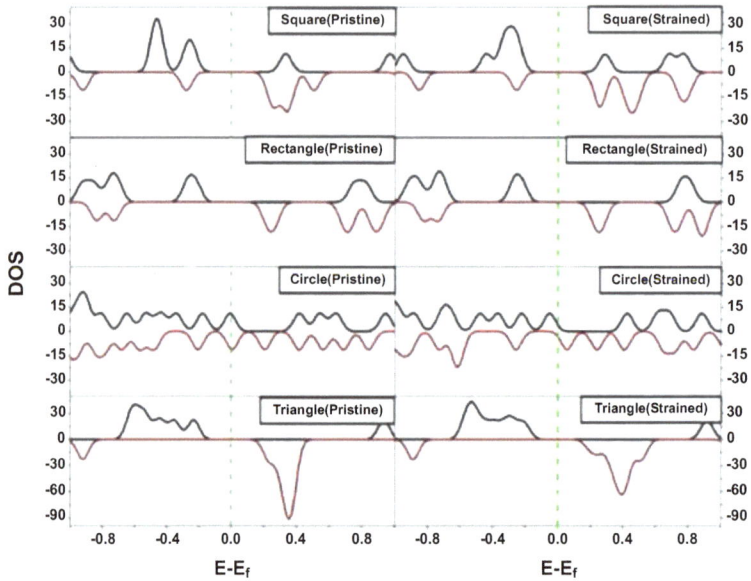

Fig. (17). Density of states for nanoribbons with applied point stress. The black and red lines correspond to DOS for spin-up and spin-down electronic states.

CONCLUSION

The structural, electronic, and magnetic properties of strained graphene nanoribbons were investigated using the spin-polarized density functional theory. Axial compressive stress was applied at two parallel edges of rectangular nanoribbons, and point stress was applied in the centre of square, rectangular, circular, and triangular nanoribbons. The C-C bond angles started to change when the axial strain was applied, and substantial rippling was detected at 6.7%. Carbon atoms formed octagonal structures along the edges where stress was applied, replacing the original hexagons. With the addition of axial stress, the electrical and magnetic properties of graphene nanoribbons altered as well. With the application of stress, the total magnetic moment changed significantly. TMM was found to vary in direct proportion to hexagonal lattice distortion and the number of zigzag edge states. The spin density maps depicted the imbalance of spin-up and spin-down states that caused nanoribbons to be magnetic. The magnetic behaviour of GNRs was also confirmed by the finite energy difference between HOMO-LUMO gaps for the spin-up and spin-down states. An increase in strain caused a shift in DOS around the Fermi level, resulting in empty states.

The application of point stress to the centre of ribbons caused magnetism in the centre of triangular-shaped GNRs, but localised moments were only seen at the edge atoms in the other ribbons. According to the predicted magnetic properties, stress causes considerable changes that might be used in spintronic applications [66]. Mechanical sensors might benefit from ripple effects in GNRs. Modifications in nanoribbon width and stacking patterns may also result in bandgap opening via quantum confinement and edge effects when nanoribbons are produced.

REFERENCES

[1]   Chen, Z.; Lin, Y.M.; Rooks, M.J.; Avouris, P. Graphene nano-ribbon electronics. *Physica E, 2007, 40*(2), 228-232.
       [http://dx.doi.org/10.1016/j.physe.2007.06.020]

[2]   Han, M.Y.; Ozyilmaz, B.; Zhang, B.Y., Kim, P. Energy band-gap engineering of graphene nanoribbons. *Phys. Rev. Lett.,2007*, 98(1-4), 206805.

[3]   Cresti, A.; Nemec, N.; Biel, B.; Niebler, G.; Triozon, F.; Cuniberti, G.; Roche, S. Charge transport in disordered graphene-based low dimensional materials. *Nano Res., 2008, 1*(5), 361-394.
       [http://dx.doi.org/10.1007/s12274-008-8043-2]

[4]   Wang, X.; Ouyang, Y.; Li, X.; Wang, H.; Guo, J.;Dai, H. Room-temperature all-semiconducting sub-10-nm graphene nanoribbon field-effect transistors. *Phys. Rev. Lett.*, **2008**, 100(1-4), 206803.

[5]   Jang, S.; Hwang, E.; Lee, Y.; Lee, S.; Cho, J.H. Multifunctional graphene optoelectronic devices capable of detecting and storing photonic signals. *Nano Lett.*, **2015**, *15*(4), 2542-2547.
       [http://dx.doi.org/10.1021/acs.nanolett.5b00105] [PMID: 25811444]

[6]   Soriano, D.; Munoz-Rojas, F.; Fernandez-Rossier, J.; Palacios, J. Hydrogenated graphene nanoribbons for spintronics. *Phys. Rev. B, 2010, 81*(1-7), 165409.

[7]   Huang, B.; Yan, Q.M.; Zhou, G.; Wu, J.; Gu, B.I.; Duan, W.H.; Liu, F. Making a field effect transistor on a single graphene nanoribbon by selective doping. *App. Phys. Lett., 2007, 91*(1-3), 253122.

[8]   Pan, Y.; Zhang, H.; Shi, D.; Sun, J.; Du, S.; Liu, F.; Gao, H. Highly Ordered, Millimeter-Scale, Continuous, Single-Crystalline Graphene Monolayer Formed on Ru (0001). *Adv. Mater., 2009, 21*(27), 2777-2780.
       [http://dx.doi.org/10.1002/adma.200800761]

[9]   Majumdar, K.; Murali, K.V.R.M.; Bhat, N.; Lin, Y.M. External bias dependent direct to indirect band gap transition in graphene nanoribbon. *Nano Lett., 2010, 10*(8), 2857-2862.
       [http://dx.doi.org/10.1021/nl100909n] [PMID: 20597528]

[10]   Yang, S.; Gong, Y.; Liu, Z.; Zhan, L.; Hashim, D.P.; Ma, L.; Vajtai, R.; Ajayan, P.M. Bottom-up approach toward single-crystalline VO₂-graphene ribbons as cathodes for ultrafast lithium storage. *Nano Lett., 2013, 13*(4), 1596-1601.
       [http://dx.doi.org/10.1021/nl400001u] [PMID: 23477543]

[11]   Novoselov, K.S.; Geim, A.K.; Morozov, S.V.; Jiang, D.; Zhang, Y.; Dubonos, S.V.; Grigorieva, I.V.; Firsov, A.A. Electric field effect in atomically thin carbon films. *Science, 2004, 306*(5696), 666-669.
       [http://dx.doi.org/10.1126/science.1102896] [PMID: 15499015]

[12]   Xu, M.; Liang, T.; Shi, M.; Chen, H. Graphene-like two-dimensional materials. *Chem. Rev., 2013, 113*(5), 3766-3798.
       [http://dx.doi.org/10.1021/cr300263a] [PMID: 23286380]

[13]   Xia, F.; Wang, H.; Xiao, D.; Dubey, M.; Ramasubramaniam, A. Two-dimensional material nanophotonics. *Nat. Photonics, 2014, 8*(12), 899-907.
       [http://dx.doi.org/10.1038/nphoton.2014.271]

[14]   Yang, G.; Zhu, C.; Du, D.; Zhu, J.; Lin, Y. Graphene-like two-dimensional layered nanomaterials: applications in biosensors and nanomedicine. *Nanoscale, 2015, 7*(34), 14217-14231.

[http://dx.doi.org/10.1039/C5NR03398E] [PMID: 26234249]

[15]   Fang, D.Q.; Zhang, S.L.; Xu, H. Tuning the electronic and magnetic properties of zigzag silicene nanoribbons by edge hydrogenation and doping. *RSC Advances,* **2013**, *3*(46), 24075-24080. [http://dx.doi.org/10.1039/c3ra42720j]

[16]   Wang, T.C.; Hsu, C.H.; Huang, Z.Q.; Chuang, F.C.; Su, W.S.; Guo, G.Y. Tunable magnetic states on the zigzag edges of hydrogenated and halogenated group-IV nanoribbons. *Scientific Reports,* **2016**, *6*, 39083.

[17]   Marin, E.G.; Marian, D.; Iannaccone, G.; Fiori, G. First principles investigation of tunnel FETs based on nanoribbons from topological two-dimensional materials. *Nanoscale,* **2017**, *9*(48), 19390-19397. [http://dx.doi.org/10.1039/C7NR06015G] [PMID: 29206255]

[18]   LiuH., XuB., ShengL., YinJ., Duan C. G. and WanX., Unexpected magnetic semiconductor behavior in zigzag phosphorene nanoribbons driven by half-filled one dimensional band. *Sci. Rep.,* **2015**, *8921*, 1-5.

[19]   Matthes, L.; Bechstedt, F. Influence of edge and field effects on topological states of germanene nanoribbons from self-consistent calculations. *Phys. Rev. B,* **2014**, *90*, 165431.

[20]   Monshi, M.M.; Aghaei, S.M.; Calizo, I. Edge functionalized germanene nanoribbons: impact on electronic and magnetic properties. *RSC Advances,* **2017**, *7*(31), 18900-18908. [http://dx.doi.org/10.1039/C6RA25083A]

[21]   Chen, C.; Huang, B.; Wu, J. Be_3N_2 monolayer: A graphene-like two-dimensional material and its derivative nanoribbons. *AIP Advances.,* **2018**, *8*, 105105.

[22]   Li, Y.Y.; Chen, M.X.; Weinert, M.; Li, L. Direct experimental determination of onset of electron–electron interactions in gap opening of zigzag graphene nanoribbons. *Nat. Commun.,* **2014**, *5*, 4311.

[23]   Pan, M.; Girão, E.C.; Jia, X.; Bhaviripudi, S.; Li, Q.; Kong, J.; Meunier, V.; Dresselhaus, M.S. Topographic and spectroscopic characterization of electronic edge states in CVD grown graphene nanoribbons. *Nano Lett.,* **2012**, *12*(4), 1928-1933. [http://dx.doi.org/10.1021/nl204392s] [PMID: 22364382]

[24]   Magda, G.Z.; Jin, X.; Hagymási, I.; Vancsó, P.; Osváth, Z.; Nemes-Incze, P.; Hwang, C.; Biró, L.P.; Tapasztó, L. Room-temperature magnetic order on zigzag edges of narrow graphene nanoribbons. *Nature,* **2014**, *514*(7524), 608-611. [http://dx.doi.org/10.1038/nature13831] [PMID: 25355361]

[25]   Kosynkin, D.V.; Higginbotham, A.L.; Sinitskii, A.; Lomeda, J.R.; Dimiev, A.; Price, B.K.; Tour, J.M. Longitudinal unzipping of carbon nanotubes to form graphene nanoribbons. *Nature,* **2009**, *458*(7240), 872-876. [http://dx.doi.org/10.1038/nature07872] [PMID: 19370030]

[26]   Jiao, L.; Zhang, L.; Wang, X.; Diankov, G.; Dai, H. Narrow graphene nanoribbons from carbon nanotubes. *Nature,* **2009**, *458*(7240), 877-880. [http://dx.doi.org/10.1038/nature07919] [PMID: 19370031]

[27]   Cai, J.; Ruffieux, P.; Jaafar, R.; Bieri, M.; Braun, T.; Blankenburg, S.; Muoth, M.; Seitsonen, A.P.; Saleh, M.; Feng, X.; Müllen, K.; Fasel, R. Atomically precise bottom-up fabrication of graphene nanoribbons. *Nature,* **2010**, *466*(7305), 470-473. [http://dx.doi.org/10.1038/nature09211] [PMID: 20651687]

[28]   Di Giovannantonio, M.; Deniz, O.; Urgel, J.I.; Widmer, R.; Dienel, T.; Stolz, S.; Sánchez-Sánchez, C.; Muntwiler, M.; Dumslaff, T.; Berger, R.; Narita, A.; Feng, X.; Müllen, K.; Ruffieux, P.; Fasel, R. On-Surface Growth Dynamics of Graphene Nanoribbons: The Role of Halogen Functionalization. *ACS Nano,* **2018**, *12*(1), 74-81. [http://dx.doi.org/10.1021/acsnano.7b07077] [PMID: 29200262]

[29]   Sun, Q.; Zhang, R.; Qiu, J.; Liu, R.; Xu, W. On-surface synthesis of carbon nanostructures. *Adv.*

Mater., **2018**, *30*, 1705630.

[30]   Narita, A.; Wang, X.Y.; Feng, X.; Müllen, K. New advances in nanographene chemistry. *Chem. Soc. Rev.,* **2015**, *44*(18), 6616-6643.
[http://dx.doi.org/10.1039/C5CS00183H] [PMID: 26186682]

[31]   Talirz, L.; Ruffieux, P.; Fasel, R. On-surface synthesis of atomically precise graphene nanoribbons. *Adv. Mater.,* **2016**, *28*(29), 6222-6231.
[http://dx.doi.org/10.1002/adma.201505738] [PMID: 26867990]

[32]   Lee, C.; Wei, X.; Kysar, J.W.; Hone, J. Measurement of the elastic properties and intrinsic strength of monolayer graphene. *Science,* **2008**, *321*(5887), 385-388.
[http://dx.doi.org/10.1126/science.1157996] [PMID: 18635798]

[33]   Cadelano, E.; Palla, P.L.; Giordano, S.; Colombo, L. Nonlinear Elasticity of Monolayer Graphene. *Phys. Rev. Lett.,* **2009**, *102*, 235502.

[34]   Bunch, J.S.; Verbridge, S.S.; Alden, J.S.; van der Zande, A.M.; Parpia, J.M.; Craighead, H.G.; McEuen, P.L. Impermeable atomic membranes from graphene sheets. *Nano Lett.,* **2008**, *8*(8), 2458-2462.
[http://dx.doi.org/10.1021/nl801457b] [PMID: 18630972]

[35]   Kim, K.; Lee, Z.; Malone, B.D.; Chan, K.T.; Aleman, B.; Regan, W.; Gannett, W.; Crommie, M.F.; Cohen, M.L.; Zettl, A. Multiply folded graphene. *Phys. Rev. B.,* **2011**, *83*(1-8), 245433.

[36]   Mathew, J.P.; Patel, R.N.; Borah, A.; Vijay, R.; Deshmukh, M.M. Dynamical strong coupling and parametric amplification of mechanical modes of graphene drums. *Nat. Nanotechnol.,* **2016**, *11*(9), 747-751.
[http://dx.doi.org/10.1038/nnano.2016.94] [PMID: 27294506]

[37]   Kim, S.J.; Choi, K.; Lee, B.; Kim, Y.; Hong, B.H. Materials for Flexible, Stretchable Electronics: Graphene and 2D Materials. *Annu. Rev. Mater. Res.,* **2015**, *45*(1), 63-84.
[http://dx.doi.org/10.1146/annurev-matsci-070214-020901]

[38]   Ni, Z.; Chen, W.; Fan, X.; Kuo, J.; Yu, T.; Wee, A.; Shen, Z. Raman spectroscopy of epitaxial graphene on a SiC substrate. *Phys. Rev. B,* **2008**, *77*, 115416.

[39]   Teague, M.L.; Lai, A.P.; Velasco, J.; Hughes, C.R.; Beyer, A.D.; Bockrath, M.W.; Lau, C.N.; Yeh, N.C. Evidence for strain-induced local conductance modulations in single-layer graphene on SiO_2. *Nano Lett.,* **2009**, *9*(7), 2542-2546.
[http://dx.doi.org/10.1021/nl9005657] [PMID: 19534500]

[40]   Huang, B.; Liu, M.; Su, N.; Wu, J.; Duan, W.; Gu, B.L.; Liu, F. Quantum manifestations of graphene edge stress and edge instability: A first-principles study. *Phys. Rev. Lett.,* **2009**, *102*, 166404.

[41]   Yu, T.; Ni, Z.; Du, C.; You, Y.; Wang, Y.; Shen, Z. Raman Mapping Investigation of Graphene on Transparent Flexible Substrate: The Strain Effect. *J. Phys. Chem. C,* **2008**, *112*(33), 12602-12605.
[http://dx.doi.org/10.1021/jp806045u]

[42]   Ding, F.; Ji, H.; Chen, Y.; Herklotz, A.; Dörr, K.; Mei, Y.; Rastelli, A.; Schmidt, O.G. Stretchable graphene: a close look at fundamental parameters through biaxial straining. *Nano Lett.,* **2010**, *10*(9), 3453-3458.
[http://dx.doi.org/10.1021/nl101533x] [PMID: 20695450]

[43]   Ferralis, N.; Maboudian, R.; Carraro, C. Evidence of structural strain in epitaxial graphene layers on 6H-SiC. *Phys. Rev. Lett.,* **2008**, *101*(1-4), 156801.

[44]   Levy, N.; Burke, S.A.; Meaker, K.L.; Panlasigui, M.; Zettl, A.; Guinea, F.; Neto, A.H.C.; Crommie, M.F. Strain-induced pseudo-magnetic fields greater than 300 tesla in graphene nanobubbles. *Science,* **2010**, *329*(5991), 544-547.
[http://dx.doi.org/10.1126/science.1191700] [PMID: 20671183]

[45]   Guinea, F.; Katsnelson, M.I.; Geim, A.K. Energy gaps and a zero-field quantum Hall effect in

graphene by strain engineering. *Nat. Phys.*, **2010**, *6*(1), 30-33.
[http://dx.doi.org/10.1038/nphys1420]

[46]   Si, C.; Liu, Z.; Duan, W.; Liu, F. First-principles calculations on the effect of doping and biaxial tensile strain on electron-phonon coupling in graphene. *Phys. Rev. Lett.*, **2013**, *111*(19), 196802.
[http://dx.doi.org/10.1103/PhysRevLett.111.196802] [PMID: 24266482]

[47]   Pereira, V.M.; Neto, A.C.; Peres, N. Tight-binding approach to uniaxial strain in graphene. *Phys. Rev. B: Condens. Matter.*, **2009**, *80*, 045401.

[48]   Lu, Y.; Guo, J. Bandgap of strained graphene nanoribbons. *Nano. Res.*, **2010**, *3*, 189-199.
[http://dx.doi.org/10.1007/s12274-010-1022-4]

[49]   Neek-Amal, M.; Peeters, F.M. Graphene nanoribbons subjected to axial stress. *Phys. Rev. B.*, **2010**, *82*, 085432.

[50]   Wang, Z.F.; Zhang, Y.; Liu, F. Formation of hydrogenated graphene nanoripples by strain engineering and directed surface self-assembly. *Phys. Rev. B Condens. Matter Mater. Phys.*, **2011**, *83*(4), 041403.
[http://dx.doi.org/10.1103/PhysRevB.83.041403]

[51]   Guinea, F.; Horovitz, B.; Le Doussal, P. Gauge fields, ripples and wrinkles in graphene layers. *Solid State Commun.*, **2009**, *149*(27-28), 1140-1143.
[http://dx.doi.org/10.1016/j.ssc.2009.02.044]

[52]   Baimova, J.A.; Dmitriev, S.V.; Zhou, K. Strain-induced ripples in graphene nanoribbons with clamped edges. *Phys. Status Solidi, B Basic Res.*, **2012**, *249*(7), 1393-1398.
[http://dx.doi.org/10.1002/pssb.201084224]

[53]   Shenoy, V.B.; Reddy, C.D.; Ramasubramaniam, A.; Zhang, Y.W. Edge-stress-induced warping of graphene sheets and nanoribbons. *Phys. Rev. Lett.*, **2008**, *101*(24), 245501.
[http://dx.doi.org/10.1103/PhysRevLett.101.245501] [PMID: 19113631]

[54]   Owens, F.J. Electronic and magnetic properties of armchair and zigzag graphene nanoribbons. *J. Chem. Phys.*, **2008**, *128*(19), 194701.
[http://dx.doi.org/10.1063/1.2905215] [PMID: 18500880]

[55]   Al-Aqtash, N.M.; Sabirianov, R.F. Spin density waves in periodically strained graphene nanoribbons. *Nanoscale*, **2014**, *6*(8), 4285-4291.
[http://dx.doi.org/10.1039/C3NR06199J] [PMID: 24615501]

[56]   Soler, J.M.; Artacho, E.; Gale, J.D.; García, A.; Junquera, J.; Ordejón, P.; Sánchez-Portal, D. The SIESTA method for *ab initio* order- *N* materials simulation. *J. Phys. Condens. Matter*, **2002**, *14*(11), 2745-2779.
[http://dx.doi.org/10.1088/0953-8984/14/11/302]

[57]   Perdew, J.P.; Burke, K.; Ernzerhof, M. Generalized Gradient Approximation Made Simple. *Phys. Rev. Lett.*, **1996**, *77*(18), 3865-3868.
[http://dx.doi.org/10.1103/PhysRevLett.77.3865] [PMID: 10062328]

[58]   Kleinman, L.; Bylander, D.M. Efficacious Form for Model Pseudopotentials. *Phys. Rev. Lett.*, **1982**, *48*(20), 1425-1428.
[http://dx.doi.org/10.1103/PhysRevLett.48.1425]

[59]   Sankey, O.F.; Niklewski, D.J. *Ab initio* multicenter tight-binding model for molecular-dynamics simulations and other applications in covalent systems. *Phys. Rev. B Condens. Matter*, **1989**, *40*(6), 3979-3995.
[http://dx.doi.org/10.1103/PhysRevB.40.3979] [PMID: 9992372]

[60]   Cooper, D.R.; D'Anjou, B.; Ghattamaneni, N.; Harack, B.; Hilke, M.; Horth, A.; Majlis, N.; Massicotte, M.; Vandsburger, L.; Whiteway, E.; Yu, V. Experimental review of graphene. *ISRN Condensed Matter Physics*, **2012**, *2012*, 1-56.
[http://dx.doi.org/10.5402/2012/501686]

[61]   Kaur, S.; Sharma, A.; Sharma, H.; Mudahar, I. Structural and magnetic properties of small symmetrical and asymmetrical sized fullerene dimers. *Mater. Res. Express,* **2018**, *5*(1), 016105.
[http://dx.doi.org/10.1088/2053-1591/aaa567]

[62]   Garg, I.; Sharma, H.; Kapila, N.; Dharamvir, K.; Jindal, V.K. Transition metal induced magnetism in smaller fullerenes (C $_n$ for n ≤ 36). *Nanoscale,* **2011**, *3*(1), 217-224.
[http://dx.doi.org/10.1039/C0NR00475H] [PMID: 20981362]

[63]   Sharma, H.; Garg, I.; Dharamvir, K.; Jindal, V.K. Structural, electronic, and vibrational properties of C(60-n)Nn (n = 1-12). *J. Phys. Chem. A,* **2009**, *113*(31), 9002-9013.
[http://dx.doi.org/10.1021/jp901969z] [PMID: 19719305]

[64]   Sharma, A.; Kaur, S.; Sharma, H.; Mudahar, I. Electronic and magnetic properties of small fullerene carbon nanobuds: A DFT study. *Mater. Res. Express,* **2018**, *5*(6), 065032.
[http://dx.doi.org/10.1088/2053-1591/aacb18]

[65]   Kaur, S.; Sharma, H.; Jindal, V.K.; Bubanja, V.; Mudahar, I. Graphene nanoribbons under axial compressive and point tensile stresses. *Physica E,* **2019**, *111*, 1-12.
[http://dx.doi.org/10.1016/j.physe.2019.02.018]

[66]   Şahin, H.; Ataca, C.; Ciraci, S. Electronic and magnetic properties of graphane nanoribbons. *Phys. Rev. B Condens. Matter Mater. Phys.,* **2010**, *81*(20), 205417.
[http://dx.doi.org/10.1103/PhysRevB.81.205417]

SUBJECT INDEX

www.ingramcontent.com/pod-product-compliance
Lightning Source LLC
Chambersburg PA
CBHW050807220326
41598CB00006B/142